nd Reinhold Company Regional Offices:
Cincinnati Atlanta Dallas San Francisco

nd Reinhold Company International Offices:
ronto Melbourne

Congress Catalog Card Number: 79-19584
42-27643-5

d in the United States of America

Van Nostrand Reinhold Company
h Street, New York, N.Y. 10020

multaneously in Canada by Van Nostrand Reinhold Ltd.

1 10 9 8 7 6 5 4 3

ngress Cataloging in Publication Data

nder title:

rocomputer control in
rocesses.

liographies and index.
process control–Data processing.
ters. 3. Microcomputers. I. Skrokov,

660.2′81′02854 79-19584
7643-5

Mini- and Microcomput
in Industrial Proce

Handbook of Systems and Applic

Edited by
M. Robert Skroko

Control Systems Division Man
Stone & Webster Engineering Cor
New York, New York

Van Nostra
New York

Van Nostra
London T

Library of
ISBN : 0-4

Manufacture

Published by
135 West 50

Published si

15 14 13 12

Library of C

Main entry u

Mini- and mi
industrial p

Includes bi
1. Chemica
2. Minicompu
M. Robert.
TP155.75.M5
ISBN 0-442-2

Dedication

I dedicate this text to world peace through technological exchange between nations. Much of this work is a distillation of worldwide experience on computer and control systems projects. In the United States, projects have been proposed, designed and/or executed on the Gulf Coast and nationwide; in Europe: Great Britain, Holland, Belgium, France, Yugoslavia, etc; in Asia: Japan, Taiwan, The People's Republic of China, Singapore, etc.; in the Middle East and North Africa: Algeria.

It is hoped that such exchanges will help lead the world to true peace as peoples of all nations get to know one another.

Preface

Many excellent handbooks on computer hardware and software have been published in the past 15 to 20 years and it is not our purpose to attempt to add one more to either listing. Most hardware texts have, naturally, described developments in electronics and have covered areas ranging from the electromechanical relay and vacuum tube to the germanium transistor, from the transistor to the silicon integrated circuit, from the integrated circuit to the medium scale integration (MSI) of the minicomputer, from MSI to the mass produced large scale integrated circuit (LSI) and to the current very large scale (VLSI) microelectronic production techniques that enable the design of a dedicated microprocessor on-a-chip.

Software texts, on the other hand, have concentrated on programming language development and network interface and communications problems. They have evolved from the machine language, boolean algebra requirement of the relay, vacuum tube and transistor systems to the current high-level languages developed for the 16 bit microprocessor and microcomputer systems.

There are, however, few computer handbooks available for applications engineers, particularly for those at work in the process industries. This handbook is essentially constructed along the lines of a distributed control system, with each of the ten chapters dedicated to a section of the plant (or aspect of system design), thus highlighting typical applications and control problems in a particular unit operation.

The purpose of this book is two-fold. First, it is hoped that it will serve as a guide to practitioners of the art of control systems engineering as it relates to computer control of processes.

Second, standardized designs of stand-alone microcomputers, linkable to host computers, for fractionating column and other unit operations control and housing up to 20 loops, are badly needed. It is hoped that our chapters on microcomputers will help push the major instrument and/or computer manufacturers toward the manufacture of such controls, as stand-alone satellite computers in distributed control systems, monitored by host computers, are the core of the control systems strategies illustrated and discussed throughout the text. As with any such effort, the standardization of languages and hardware systems design must be accomplished through the technical societies. The Instrument Society of America (I.S.A.) and the American Society of Mechanical Engineers' (A.S.M.E.) automatic controls division have begun such efforts.

Typical, large petrochemical processes are analyzed by breaking down these world-sized plants into discrete unit operations (since these projects are generally huge and may aggregate from 100 million to 500 million dollars in total projects costs.)

This handbook has grown from the editor's and contributing authors' industrial applications experience, covering more than 25 years in control systems. Some of the material has been extrapolated from published papers and talks. Marvin Weiss, the author of Chapter 6, is well known in the I.S.A. for a long history of talks and papers on process analysis instrumentation. He has worked on some of the first, pioneering applications of continuous analyzers to industrial process control. In this text, he has added computer controlled analyzers to his list of credits. Dr. Hoo-Min Toong has expanded his "Microprocessor" article, published in the September 1977 issue of *Scientific American,* into the chapter on microprocessors for industrial control. Dr. Toong's research at M.I.T. on distributed computing systems, as well as his work as head of the M.I.T. Digital Systems Laboratory, are the basis for his experience with the "triad" development system described in his chapter.

Merrill Thor based his chapter on specification design for process control computers on his experience with instrument/computer manufacturers and engineering contractors on projects constructed in Europe, as well as in the U.S. His chapter on networks and system configuration is drawn partly from problems solved on those projects. Aaron Kramer contributed the long chapter on fractionating column control and the chapter on compressor control based on his experience in control systems with an instrument manufacturer and long tenure as a professor at the State University of New York, Maritime College, as well as his work as a consultant at Stone & Webster Engineering Corporation for the past five years, where he has helped solve some difficult control systems problems through simulation on computer.

Morris Leitner, in Chapter 5, has applied sequence control systems to power plant projects and large petrochemical unit operations. Steve Cocheo, the collaborator on this chapter, is an engineer in the Control Systems Division of Stone & Webster and has previous experience with instrument and computer manufacturers. Ralph Bothne, the President of EMC Controls, is the author of Chapter 10, on systems house interface to computer project execution.

This work has been a part-time effort for all of the contributing authors and the editor. All of us hold full-time, responsible positions in some aspect of the process industries: in engineering contracting organizations, in systems organizations or in educational institutions. The contributors to the text are thus active practitioners in this field and have extensive experience in the design of plant process control systems, augmented by operation and start-up experience. Most have worked in the field since the early 1950's and have witnessed the changes in control instrumentation hardware and systems as it progressed from the large-case instrumentation hardware stage of the 1940's and 1950's to the large, full-graphic and semi-graphic panels of the 1960's and to the current "TV" screen concept of computerized operator consoles. They have thus been witnesses to the history of instrumentation miniaturization and the beginnings of electronic instrument application and its evolution.

It has taken almost two full years to complete this volume. Chapters were outlined on airplanes over the Atlantic and Pacific Oceans on long flights to Europe and Asia for business travel, during commuting train rides between New York City offices and various surrounding suburbs and, largely, in home-based office space.

We begin with a section on the history of industrial computer control and microelectronics. With this as background, the requirements for a computer project in the process industries are outlined as a prelude to later chapters on unit operations control, such as fractionating column control and sequence control systems. The chapters on computer controlled analyzers and systems hardware configuration represent designs of sub-systems within the overall project. Microprocessor hardware and software development is covered early as a prerequisite to the distributed control concepts used throughout.

The ethylene process is chosen as the basis for discussion of distributed control strategies because it embodies the complexities of most other petrochemical processes. The front-end furnace section is probably the most complex of such furnace processes, while the fractionation train is representative of a typical train in any petrochemical process or petroleum refinery.

I want to acknowledge my wife and family, who were very patient during the many evenings and weekends that I devoted to this work and to thank Mrs. Ruby Vlaun Farley and Mrs. Barbara Vetter, who typed the manuscript. Thanks, too, to Joseph Greenspan, who did some of the illustrations.

Acknowledgment is also due to the various computer and control systems manufacturers who allowed the reprinting of some photographs and text material, such as The Foxboro Corporation, the Bristol Company, the Taylor Instrument Company, Honeywell, Inc., I.B.M., EMC Controls, Inc., Metromation, Inc., and Digital Equipment Corporation.

It is hoped that the text produced by this combined effort will be found useful to practicing control systems engineers, process engineers, technology managers and others interested in digital computer control applications in the process industries.

Robert Skrokov
North Plainfield, New Jersey

Contents

1
Introduction

M. Robert Skrokov
Chief, Control Systems Division
Stone & Webster Engineering Corporation
New York, N.Y.

HISTORY: COMPUTER PROCESS CONTROL

Digital computer control in the chemical process industries has been studied, specified, economically justified and widely discussed in the industries and the technical societies for almost 25 years.

Some of the earliest chemical industry digital computer applications are traceable to the mid-1950's, when attempts were made to capitalize on post-war, national defense, electronic industry developments with the injection of some of the technology into the chemical process industries. One such effort was a joint venture between several major operating companies and a small computer manufacturer in 1955, which resulted in more than a dozen successful applications in such processes as ethylene, ammonia, vinyl chloride, styrene butadiene rubber and industrial alcohol, among others. Initially, the small computer company used the impetus it had received from its national defense work in electronics to promote joint economic and technical evaluations of the digital computer for process monitoring and control. These evaluations pinpointed a number of unit operations in each industry best suited to computer control and led to projects that later resulted in some 19 successful applications in the early 1960's. The applications were based on the available hardware of the day, which was slow in speed because of the drum-type memories employed, and was massive in size when compared to current microelectronic based hardware.

The primary purposes and strategies of these projects included performing control calculations on conversions and yields, logging operating records, alarming off-normal variables and using advanced control strategies, such as adaptive control on plant simulation models in order to improve plant productivity and profitability.

The initial experience gained in these early years led the participating operating companies into a development of their own internal computer technology. Any profitable successes resulting from these efforts were not widely publicized. However, some other fairly large projects that did not succeed for one reason or another were widely discussed in the trade journals of the time. In the mid- and late 1960's, a number of projects were begun in the petroleum, steel, pulp and paper and other process industries. Many of these projects became showpiece installations, with little, if any, financial return on investment. They were based on single, large, main-frame computer hardware, centrally located in control rooms and necessitating long runs for input/output signals to process instrumentation and control valves. Proper interface design for such systems was difficult, as were installation, start-up and maintenance.

The small computer company was eventually bought out by a larger company. The large computer companies began to sell hardware to the major operating companies at top management levels during this period. Computer projects were then created around this machinery without adequate systems analysis.

All of these factors contributed to the failures publicized in the trade journals. Thus, the propagation of the information about project failures, coupled with the secrecy of the successful computer users, tended to inhibit the growth of computer applications. One by one, the large computer manufacturers dropped out of the process control field, concentrating instead on commercial data processing systems in other business areas.

Therefore, it may be said that a general and widespread application of computers to process control is only just beginning after an initial start dating back almost 20 years. Until recently, most process plant designs have been conceived without regular consideration given to computer application.

Today, almost every major project considered for the process industries seems to have some requirement for digital computers written into the scope of the project as a matter of routine. Thermocouple and signal multiplexing, computer-driven stream analyzers and sequencing control systems are included in most projects as a minimum requirement. On a unit control basis, closed-loop control of distillation columns and cracking furnaces by computer is specified, while on a plantwide basis, off-line or on-line optimization is often stipulated.

The current revolution in microelectronics has helped spur renewed interest in computer application to the process industries. Resultant orders-of-magnitude reduction in microcomputer systems hardware cost over the past few years is but one impetus for this interest. Microcomputer hardware also reduces control board size by similar orders of magnitude while enabling use of more sophisticated control strategies on a unit basis. It also makes the design of distributed control systems practical and realistic (i.e., the placement of unit control hardware in the field closer to the points of measurement and control), thus providing better redundancy and availability.

An understanding of recent trends in process control instrumentation and computer application may be obtained by an examination of the evolutionary phases through which computer application has progressed over the past 20 to 25 years. Initial development of process industry computer systems application proceeded through approximately five phases, as discussed below.

The first phase was the attempt of a small computer company (the company cited above) to work directly with operating company personnel to jointly evaluate the economic and advanced control possibilities of computer control in certain areas. This phase succeeded in defining the principal areas in which computer control would be successful.

The second phase consisted of the development of industrial systems by the computer manufacturers based on their maxi-machinery. The style was essentially one of making the application fit the available hardware. Being aware that many potential process applications existed for their machinery, and having large and aggressive sales forces available, they launched a number of nationwide sales application demonstrations to various segments of the process industries.

One such series that may be recalled involved extensive presentations to the pulp and paper industry in the early 1960's. These exhibits were based on large mainframe systems which lacked adequate front-end hardware and software for process instrumentation interface, and systems strategies for distributed plantwide control as a redundancy mode in case of failure. These sales campaigns were successful in forcefully injecting the maxi-computer of the age into some rather unprepared environments. Many process operating company staff engineering departments still had limited or inadequate computer experience; on the other hand, computer manufacturers lacked staff experienced in the idiosyncrasies of the processes they were attacking, as well as experience with instrument interface problems.

In many instances, large main-frame hardware was sold to, or authorized by, operating company management who delegated the hardware to their central engineering departments for process application without a concomitant authorization for prior systems analysis. As a result, effective real-time process computer control failed in a number of applications, with each failure creating psychological barriers to further application and development.

Coincidentally, development work on electronic instrumentation was progressing through an early design phase. Standardized signal levels were being developed through the engineering societies as well as by the instrument manufacturers. The injection of large computer systems into this environment only added to the confusion. Wide acceptance of computer systems at that time was thus highly improbable.

The third phase was one in which more of the major operating companies began to build internal computer capability based on their process technology. They also began, at this time, to develop the software necessary to more realistically apply computers to process control. However, the style was still one of an attempt to apply one large system for process optimization and management information as well as for local unit operations process control. Most instrument sensing and control signals were run to and from the central processor, which was not only technically difficult for computer interfaces, but was expensive as well. In addition to instrument signal and grounding problems, operator communications and training were lightly addressed, as was the development of conversational-level programming languages and redundancy strategy, to mention but a few of the key obstacles to wide success.

Thus, having had the experience of these early failures, the operating companies began a more concerted effort to apply computer control using systems engineering techniques. Some organizations produced their own process control languages, others designed specialized operator communications consoles oriented to their processes, while still others joined forces with suppliers to exert a strong influence on total hardware and software development.

The fourth phase focused on the introduction of complete hardware and software systems by the major analog instrument manufacturers based on main-frame concepts of their own. These systems were better suited to interface with real-world analog instrumentation. Redundancy was also designed into these systems. However, redundancy was accomplished through the use of two large central processing units in one design, and many instrument signals still had to be wired back to these machines. Distributed control had not appeared on the horizon as yet. Direct digital control (DDC) was accomplished through the master computer, as was set point control (SPC) on key process control variables. Failure of one system did, however, allow the other to take over.

Lately, another phase has developed in which engineering contracting organizations, with a computer control staff and proprietary process knowledge, are being called upon by users to offer turn-key responsibility for process computer systems. This may partly be the result of some shift in engineering manpower from operating company staffs to contracting engineering staffs, due to the cyclic employment situation in the engineering profession in the last 15 years. Another possibility is that it resulted from market pressures exerted on the engineering contracting organizations to produce a total plant design that illustrates a high level of capability in a competitive market. Whatever the cause, there is a noticeable tendency for operating companies to include a computer control package as part of most new plant projects given to engineering contractors. The process control computer is beginning to be treated as just another element in the total control system.

Another result of these evolutionary phases is that they may also have helped create a new group of service organizations for process computer application. These are the systems-oriented organizations that are geared to provide a viable marriage of both hardware and software to a particular process industry by virtue of their knowledge of that industry. The need for a systems "middleman" has probably emanated from both the operating companies (generally lacking a large programming staff) and the contract engineering organizations (also without extensive programming staffs). The systems organization, then, is this "middleman" between the primary hardware manufacturer and the end user.

A systems house will take a main-frame central processing unit (CPU), design a custom front-end and hardware package suited to its own distributed control concepts, and sometimes also provide a process control language. These organizations usually limit their work to particular process industry areas, such as petroleum refinery or olefins plant technology. In designing software, they provide the "executive" programming necessary to process information within the system, as well as the "applications" level programming pertinent to the process being controlled. Both the ultimate user and the contract engineering organization want systems responsibility to be supplied as part of the computer system. The users no longer want to purchase a large machine as "bare-bones" hardware; the contractors want to use available staff to expeditiously execute large process plant designs, and cannot be concerned with the many marriages of hardware, software, peripheral interfaces and communication links. All of this must be supplied as a package and guaranteed to operate as such within the context of a large, probably world-size, process complex.

An ability to produce systems based designs has thus brought the following organizations into focus:

1. The control engineering or research department of the major operating companies specializing in computer control of proprietary processes.
2. The systems organization with a reputation and knowledge in a series of related processes (such as ethylene or ammonia production).
3. The major instrument manufacturing companies able to supply both computer systems and instrumentation.
4. The major computer manufacturers with systems staffs knowledgeable in process control.
5. The large engineering contracting organizations with the ability to design, specify, purchase and supervise the field installation of computer systems as part of a complete plant design.

The historical development outlined thus far has produced a library of field-proven hardware and software systems available through all, or through a combination of, the organizations mentioned. This library is currently in the process of standardization.

Another phase, then, in this development is an attempt at standardization. This phase is still developing and is incomplete. Currently, all suppliers are aiming at producing standardized systems under marketplace pressures. Distributed control concepts, redundancy and high-level process control languages are some of the products of standardization beginning to appear. As an example of software evolution, standardized languages were developed from variations of Fortran IV and BASIC that finally evolved into "conversational-level" programs usable by control systems engineers without extensive professional programming aid. Machine and assembly language based programming systems were too cumbersome to use in the world of process control.

Distributed control is appearing in both computer and instrumentation systems designs. The development of the microprocessor has made possible the location of microcomputers closer to the point of final control and the redesign of conventional analog instrumentation control boards and systems. Variations in both of these are beginning to appear on the market.

As with any new development, it takes time for the better solutions to settle out of the mix of possibilities. It has taken time to reach even the present maturity in process control computer application. Recent developments in the electronics industry have helped to quicken this maturity. With the revolutionary development of large scale integrated circuitry (LSI) and, currently, the very large scale integrated circuitry (VLSI), hardware costs have been drastically reduced, reliability has increased, panel board space can now be compressed and advanced control technology is more readily available to the solution of everyday problems. Furthermore, this revolution in electronics has spread throughout our culture, breeding familiarity as it goes, in the form of digital wristwatches, home and auto microcomputers and the ever-present hand calculators. This technological assimilation has had a positive effect on the process control industry in that almost all major control instrument vendors are now offering a line of plant control instrument systems incorporating some form of microprocessor hardware. Major computer manufacturers are also rapidly designing distributed systems.

What important lessons were learned in the last 15 to 20 years? At least two may be highlighted as outstanding: the need for a systems engineering analysis and specification stage prior to hardware and software acquisition, and the need for a unit operations (or distributed control) strategy of control systems design.

The development of microcomputer hardware has enabled the placement of advanced digital control technology at key unit operations throughout the plant. In petrochemical complexes, distributed control allows for local computer control of the furnaces, the fractionating towers, the cracked gas compressor and other downstream units under the supervision of a host computer. Control sophistication not previously practical is now possible. Plant optimization and management information is handled by the host computer with simultaneous monitoring of the distributed microcomputers.

Each "micro" can maximize local control strategy while minimizing energy usage. The host can further improve these local targets based on a plantwide analysis, while maximizing profit and/or product throughput and minimizing plantwide energy usage. Complex optimization is now possible in the real world, concurrent with distributed control.

MICROELECTRONICS: THE IMPACT

As some understanding of the history of computer systems in process control is essential for a study of the state-of-the-art, so is a rudimentary understanding of the revolution in microelectronics that has occurred over the past 10 to 15 years.

A microcomputer on a single chip, comparable in performance to some of the early minicomputers, is possible only because of the development of LSI metal oxide semiconductor (MOS) technology.

Microelectronic development has progressed exponentially over recent history. It has sprung from the earlier development of the germanium transistor, to the basic silicon integrated circuit, to the metal-oxide semiconductor technology of the MSI (or medium scale integrated circuit) to the current LSI and, lately, the VLSI.

Terms such as NMOS, PMOS, CMOS, TTL, I^2L, RAM, ROM, etc., are technologies and software that may all be incorporated in a single microcomputer chip. (Since a glossary defining and clarifying such technology is essential to a fuller understanding of new technological developments, it is hoped that the glossary in this book will make the text more usable and valuable to practicing engineers and students of computer control; it has been compiled in an attempt to collect all pertinent terminology in one reference area.)

A microprocessor is the central arithmetic and logic unit (ALU) of a computer. The "micro" portion of the term indicates that the processor is fabricated on a single chip of silicon. A typical such chip is approximately 0.5 cm on a side and may contain up to 20,000 equivalent transistor functions.

A single chip microcomputer is shown in Figure 1-1. This illustrates the Intel 8748 chip, which is approximately 6 × 7 mm in size. This one LSI chip houses the central microprocessor as well as the program memory and all auxiliary functions such as input/output and timing circuits necessary to accomplish the functions of a full minicomputer of the 1960's.

The cost of the microcomputer on-a-chip is approximately $250 in quantities of 25 or more. However, it must be programmed in machine language since it is microprogrammed for a dedicated application. Such an application might center on the optimization of an automobile engine for improvement of fuel economy or be dedicated to appliance or machine control in the home or office.

Special purpose controllers have been developed for industrial process control but

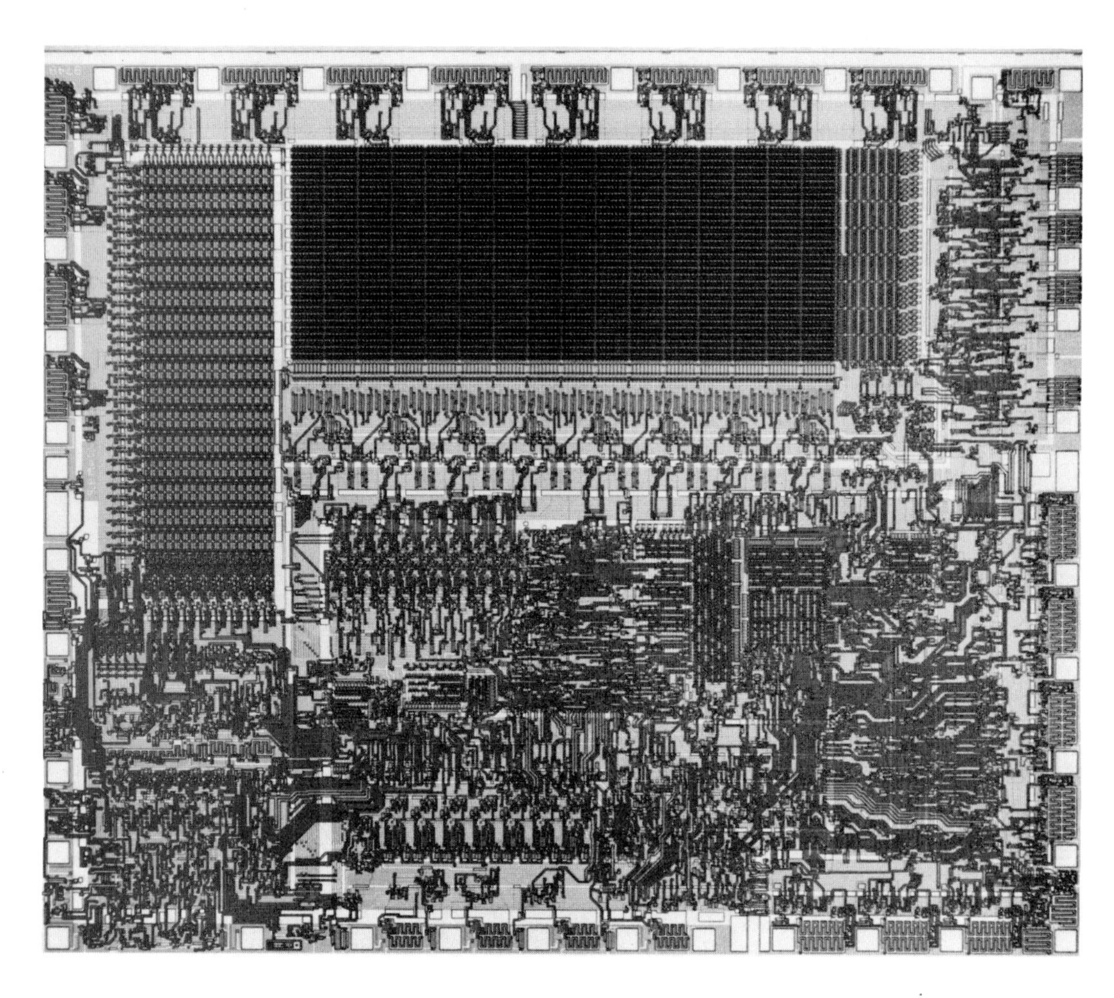

Figure 1-1. A single chip microcomputer: the Intel Corp. 8748. (*Courtesty of Intel Corp.*)

not, as yet, in the form of a microcomputer on-a-chip. Proposals have been made for such a development to the instrumentation and computer industry and specifications are being designed for a system based on local (plant) distributed microprocessors monitored by, and downloaded from, a host computer. (See Figure 2-5.)

A microcomputer on-a-chip dedicated to distillation column control and operating in a distributed mode is based on local heat and material balance strategy for the maximization of a desired product composition and/or throughput with minimal energy usage. Such "micros" are designed for control of specific towers in petroleum refineries or petrochemical plants such as deethanizers, demethanizers, ethylene towers or fluid catalytic cracking units.

For the electronics industry, special chip manufacture is a function of production volume. The question remains whether a large enough market volume for such applications exists in a given production year to justify the "masking" of special chips dedicated to column optimization. Tower loading, climatic conditions, tuning etc., are different for each type of tower. Each tower must be studied and evaluated for standardization of input signals, loading and purity range of the product desired.

Currently, column optimizer controllers can more easily be developed from the state-of-the-art hardware around the single board computer concept such as the Intel Corporation's i SBC 86-12 or Digital Equipment Corporation's LSI-11. A single board computer on a card about the size of this page houses the CPU, I/O (input/output) memory and interface functions, each on separate microprocessor chip on the one board. (See Figure 1-2 for the Intel i SBC 86/12.)

A mass-produced (or standardized) SBC such as this could be purchased for less than $300, in 1979.

Since process industry end users and engineering contract organizations are necessarily more concerned with overall control systems design and do not have man-hours allocated for elegant information processing within computer hardware, they tend to concentrate on applications level programming for proprietary process systems. Non-proprietary programming is functionally given to the systems organi-

Figure 1-2. A single board computer: the Intel Corp. iSBC-86/12. (*Courtesy of Intel Corp.*)

zations, required (by specification) to supply a complete package incorporating the end users' proprietary applications software.

A typical state-of-the-art distributed system might appear as shown in Figure 2-5, which illustrates Stone & Webster's standard design for a distributed system.

The distributed configuration eliminates the expense of wiring each signal back to the control room. A host minicomputer communicates with a number of satellite microcomputer systems. A computerized gas chromatograph system is used for analysis data.

Two identical operator positions are shown. Each position includes two CRT displays and a keyboard. An instrument panel board is optional. Two typewriters are shown—one for logging operations and one for alarm functions.

A communications microprocessor with redundant back-up supervises the flow of data among the satellites, the CRT devices and the host computer.

The host minicomputer performs process monitoring and control functions using data acquired at the satellites and sent over the high-speed communications line. Special functions such as furnace de-coke, furnace yield automation and overall plant optimization are also performed in the host. All programming is done at the host using the programmer's CRT terminal and the line printer. Programs for the satellite micros are down-line loaded (sent over the communications link from the host to the satellite).

Each satellite microcomputer system is fully redundant for reliability. Process input/output equipment is included in the microcomputer system. Eight microcomputer systems may be used in a typical olefins plant design.

A portion of Chapter 2 is dedicated to a discussion of such a distributed, microprocessor based computer system for industrial process control. The advantages and disadvantages of a distributed system over a conventional large CPU system and over conventional analog instrumentation, operator communication, control board design and size reduction, interface and redundancy problems are introduced and expanded in Chapter 8.

2.
Planning and Executing The Computer Project

M. Robert Skrokov
Chief, Control Systems Division
Stone & Webster Engineering Corporation
New York, N. Y.

INTRODUCTION: ASSUMPTIONS AND THE PROJECT PLAN

Control systems design projects based on the digital computer are fundamentally not very different from those incorporating analog instrumentation hardware and control. Assumptions made regarding project goals, project organization and staffing, the need for a systems engineering analysis prior to scope definition, careful documentation procedures, adequate on-site and vendor factory hardware testing, tuning and personnel training procedures are required for both analog and digital systems projects. However, projects based on the digital computer require a more thorough and detailed systems engineering effort, monitored by a tight and competent project organization and followed by rigorous testing of both the hardware and software.

More elaborate training procedures for operating personnel are also necessary in order to realize projected goals. Digital systems projects require such efforts because of hardware and software complexity and because these systems usually control a larger number of process units in one control scheme than is possible with a conventional analog instrumentation loop.

The microcomputer, when used as part of a local closed-loop system, should be treated as just another block in the control loop as regards flowsheet design and specification production during the system design phase. It is, after all, functionally just another control instrument system, although it may (and most probably will) interface into a communications network with higher level computer hardware. Thus, the local microcomputer may or may not have a "front end' for this communications link. If the "micro" is part of such a system, it obviously requires a "front end" designed to interface with a CRT (cathode ray tube) screen and keyboard for manipulation of control commands and feedback response indication. If it is designed as a stand-alone controller, it may incorporate only a manual, or remote, means of set point manipulation (as shown in Figures 2–1 and 2–2).

One of the objectives of this text is to clarify and illustrate by example, wherever appropriate, those unit operations of the typical process plant that are best suited to computer control. Another—and perhaps as important—objective is to highlight any special organizational requirements that are unique to projects incorporating digital computer systems. Therefore, this chapter begins with a discussion of the requirements for a computer project organization

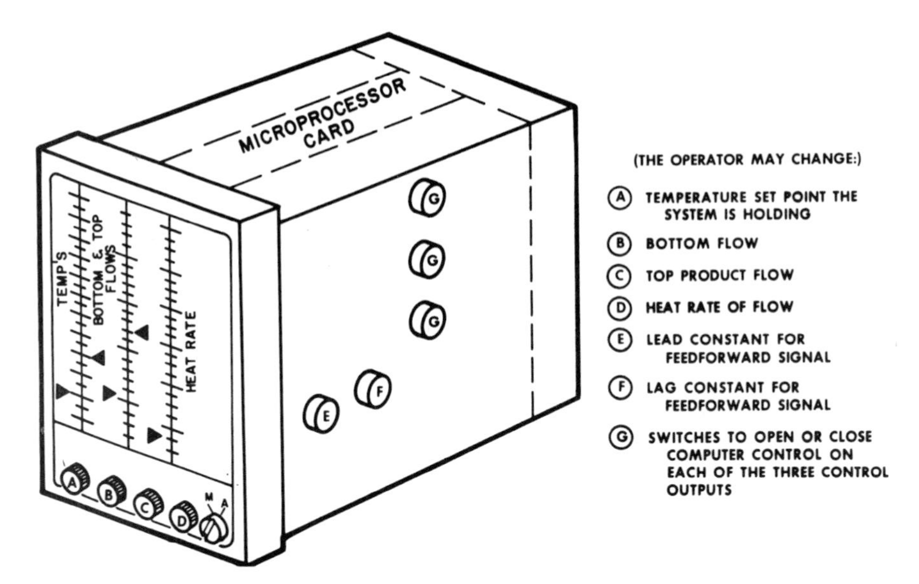

Figure 2-1. "Column optimizer" (Conceptual): local microprocessor computer.

and execution plan, lists some necessary clarifying assumptions and proceeds into a discussion of scope definition and system design. The importance of an early systems engineering analysis is discussed, as are some typical specification requirements for hardware and software. (Chapter 7 is devoted to a full treatment of specification design and content.) Some human engineering aspects of man-machine interface are introduced, as well as some typical electronic common mode rejection problems regarding interface with sensing and control instrumentation.

Final sections of the chapter begin with a discussion of operating personnel training requirements, start-up and systems staging procedures, and end with an outline of typical process unit operations in the chemical process industries for which economic

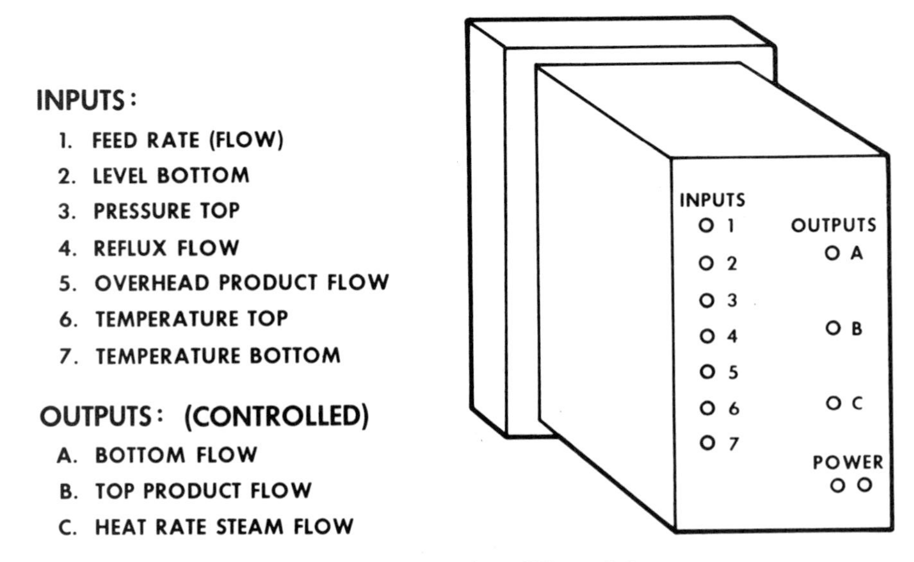

Figure 2-2. Rear view of Figure 2-1.

benefits of computer control have typically been demonstrated or are likely.

Chapter 2 is, therefore, an overview of some important assumptions, requirements and considerations for the planning, execution and completion of a successful computer control project in a chemical process plant. Problems relative to the use of systems based on the digital computer for process control are examined as an introduction to later chapters which provide detail on hardware and software systems and specification design. Problems relating to hardware and software design are thus, of necessity, discussed, but only as they are applicable to the successful implementation of *systems* using these elements. The design of internal computer circuitry, bread-boards, micro-"chips" or the elegance of various programming techniques and languages that make possible more efficient "bit-processing" within the machines are neither within the scope of this chapter nor the text in general. Rather, by presentation of proper systems design techniques and illustrations of typical process applications, it is hoped that the result serves as an aid to the reader in understanding the development and execution of computer projects in the process industries.

Assumptions and Requirements for the Successful Application of a Process Control Computer Project

Various problem-solving techniques (e.g., the method of Kepner and Tregoe[1]) and the first phases of the "scientific method," when applied to engineering projects, are the basis of so-called "systems engineering" methods. The systems engineering techniques, if properly applied during the conceptual phases of a computer project, provide a means of clear definition of project goals and serve as important tools in cost-benefit analysis to help ensure successful implementation and completion of the project.

Such techniques call for a clear and systematic outline of assumptions and problem statements prior to any attempts at solution. Therefore, let us first list some key assumptions particular to computer control projects in the process industries. This technique will lead (almost naturally) to pertinent questions and, eventually, to a set of concise statements regarding the scope, system design techniques, staffing requirements, etc., of a process industries computer project.

As noted in the previous chapter, a number of large process control computer projects failed to produce expected benefits and savings during the early history of computer control in the process industries (during the 1950's and 1960's), partly because of a low priority given to preliminary systems engineering studies prior to any attempts at final design and implementation. Therefore, an analysis or feasibility study effort during—or prior to—the early phases of such projects is vitally important to a successful installation, (as is a benefits-accrual phase at the completion of the project). Systems engineering methods should begin before the official "kick-off" date for the project and should continue to be used throughout the detail design phases, as well as for the final evaluation analysis at the completion of the project.

What, then, are some important assumptions concerning the design of a computer project that should be addressed prior to specification and systems design?

Before a cataloging of pertinent assumptions can be arrived at, a listing of generalized questions concerning these projects should be made. (A Socratic question and answer problem-solving technique is a useful method for pinpointing important aspects of a problem. It is being used here in order to outline pertinent background information on the design of process control computer projects discussed later in the text. The answers gleaned through the use of this technique give the reader a series of concise statements on project magnitude and scope, computer systems hardware and software specification, and project execution requirements.)

Questions on Key Process Control Computer Project Requirements

1. What organizations typically undertake process control projects utilizing computers in the process industries?
2. What is the type, magnitude and size of such projects?
3. What are typical time and manpower schedules for an olefins or other chemical, petrochemical or other process computer project?
4. What are the project staffing requirements for each type of organization listed in Item 1?
5. What types of computer hardware are being considered for the project (i.e., a large scale maxi, a mini and/or a micro? or a combination of these)?
6. Which software systems are used? What are their sources?
7. What should the failure mode be for effective process control (in case of computer failure)? Should the system revert to a redundant computer or to conventional analog instrumentation?
8. If analog instrumentation is used as back-up for the failure mode, what types of conventional instrumentation are used on such projects and what is the interface hardware to the computer?
9. What are the man-machine (human engineering) interface requirements for the particular project? What will be the conceptual control room and control board layout for the entire control scheme?
10. How are such systems tested and where? In other words, what are the vendor factory test procedures, where will the system be fully staged, and which parties are responsible for what tests?
11. What are the personnel training responsibilities for each organization involved (i.e., the ultimate owner, the systems organization, the contractor, etc.)?
12. What are the usual start-up procedures, problems and requirements?
13. How is documentation of hardware and software accomplished? What party provides which documents?
14. What are the spare parts requirements for the project? Is the project being constructed in a remote area of the world, such as the Algerian desert or the northern reaches of The People's Republic of China? (The location and the parts' plant accessibility must be considered here.)
15. What are the follow-up responsibilities of each organization listed in Item 1?

Assumptions on Requirements for Process Control Computer Projects

The following assumptions now arise from an examination of each key item given above.

1. *Organizations undertaking large computer projects in the process industries.* It is assumed that the project is being executed by one of, or by a combination of, these organizations:

- A chemical, petroleum or petrochemical process operating company or organization of companies.
- A contracting engineering organization.
- A control instrument/computer manufacturer.
- A digital computer manufacturer.
- A process-knowledgeable systems organization.

Some examples of each category* include the following.

- Large, and typical, U.S. operating companies able to undertake olefins complex (or other process) computer projects: Exxon Engineering and Research Company, the Texaco Corporation, Mobil Chemical Corporation,

*The Appendix at the end of this chapter lists names and addresses of some U.S. corporations in each category.

etc., in the petroleum/petrochemical industry; Dupont de Nemours & Company, Monsanto Company, American Cyanamid Company, etc., in the chemical/petrochemical industry.

- Overseas operating companies: Sumitomo Chemical (Japan), Showa Denko (Japan), Sonatrach (Algeria; oil and gas), Petrochim (Belgium), Ato Chimie (France), Shell (Netherlands), H.I.P. (Hemijska Industrija Pančevo; Yugoslavia), B.P. (British Petroleum), I.C.I. (Imperial Chemical Industries; Great Britain), etc.
- U.S. based contracting engineering organizations: Stone & Webster Engineering Corporation, Foster-Wheeler Corporation, the Lummus Corporation, M. W. Kellogg Company (division of Pullman), etc.
- Overseas contractors: J.G.C. (Japan Gas Corporation), Comsip (France), C.N.T.I.C. (Peoples' Republic of China) Davy Power Gas (Great Britain), etc.
- Process-oriented U.S. control instrument/computer systems manufacturers: the Foxboro Corporation, Honeywell, Inc., Taylor Instrument Company, Fischer & Porter Company, Fisher Controls, Inc., the Bristol Company, etc.
- Overseas control instrument/computer manufacturers: Siemens (West Germany), Yokogawa (Japan), Kent (Great Britain), etc.
- U.S. digital computer manufacturers: IBM (International Business Machines Corporation), D.E.C. (Digital Equipment Corporation), Control Data Corporation, Hewlett-Packard Corporation, Varian, Modcomp, Inc., Data General Corporation, etc.
- U.S. systems houses knowledgeable in olefins processes: Biles and Associates, C.A.T.C.O. (Control Automation Technology Company), EMC Controls, Inc., Setpoint, Inc., etc.

2. *Project size and magnitude.* Continuing with our listing of assumptions, the next point is one of size and type of the computer project usually undertaken in the process industries (Item 2 in the list of project requirements).

It is assumed that the magnitude of the computer project is in the order of 1 to 5 MM U.S. dollars and that it encompasses a complete petrochemical complex. An ethylene project is taken as the basis because it represents all unit operations usually found in a petroleum refinery or chemical plant, with the added technological complexity of the large cracking furnaces. Ethylene projects of the order of 1 billion pounds/year can require approximately $6 MM U.S. dollars, of total instrumentation, (including control boards, analyzers, etc.) A plantwide computer system may add an additional 3 million dollars, more or less, to this cost.

3. *Project schedule.* The next assumption concerns the project's working time schedule and typical "critical path" controls and "milestones." (See Item 3 of the project requirements list.)

As shown in Figures 2–3 and 2–4, a typical project dedicated to computer control of a petrochemical complex, or its equal, can run through two chronological years, from initial project approval and organization to final field acceptance testing of the system.

4. *Project staffing.* How should a computer design project be staffed? In other words, who are the usual participants in projects dedicated to process monitoring and control, and what are the responsibilities of each of the participants?

It is assumed that the project is assigned to a knowledgeable process engineering/control systems computer team in the central engineering department of an operating company, *or* that it is assigned to a similar team on the engineering contractor's staff, *and/or* that a systems project management team is assigned if a control instrument/computer manufacturer is given a portion of the project. Similarly, a systems management project team is assigned with a "systems house" organization, if a systems organization is used on the project.

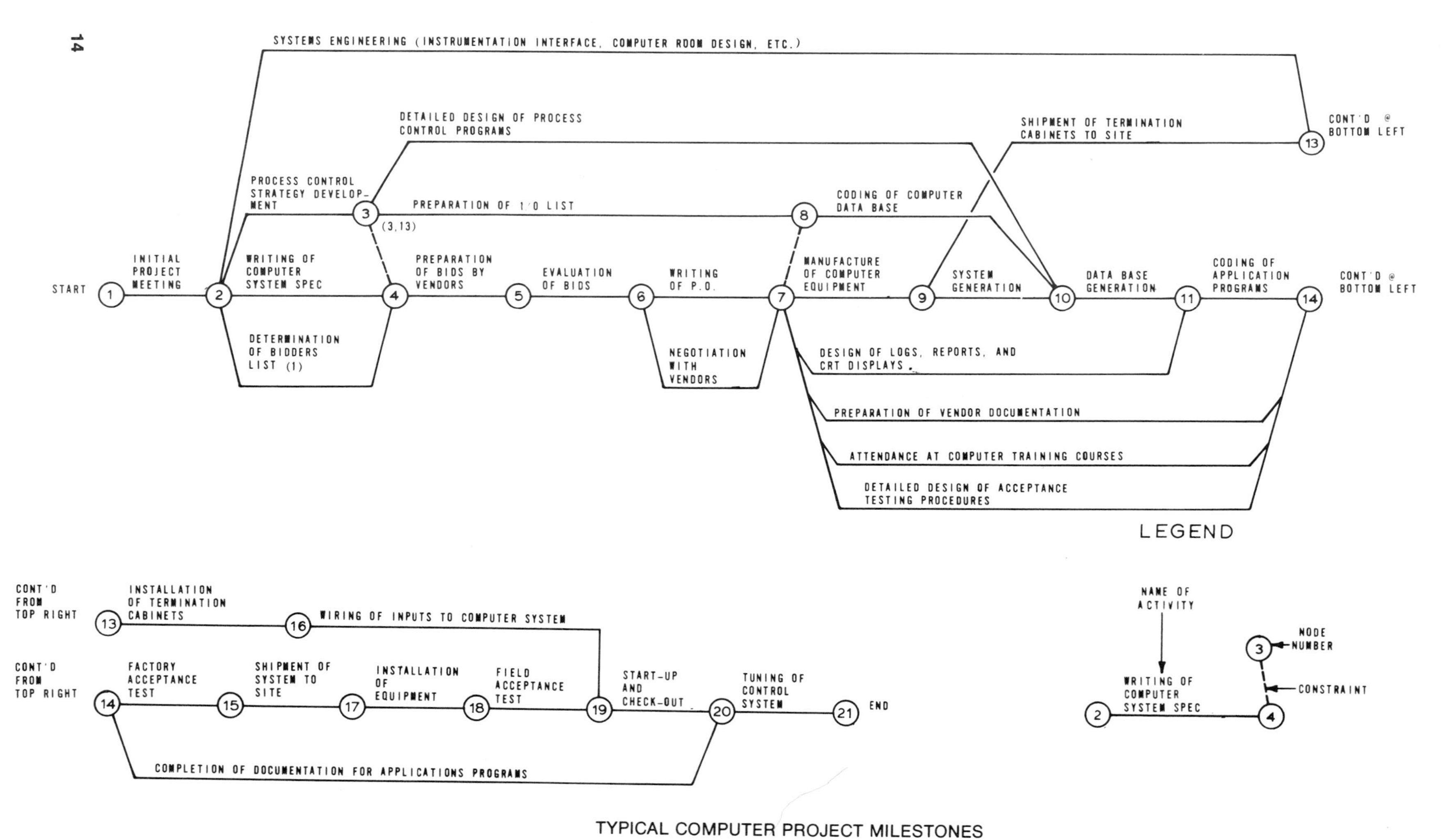

Figure 2-3. The critical path.

Below are typical milestones for computer project. Elapsed times are given with respect to the award of contract date.

Milestone	Elapsed Time
1. Completion of detailed hardware and software specifications.	3 months
2. Bids received from computer vendors.	5 months
3. Bids evaluated.	6 months
4. Purchase order sent to selected computer vendor.	7 months
5. Completion of manufacture of hardware.	12 months
6. Completion of programming.	18 months
7. Completion of factory acceptance test.	19 months
8. Arival of system on site.	20 months*
9. Computer system in place	22 months*
10. Field acceptance test.	23 months*
11. All inputs wired to computer	24 months*
12. Computer system in control of plant.	25 months*

*These dates do not take into account plant construction schedules. They are the earliest dates at which the milestones can be reached.

Figure 2-4. Typical project time schedule. (Elapsed times are given with respect to the contract award date.)

5. *Computer hardware.* What are some of the possible hardware configurations, based on proven system concepts? It is assumed that the computer or computer network is part of a process monitoring and control system and that, as such, it is tied to sensor based instrumentation. It is also assumed that the system operates in real time, and that the system is designed along distributed control concepts—i.e., that local control is accomplished by microprocessor based controllers and that some multi-level (or hierarchy) of control exists (similar to that illustrated in Figure 2–5).

A distributed control configuration using satellite computers, located in the field, eliminates the expense of wiring each signal to the control room and enables local optimization control of unit operations with none of the extensive software requirements of file transfer and command/feedback communication required in a non-distributed design. In this distributed network, the host computer also performs the alarming, logging and set point (regulatory) control functions. Data acquired by a satellite is sent over a high-speed redundant data link to the host. Set points calculated in the host are transmitted to the satellites via the data link. If the host should fail, first-level computer control continues. Additionally, two identical operating positions are located in the control room, thus allowing two operators to communicate simultaneously with the system. (Further discussion on the advantages and disadvantages of such systems is given in Chapter 8.)

In general, the host computer is of the

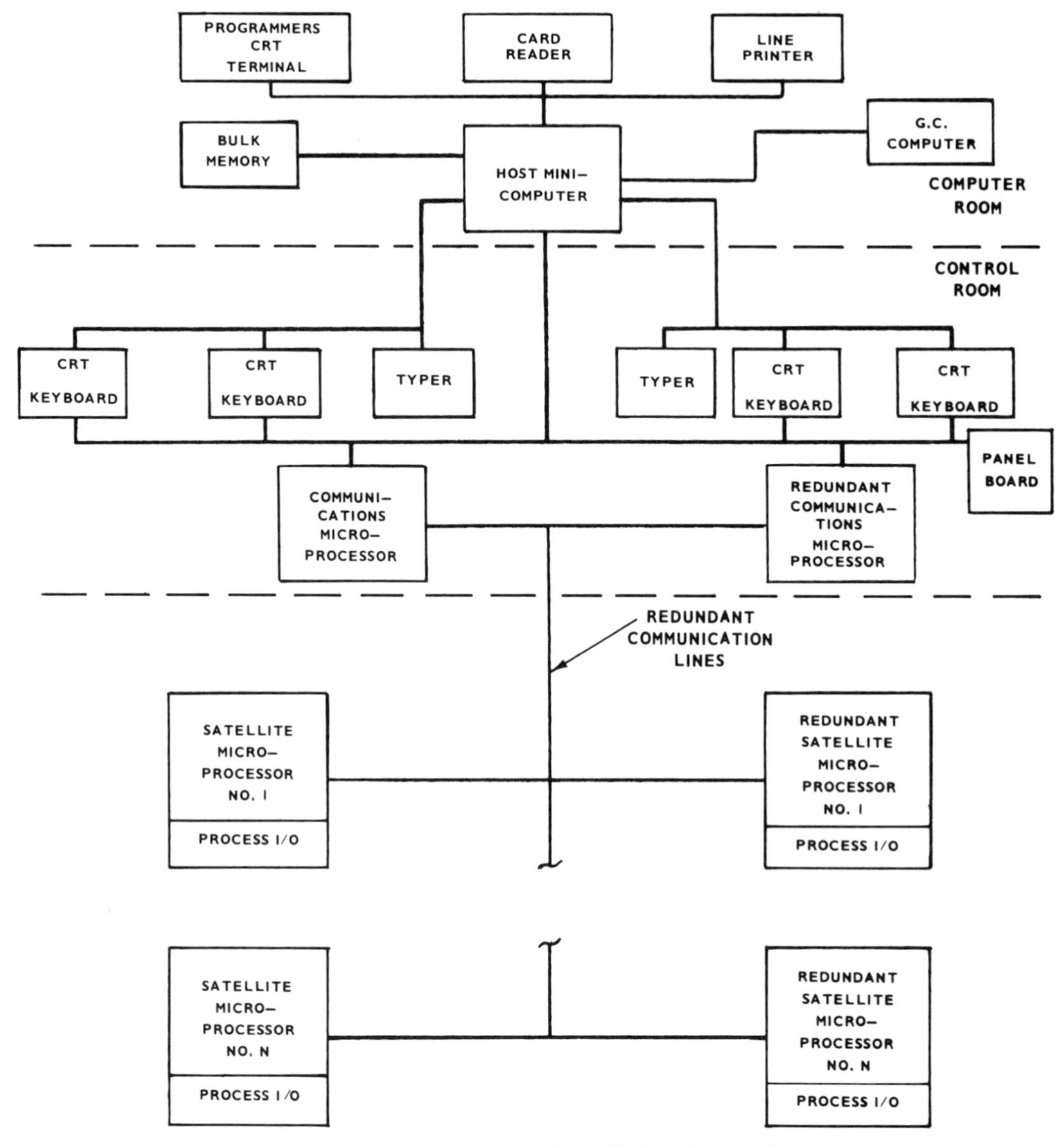

Figure 2-5. Hardware configuration for a distributed control system.

type provided by Foxboro in their "Fox-1" model, or the Digital Equipment Corporation Model PDP 11/34. Local microcomputers are envisioned as equivalent to D.E.C.'s LSI-11 or the Intel Corporation's 8080 system—the "i SBC 86/12 16 bit Single Board Computer" (see Figure 1–2 in Chapter 1). Other sources are described later in this chapter, under "system design."

6. *Software systems*. It is assumed that operating software systems are procured and specified as part of the system "package", that the software is a high-level, conversational mode, process control language based on Fortran IV or the newer Fortran VII, and that the system is supplied with all operating "standard" software (that is, that a standard data base be available for the processing of information within the host computer system; that alarm limits and summaries, commands, etc., can

readily be generated in the available software in conversational mode).

Conversational mode (or process control language "packages") is an important facet of computer projects undertaken in the C.P.I., since time, cost and manpower budgets for the computer system must mesh with the usual rigid schedule for the overall plant. Thus, development systems, requiring new language format and structure, cannot be tolerated. In fact, all computer hardware, software and interface systems should be predesigned and precoded. For example, any system under consideration for an active project should ideally have a data base incorporating all necessary addressing and reporting routines, ready for final generation in conversational mode.

Table 2–1 lists some process control languages usually used in real-time, process control systems applications.

Table 2–1 is intended only as a partial listing of some of the process control languages available from a number of industrial sources, namely, computer systems organizations (such as Biles and Associates, Copeland & Roland and Metromation, Inc.), instrument/computer systems manufacturers (such as EMC Corporation, Foxboro Corporation, Taylor Instrument Company, Honeywell, Inc., Leeds and Northrup Company, Hitachi Corporation and Westinghouse Corporation) and operating companies in the process industries (such as DuPont Corporation and Merck, Inc.).

7. *Failure mode.* As illustrated in Figure 2–5, the failure mode of the host computer is assumed to revert to a distributed network of local microcomputers, wherein each "micro" will continue to operate the local unit operation, either directly or through a localized group of equivalent analog control loops, in order to continue to achieve local goals, after failure of the host. For example, a demethanizer tower could be assumed to operate on local control to minimize reflux and energy used, while still maximizing product throughput and quality. Methods for the latter control scheme are described fully in Chapter 4.

8. *Conventional analog instrumentation back-up.* It is assumed that control systems designs of the 1980's will incorporate distributed systems of analog-equivalent digital instrumentation for large chemical projects similar to those currently on the market, such as Honeywell, Inc.'s "TDC-2000", Foxboro Corporation's "VIDEO-SPEC" and Taylor Instrument Company's "MOD III"; illustrations of each type are shown later in the section on systems design. These are all CRT screen display based systems, backed by digitized hardware equivalents of conventional analog

Table 2–1. Examples of Process Control Languages.

Biles and Associates' "AIM"
Foxboro Corporation's "IMPAC"
Copeland & Roland's "CRISP"
EMC Corporation's "ECL"
Fisher Controls Corporation's "PROVOX"
Honeywell, Inc.'s "BICEPS"
*Metromation, Inc.'s "FLIC"
Taylor Instrument Company's "POL 3"
Staff Corporation's "PRO"
Leeds and Northrup Company's "CODIL"
DuPont Corporation's "INFOTROL"
Merck, Inc.'s "AUTRAN"
Hitachi Corporation's "PCL"
Westinghouse Corporation's "PROTEUS"

*Licensed from the Union Carbide Corp.

instrumentation and linked by means of a communications network to the central display console(s).

9. *Man-machine interface.* It is assumed that the primary man-machine interface of the 1980's will be via CRT screen using the hardware described above and that chemical operators will have approximately three such screens available: one for *process overview* (equivalent to the older graphic panel concepts), giving flowsheet pictorial representations of the various subunits of the process, as well as full process flowsheet views, if required; one dedicated to *alarm summaries, unit reports,* etc.; the third dedicated to *status reporting* of individual loop controls, giving status of set point, process feedback and final control element position. (A discussion of the history of man-machine interface design, human engineering aspects and control room and control board design is given later in this chapter.)

10. *System testing.* It is assumed that the system is tested at three points during the course of the project. First, it is assumed that there must be a test by the vendor of all hardware and software at the vendor's factory, when he feels that his manufacturing process is completed and ready for this test. Second, a "staging" of the entire system may be required at the contractor's (or client's) option. Third, the entire system must be tested on the job site, tied to all process inputs and outputs. ("Staging" and other tests are described further later in this chapter and in Chapter 8.)

11. *Personnel training procedures.* It is assumed that all personnel involved in the project (from the operating company shift personnel to project and control systems design engineers) will take training at the computer vendor's facility in order to become familiar with hardware and software requirements for the particular system. This may take up to six people as much as a month each to learn. (Overseas companies sending personnel to U.S. vendors' facilities may require even more time and numbers of people because of the remoteness of some areas, so that more information must be absorbed during one stay.)

Operating personnel must learn various addressing routines—how to call for alarm and logging reports, how to call up various loop configurations for status, etc. Process superintendents and shift supervisors require the ability to call logging reports, whereas owner programming personnel must learn data base generation techniques, as well as the particular language structure and format, so that control loop or other configurations may be changed if required during start-up, or later, after the system has been finally accepted.

Maintenance personnel must become familiar with upkeep requirements and equipment replacement techniques.

The contractor engineering personnel must take courses relevant to system configuration and language and data based generation, and on format and system operation.

If a systems house process control language is being purchased as a separate package to be inserted into the operating system software, obviously all parties involved must become familiar with that language in addition to the one standard on the machine in question.

12. *Start-up procedures.* It is assumed, again, that personnel from the owner organization and systems people from the contract organization will be present at the job site, along with representatives of the computer manufacturer. Start-up may take all participants into a minimum six week duration for a large plantwide computer control project. If overseas, this period may be much extended, depending on the thoroughness with which the acceptance testing procedures are carried out prior to shipment and also on how well installation and check-out procedures are executed at the overseas location.

13. *Systems documentation.* It is assumed that the computer manufacturer will provide a certain minimum number of copies of complete software documentation (if a complex system has been designed and

various secrecy agreements have been signed, this will include a program listing).

The "systems house" will, similarly, provide documentation on the system if it was procured through such an organization.

The prime engineering contractor will provide manuals on the process, detailing all control systems concepts, and giving a hardware listing and any applications-software documentation on such software proprietary to his efforts.

14. *Spare parts requirements.* Spare parts requirements are a function of the following:

- Computer, analysis instrumentation interface and back-up hardware complexity.
- Wear and tear expected in the climate destined for the equipment. (For example, sand storms in the Algerian desert, if such is the destination, should be allowed for, as sand has been known to penetrate into control room atmospheres.)
- Remoteness of the geographical areas. (Remote areas require that a higher percentage of spare parts be allocated. Most contractors allocate, as a minimum, 20% or higher spares for remote areas of the world, as against 10% in the U.S. or Western Europe, for example.)
- Personnel training. (Untrained installation personnel may damage an excess percentage of replacement parts during servicing procedures. Untrained start-up or operating personnel may, similarly, create a parts shortage.)

The more complex the equipment package is, the more "complete" the spares package that may be required. For example, continuous, automatic analyzer sampling and programming systems require delicate handling by trained personnel if such systems are to operate reliably and continuously during the transmission of meaningful product or stream composition data to the local, or host, computer. In remote areas, complete spare analyzer sample conditioning or programming assemblies may have to be provided as part of the system "package."

15. *Follow-up responsibilities.* The prime contractor for the computer project may have the responsibility, by contract, to continue to debug the system long after the initial start-up phase, if the system is complex or extensive (if, for example, it is a plantwide system based on computer control of an entire olefins complex). In many instances, the prime engineering contractor on an olefins project has probably supplied the basic process technology, with guarantees, and must run plant evaluation tests long after start-up, in order to prove that the facility meets certain stipulated product throughput rates in both quantity and quality. In doing this with a computer project, he would most likely be required to use the computer for data gathering and be required to show that the computer hardware and software system is operating in a reliable and proven fashion, if the data being presented by the system is to be useful in the plant acceptance phase.

Any sub-contractors (such as systems organizations that have supplied a portion of the hardware, software or process technology) would likewise be required to meet their sub-contract requirements at this point and thus be required to field-prove their portion of the system, "meshed" into the overall plant control scheme. This might require the debugging, redesign or replacement of certain sub-systems long after the initial start-up phase has passed.

PROJECT SCOPE

From the listing of assumptions previously developed and analyzed, a broad-based project scope document may be written that incorporates all of the important considerations and requirements for a large-scale process control computer system. Such a project scope document should detail what is expected to be achieved by the project. It should indicate the overall purpose of the project; i.e., what plantwide goals are expected to be met. In doing this, local targets should also be clearly indicated. For example, what are the tangible and intangible goals that might be expected to accrue from the project?

Tangible Benefits

Increased plantwide production rates. These are usually achievable by setting computer system targets closer to unit capacity limits. (It is known, for example, that chemical operators will, if allowed, set distillation column reflux rates excessively high in order to meet product yield requirements, since with ordinary instrument control systems, it is difficult, if not impossible, to judge close operating targets while allowing for all of the process and ambient variables that mitigate tight operation to tower capacity.)

Improved product yield. For a given feedstock slate (or mix), the computer may be able to improve the product mix. That is, it may operate so as to increase the throughput of an expensive (or desired) product, while reducing the throughput of a less desirable (or less expensive) product. This yield production and control aspect of computer systems technology is becoming more important as these systems begin to spread across widely divergent geographical areas. Obviously, Gulf Coast (U.S.) feedstocks vary radically from those of the Algerian desert or those of the northern reaches of the Peoples' Republic of China.

Improvement of product quality. Product quality may be improved with a computer system through closer control of on-line product composition by computer-monitoring and control of the product analyzer instrumentation.

The maintenance, calibration and operation of on-line analysis instrumentation has always been troublesome, although calibration, operation and maintenance of gas and liquid Chromatograph analysis instrumentation, infrared analyzers and the like are essential to the realization of composition targets for desired products in any plant. Computer control allows for automatic calibration (to known chromatogram standards), automatic setting of elution times for each component of interest, and proper maintenance through the elaborate alarm reporting systems that most large computer systems utilize. (A full discussion of computer controlled analyzers is given in Chapter 6.)

Improved use of raw materials. This is also possible with the computer yield control program. Not only is the product slate improved, based on the given feedstock slate, but the energy saved may be appreciable through decreased use of fuels such as steam or fuel oil. In catalytic processes, catalyst usage may be drastically improved, as well.

Reduction in equipment maintenance. This is possible through reduction in emergency shutdown operations through reduction in wear and tear on all elements of the control system (control valves, etc.) and through the possible use of planned, or preventative, maintenance techniques made possible by the elaborate reporting and display systems of the computer.

Improved automation. A reduction in the number of operating personnel may also be a possibility. Each shift operator is able to scan and monitor an increased number of plant units with the computer than with conventional analog instrumentation. The combination of a larger number of process variables (such as temperature, pressure and flow rate) into more sophisticated control loops based on, for example, feedforward/feedback-corrected algorithms and other such advanced control strategies such as adaptive control, moves the operator upwards one or more steps in the control hierarchy ladder for the plant. This increased sophistication of control places fewer operators at the control console(s) and requires that each such operator now maintains the plant at management-type objectives and within closer target limits for each objective than is possible with conventional analog instrumentation. (Further discussion on console design and man-machine interface—the human engineering aspects of such design—is detailed later in this chapter.)

Capital cost savings in original equipment. A computer system may eliminate the requirement of an added unit through the on-line maintenance of that unit. For example, a stand-by cracking furnace might be eliminated in an ethylene complex through the use of zone, on-line, de-coking procedures operated by means of the computer system. Ethylene cracking furnaces produce coke build-up in the hydrocarbon tubes during the life of a given "run." Therefore, one of (for example) eight furnaces is usually designated for stand-by operation while one of the other seven furnaces is de-coked or cleaned off-line. The elimination of one such furnace in an ethylene complex may easily pay for the cost of a large-scale computer system, as such a furnace may approximate a cost of 10 million (U.S.) dollars, including controls and accessories.

Intangible Benefits

Improved accounting and engineering data analysis and gathering. The computer has the ability to convert direct variables such as flow rate units, temperature units and pressure units to the desired accounting units and provides a much more accurate date base and record-keeping system for such vital data. The computer also may provide more accurate data on indirect process variables, such as product composition and quality, through analysis of available data from on-line analysis instrumentation.

An increase in process knowledge. This may accrue from a number of phases in the computer project. Early systems engineering studies necessitate a clear definition of the process and all of its idiosyncrasies, thus partly improving the existing base of process knowledge. Later, as the data-gathering process begins (and continues), a historic base is built upon which further process control strategies can be based, thus also improving process knowledge.

Automatic start-up and shut-down routines, sometimes only possible with a computer, may provide tangible savings and benefits as well as intangible benefits. Faster start-up times for cracking furnaces may, for example, produce fuel savings as well as longer furnace run-times between shut-down sequences because of the automation procedure, while also increasing the total cracked product throughput rate.

Communicating Project Goals

Once technical and economic goals are clearly defined, they must *also* be effectively communicated in all relevant directions within participating organizations, during both the design and implementation phases. Desired project objectives must be understood by operations management as well as by line (shift) operators if these goals are to be realized.

Project scope can only be set and fully determined by the ultimate user of the computer system (the operating company or client). Tangible and intangible benefits and project purpose (or intent) lie primarily in the province of the ultimate user. The ultimate purpose of the project may involve proprietary sales strategy, economic goals or technical objectives which may not be revealed without jeopardizing the position of the operating company within its market sphere.

System designers, located in the contract engineering organization, the computer manufacturers' organization or the "systems house" organization may not (and probably should not) be privy to the ultimate project purpose. They should, however, be given a full scope document on the project, and it should outline all important aspects and requirements for control of the process in question. It is thus obvious that the scope document for the computer system should be (and most usually is) written by the ultimate user of the system.

A typical computer system "scope" document might be written as indicated in Chapters 5, 7 and 10.

THE CRITICAL PROJECT MILESTONES

The critical milestones, Figure 2–3, is designed from the ultimate user's point of view (the owner or final client) and from the prime contractor's point of view (the engineering contractor, who may be assigned the full engineering task for the project, the process, and the computer and control systems design).

As shown in Figure 2–3, the key functional activities in the critical path diagram are the following.

1. *Initial project meeting*. Initial project assignments are made and a kick-off meeting is held to identify participants, to set (or clarify) project goals and to make assignments and issue time and manpower schedules.

2. *Systems engineering*. The systems engineering activity parallels key sub-activities between nodes 2 and 13, consists of systems analysis and study, instrument interface design, control room and console(s) design, etc.

2–4. *Computer system specification design*. This consists of the drafting and writing of the computer system specification (details of which are given in Chapter 7 and later in this chapter).

2–3. *Process control strategy development*. Development of strategy for all units may take 3 weeks (early finish), if solidified from previous experience, and 13, or more, weeks (late finish), if not.

3–10. *Design of process control programs*. This is the software "umbrella." It covers all activities, from conceptual flowcharting and logic diagram construction to final coding of the data base, and may take over 10 weeks of project schedule time, depending upon the complexity of the project.*

*Note: A large olefins project may require up to 6 months.

3–8. *I/O list preparation*. The input/output list documents all information coming to the computer, as well as all outputs (whether command or display) emanating from the computer. Figure 2–6 illustrates one page from a typical I/O listing for a petrochemical process control system. Duration is approximately 5 weeks or longer because of other activity interface such as bid evaluation or purchase order writing.

8–10. *Coding of computer data base*. Process alarm summaries and report designs, operator interface techniques, methods of addressing specific control loops, changing control modes, etc., are all part of the data base. This effort may take 6 weeks to complete because of constraints such as system generation (9) and manufacture of hardware (7).

10–1. *Data base generation*. The implementation of the coded data base onto the purchased hardware may take, as a minimum (with proper training on the particular machine), one week of the project schedule.

7–11. *Design of logs, reports and CRT displays*. The custom design of desired reports for hard copy and for CRT screen reporting may take 10 weeks of elapsed time and may be started immediately upon completion of purchase of the hardware system. These are reports peculiar to the industry and client for which the system is intended. Figures 2–7, 2–8 and 2–9 respectively illustrate three such reports: the "status report" (printed every hour, or on demand); the "data log printout," a report produced once per shift (or on demand); and the "alarm scan printout" (a typical report for an ethylene furnace control system is shown).

11–14. *Coding for applications programs*. Applications programs are those software systems devoted to, and specifically designed for, the implementation of process control algorithms, optimization routines, custom logs, reports, etc. This activity is a key one for control system strategy generation and is thus

ETHANE
NAPHTHA FURNACE

TAG NO.	POINT IDENT.	DESCRIPTION	PURPOSE	SOURCE	TERMINATION NO.	SCAN RATE	ENGINEERING UNITS	ALARM LIMIT		RANGE		SIGNAL TYPE	MUX ADDRESS	TRIGGER LIMIT	
								HIGH	LOW	TRANSDUCER	ENGINEERING			HIGH	LOW
FT-102 A-G	322	FLOW TRANS	NAPHTHA FEED	FIELD	A-6	1/SEC	BPH	–	–	4-20MA	0-200%	CURRENT	A-6	-	-
FIC-102	323	REMOTE SET CONTROLLER	NAPHTHA FEED	PANEL	A-7	1/SEC	0-100%	–	–	4-20MA	0-100%	CURRENT	A-7		
ZT-102 A-G	324	NAPHTHA VALVE POS.	NAPHTHA FEED	FIELD	A-8	1/SEC	0-100%	–	–	4-20MA	0-100%	CURRENT	A-8		
FT-103 A-G	325	FLOW TRANS.	DILUTION STEAM FLOW	FIELD	A-9	1/SEC	LB/HR	–	–	4-20MA	0-25K	CURRENT	A-9		
FIC-103 A-G	326	REMOTE SET CONTROLLER	DILUTION STEAM	PANEL	A-10	1/SEC	0-100%			4-20MA	0-100%	CURRENT	A-10		
TT-119 A-G	327	TEMP TRANS.	MODULE 1 COIL OUT TEMP	FIELD	A-11	1/SEC	°F	1950		4-20MA	400°F-2000°F	CURRENT	A-11		
PT- A-G	328	PRESSURE TRANS.	MODULE 1 COIL INLET PR.	FIELD	A-12	1/SEC	PSI	50		4-20MA	0-100%	CURRENT	A-12		
AR-	329	ANALYZER	CRACKED GAS EFFLUENT	FIELD	A-13	1/SEC				4-20MA	% RATIO	CURRENT	A-13		

PAGE __ OF __

Figure 2-6. A typical computer input/output list.

usually designed and implemented by either the ultimate user of the computer system or the prime engineering contractor. A portion of this activity may also be provided by a process-knowledgeable systems organization. Twenty-four weeks duration for this activity is not unreasonable for a large petrochemical complex.

14–15. *Factory acceptance testing*. This activity is important for the de-bugging of software and the checking of all the included electronic systems (sample loops of analog control instrumentation

STATUS REPORT (Printed every hour - or on demand)

DATE, **TIME** **FURNACE NO.** **MODE** (Cracking
(Decoke
(Standby, etc.

Hydrocarbon Feed Rate ____________

Steam Feed Rate ____________

Severity Function ____________

Coil Outlet Temp. ____________

Coil Outlet Pressure ____________

Selectivity Function ____________

Steam Production ____________

Type of Hydrocarbon Feed ____________

Latest Maximum Tube Metal Temp. ____________

Estimated Time to End of Mode (Hrs.) ____________

Figure 2-7. A typical computer status report.

DATA LOG PRINTOUT (Once per shift or on demand)

DATE	TIME	FURNACE NO.	MODE	
			TOTAL	
		Furnace On-Line Time	______	
		Exchanger On-Line Time	______	
			TOTAL	AVERAGE/LAG
		Hydrocarbon Feed	______	______
		Steam Feed	______	______
		Steam Production	______	______
		Fuel Consumption	______	______

Optional Print Out of all Inputs to Computer or any selected list, in same format as I/O list titles

Figure 2-8. A typical computer data log print report.

also are usually wired and tested with the system at this stage). A "staging" test is sometimes carried out at the contract engineer's facilities or at the prime systems organization's facility. The "staging" process provides a point in the test cycle at which all relevant hardware and software systems can be linked and operated as a unit. (Chapter 8 covers this point in further detail.)

17–18–19. *Field installation and acceptance testing*. This activity takes us to the final linkage of the computer system into the process control chain. This is the mechanical and electrical installation phase of all equipment, when input analog process variable signals are wired into the system, as are the output control signals to the control valves and other final control elements. The control room installation of wiring and peripheral equipment is also completed at this point, for a minimum duration of 4 weeks. Acceptance testing is begun and carried out during this stage. (Acceptance testing as outlined by the Instrument Society of America Standard RP 55.1 is discussed later in this chapter.)

DATE: 9-26-77 SCAN TIME: 10:30

TAH 101A	HIGH FLUE GAS TEMP. FURNACE A
ALARM	10:11:30
RESPONSE	10:12:15
ASHL 102G	HIGH FLUE GAS OXYGEN FURNACE G
ALARM	9:53:20
RESPONSE	9:54:30
CLEARED	10:30:00
TAH 110F	HIGH TLX TEMP. OUT FURNACE F
ALARM	10:10:17
RESPONSE	10:12:15
PAL 101E	LOW NAPHTHA FEED PRESSURE FURNACE E
ALARM	8:40:14
RESPONSE	8:42:12
PDAH 102F	HIGH DELTA P MODULE 1 RADIANT SECTION FURNACE F
ALARM	8:53:12
RESPONSE	8:54:13
CLEARED	8:55:17

Figure 2-9. A typical computer alarm scan printout.

19–20–21. *Start-up, check-out and tuning of the control system.* The final phase of the project involves the de-bugging of software, the tuning of control algorithms and the checking of electronic circuitry for the detection of poor grounding or wiring errors.

Of course, the training of all operating and start-up personnel has proceeded in parallel with all these efforts—during activity nodes 7–14 and probably continuing beyond this point. Similarly, the documentation of applications programs was begun at activity node 14 and is presumably completed before final tuning at node 20.

STAFFING THE PROJECT

The success of most human effort is usually a function of the caliber of person(s) undertaking the task. Given the high-technology complexities of a project dedicated to a closed-loop process control computer system, the staff structure, personnel qualifications and duties required of participants are vitally important to the success of the venture (particularly one intended for a world-sized petrochemical complex). It seems appropriate, therefore, to detail the organizational staff structure required for such work, along with the qualifications and duties of staff members.

A generalized staff structure for such a project might be constructed along the lines indicated in Figure 2–10.

The Computer Project Staff

The Project Manager

The computer project manager should be an individual with up to 15 to 20 years of petrochemical process control systems design experience, largely centered on projects involving sophisticated control systems hardware and software. This background should include a stint in the management and/or design of projects in

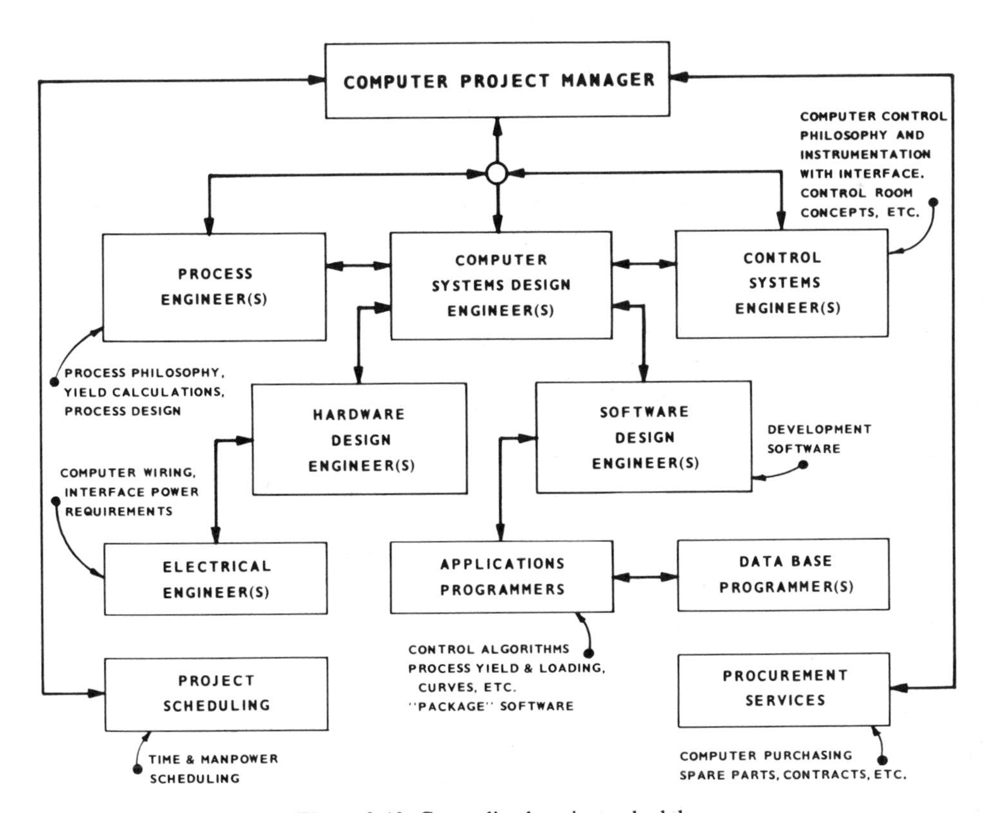

Figure 2-10. Generalized project schedule.

the chemical process industries, wherein electronic instrumentation and process analyzers were interfaced with a real-time, closed-loop, process control computer; it should also include the supervision of the design of the systems hardware and software and the instrument interface hardware and wiring and experience in the acceptance, field testing and installation phases of such projects.

The project manager is the chief executive of the project. As such, he must coordinate all contributory efforts and help set the project guidelines and policies on technical as well as production (and possibly financial) aspects. He must help design manpower allocation and time and cost budgets and schedules, and, in doing the latter, he must be responsible for the effective execution of the work within the budgets and schedules set for the project.

He may also be called upon to contribute to some of the technical aspects of the project. For example, he might help set the control strategies for some of the units, he might help set the software design philosophy or he might assist in the purchase of outside "packages" for some of these functions.

Once philosophies are set, he might be called upon to arbitrate on changes in technology, method of work execution, etc. In essence, he is the project leader, responsible for goals and schedules, and must therefore delegate assignments and follow work progress to ensure that the project is completed technically, and within these budgets and schedules.

If the computer project is but one phase of a large plant design, he must also execute his responsibilities so that the computer portion meshes properly with the rest of the plant—through the design phase and on into field installation and start-up.

The Process Engineer

The process engineering input to petrochemical plant computer systems design requires a staff of heavily experienced personnel, each with 10 years or more of background in process unit design and product yield calculation methods, as well as the usual experience in economic, heat and material balance calculations and extensive field experience in unit operation. This field experience should include process guarantee test runs on complete plants (as current computer projects must, in many instances, be included in such plant guarantees, if improvement in plant operation is claimed for the computer portion of the complex).

The process engineer(s) may be required to contribute to plant modeling and off-line computer simulation of sections of the plant in order to help develop optimization level (or other) control strategies and to predict or prove product throughput and quality guarantees on the particular process. The latter effort usually involves an off-line LP (linear programming) optimization mathematical solution for a matrix of feedstocks, which may later serve as the basis for an on-line optimization control of product throughput or plant profit for the process control computer.

Given a feedstock slate, the process engineer is charged with the design of the plant product mix, based on a set of desired market strategies. He is also usually responsible for setting the desired flow rates of utilities and hydrocarbon streams for the plant, as well as for other process variables such as key operating temperatures and pressures.

The Control Systems Engineer

Generally, the control systems engineering job description may be broken down into two basic functions—one requiring the design of control philosophy for all aspects of the plant and the other requiring the design of detailed hardware (instrumentation) for the execution of the control philosophy. The first effort requires a process control flowchart design (see Figure 2-11 for a typical process control diagram format) and description, illustrating the control strategy for each unit operation in the plant. The

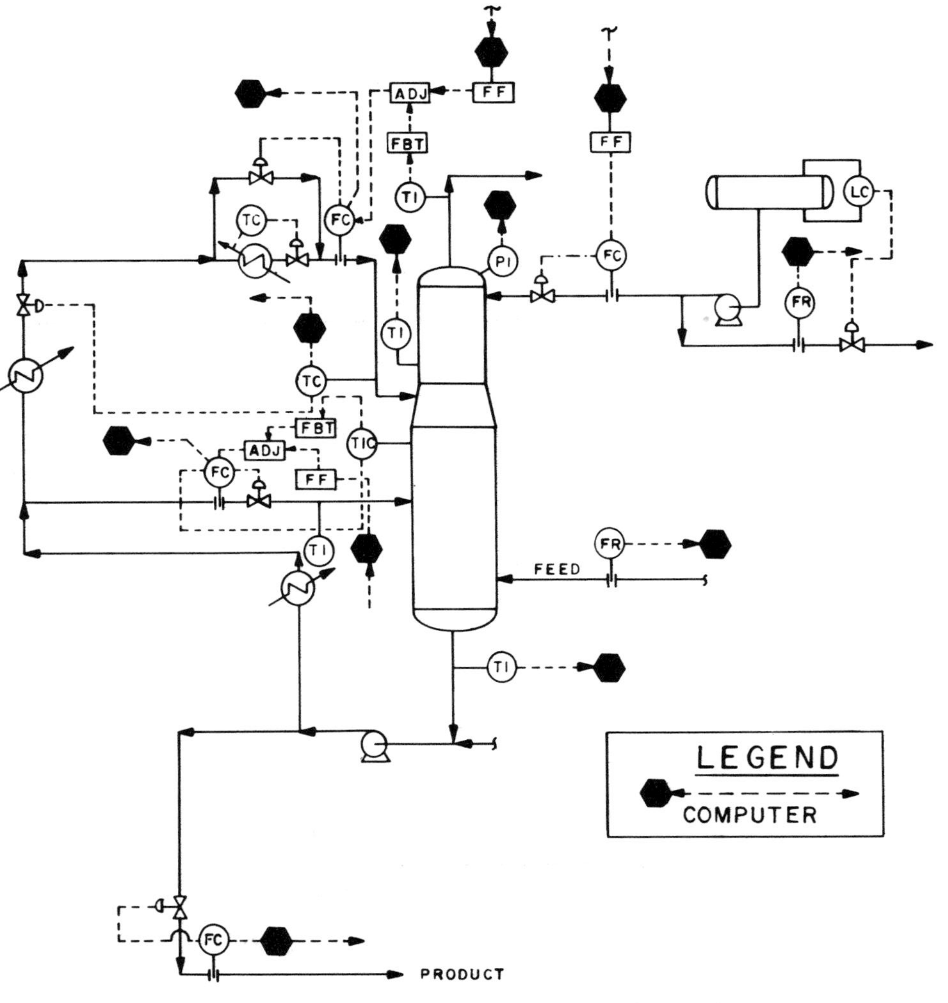

Figure 2-11. Flowsheet indicating computer input/output (PCD).

second work effort involves the production of specifications, data sheets and purchase requisitions (and the supervision of detailed installation drawings) for all control and monitoring instrumentation, control valves, analyzer instrument systems, etc.

In the case of a computer project, this engineering function helps set the computer control philosophy for the process in question and designs the instrumentation interface, as well as the control, the computer console(s) (see Figure 2-12 for a typical console design) and panels and the computer and control room layouts.

This effort requires a detailed knowledge of physics, for the control and monitoring instrumentation, and mathematics, for key control schemes in the plant. It may also include the modeling and simulation, on computer, of the more difficult control strategies for the process. (This simulation may be used later for the training of operating personnel on the computer system.)

The control systems engineering effort, therefore, requires a staff of experienced engineers and designers capable of producing a total control system, married to the computer (and which will operate on computer failure), while also contributing to the computer system design philosophy, specification, modeling and simulation tasks required to ensure success on closed-loop control for the computer system.

The control systems staff should be headed by a lead (project leader type) control systems engineer whose background covers extensive process control systems design, supervision and start-up experience in conventional electronic and pneumatic instrumentation, control valves, process analyzer systems and other such hardware, on large-scale C.P.I. projects, preferably having incorporated closed-loop computer systems. This experience should be broad-based in the process industries, covering a minimum of 10 years and including the field testing and start-up of such systems.

The staff under the lead control systems engineer should be a group of specialists, each one (ideally) experienced in some facet of control hardware specification and theory. (A process analyzer specialist is usually required to ensure the proper design of such instrumentation and the sampling systems that are part and parcel of such systems. Similarly, knowledgeable instrument interface designers and control valve modeling and simulation specialists are required as well.)

The Computer Systems Design Engineer

The computer systems design engineering function is the "workhorse" function of the computer project. The lead computer systems design engineer must help set the computer control philosophy and supervise final computer project system specifications for both hardware and software; that is, he must help determine the final hardware configuration that is to be used on the project (will it be a single CPU, a dual, redundant CPU or a totally distributed network system of computers?). (Refer to Figure 2-5.)

Ultimate decisions on the configuration of the system are, of course, usually made by the ultimate user of the system—the process operating company. If the computer systems design engineering function, as described here, is on the staff of the ultimate user, then it may be consulted in the choice of the projected configuration for the system. The final configuration may change during the bid evaluation procedure, as new ideas are presented by vendors and are accepted or rejected.

The computer systems design engineer is the key technical consultant on any contacts with computer vendors. He must lead, or, at the very least, heavily partake in, technical negotiations and evaluation meetings with computer vendors. Once the specification of the system is completed, he and his staff must join in almost all of the activities of the project. Referring to the "critical path" (Figure 2-3) for the project, the computer systems design func-

Figure 2-12. Computer console for a process monitoring and control computer system. (*Courtesy of EMC Controls, Inc.*)

tion is involved in all of the activities shown except for activity 3–10, design of process control diagrams.

Once the hardware is selected, the software staff must begin to work on the applications programming and the building of the data base for the system. The staff must generate the I/O list (Figure 2-6), code the data base, generate the data base and design specific logs and reports for the system (Figures 2-7, 2-8, 2-9) and CRT displays.

This group must, therefore, be staffed with personnel heavily experienced in proven hardware and software dedicated to real-time process control. Applications programmers must have had experience with process control language development and usage as listed in Table 2-1. Data base programmers must be Fortran IV or VII experienced and have a machine assembly language background on IBM or D.E.C. hardware, as an example.

Hardware systems designers should have up to 10 years experience on the design, and on the configuration of process control computer systems centered on hardware such as D.E.C.'s "PDP-11 Series," Foxboro's "Fox 1," Modcomp's "Classic" and EMC's "EMCON D.," etc.

The Electrical Engineer

The electrical engineering function here is concerned primarily with the power distribution requirements of the total control system and the cabling and wiring of the instrument interface to the computer. Working in close cooperation with both the control systems engineer and the computer design engineer(s), and given the various signal levels (voltage, current, impedance) that must be considered in the design, a compatible power and wiring system must be developed that will properly interface to the power distribution and wiring system of the rest of the plant.

Considering the job aspects as generally outlined above, the electrical engineer(s) on the project should have up to 10 years experience in electrical systems design on control systems projects in the process industries. The experience should include designs incorporating multiplexing systems of thermocouple and signal wiring, analyzer instrumentation, programmers, etc. This group should also be knowledgeable in "common mode rejection" circuit design on the control instrumentation hardware used for interface with the computer, and should have some experience with the communications signal levels used in the particular computer network.

Interactivity Between Disciplines

Figure 2-10 (p. 25) illustrates interactivity between the disciplines involved. Activity lines are given, but these lines are by no means complete. The diagram is a generalized (conceptual) pictogram and not a definitive line-and-staff organization chart.

It may be obvious to the reader, from the above discussion, that there is much interactivity between disciplines on such projects, as well as some gray areas of responsibility and overlapping of the various functions depicted. As an example, a "split of work" between process engineering and control and computer systems engineering might be illustrated as shown in Table 2-2, but this too might vary, depending on whether these disciplines were part of the ultimate user's organization, the contractor, the computer manufacturer, the instrument/computer manufacturer or the "systems house."

Table 2-2 does, however, serve as an illustration of some of the important gray areas of responsibility of such projects.

Definitions of Tasks

Interface with vendor. Meetings and correspondence; vendor schedules.

Process strategy. How computer control should be applied to improve plant performance (yield, etc.); feedforward loading curves; LP optimization models; benefits and savings.

Process control philosophy. Implementation of process strategy; development of control strategy; PCD's; commputer control diagrams; simulating control systems; etc.

Data base programming. Filling out the data base input forms.

Developing and testing of new process strategies. Developing new strategies not supplied by the computer vendor (such as distillation column feedforward techniques

Table 2–2. Computer Control Project "Split of Work" (CSD/CSE Process).

TASK*	PRIMARY RESPONSIBILITY**†
Interface	CSD
Process strategy	P
Data base programming	CSE
Data base programming	CSD
Development and testing of new process strategies	P
Creation of logs, reports and displays	CSD
Acceptance test procedures	CSD
Staging of computer system	CSD
Project management	CPM
Field start-up	CSD
Systems engineering	CSE

*See "Definitions of Tasks," below.
**CSD = control systems design function.
CSE = control systems engineering function.
CPM = computer project manager.
P = process engineering.
†Tasks for which CSE and CSD have *primary* responsibility, with input from process and *vice-versa*.

and furnace optimization) and the testing of these strategies on the computer.

Creation of new logs, reports and displays. Specifying logs, reports and displays by filling in forms or by other methods used by the display builder; report generation programs.

Staging of the computer system. Residence at staging site for loading the programs into the system; interface with the vendor.

Acceptance test procedures. Writing specification for field and factory acceptance testing.

Project management. Lead responsibility for scheduling and man-hours for the computer portion of the overall project.

Field start-up. Residency on site for 6 to 12 months.

Systems engineering. Sizing of computer room, electrical and air conditioning requirements; instrumentation interface; computer signal input/output lists; etc.

Personnel Training

There are a number of training courses available from computer manufacturers, instrument/computer manufacturers, systems houses, etc., that should be acknowledged and used throughout the course of a computer project.

The list in Table 2-3 indicates some typical training courses available from an instrument/computer manufacturer. Similar courses are given by the large computer manufacturers. Many are designed for various levels of experience, while others are aimed at specific functions such as electronic maintenance or assembly language programming.

During the course of the computer pro-

Table 2-3. Typical educational services available from instrument/computer manufacturers.

INSTRUMENT COURSES
Fundamentals of mechanical and pneumatic instrumentation
Electronic measuring devices
Electronic maintenance
Instrument systems maintenance
PROCESS CONTROL COURSES
Fundamentals of control
Instrument and control technology
Control systems engineering
Distillation control systems
Industrial pollution control systems
Energy conservation control systems
COMPUTER SYSTEMS COURSES
Introduction to process computers
Advanced computer process control
FORTRAN IV programming
BASIC programming
Digital logic fundamentals and applications
"PCL"* LANGUAGE
Assembler language
Programming
Operational maintenance

*"PCL" here denotes any of the manufacturers' process control languages listed in Table 2-1.

ject, the ultimate user should expose his plant operating shift personnel, design and project engineers (and managers) to particular courses intended for that level on the computer chosen for the project. Similarly, the contractor's engineers and key design personnel should attend the same courses as the job progresses.

Training on the particular machine (the computer chosen for the project) is a key ingredient in the success of such projects and should be included in the manpower, time and cost budgets and schedules for the project at its inception.

DESIGNING THE SYSTEM

The previous sections of this chapter were devoted to a discussion of some aspects of the organization, planning and execution phases of the computer project. In this section, the "production" or detail design phase of the project is outlined.

The process of design should begin with an analysis or conceptual effort prior to any attempts at the production of detail systems specifications. In that sense, a computer project is not unique. Any control systems design project must begin with the production of conceptual flowcharts outlining the control strategies to be used in the system (see Figure 2-11). A typical such flowchart is the PCD (process control diagram).*

The PCD is a document used by a number of engineering contractor organizations in order to delineate, in a clear and efficient manner, all of the control strategies designed for a particular unit operation. For a process control, closed-loop computer system, the strategy may become too complex to rely on this diagram alone, but it is a document that illustrates *all* key control and monitoring points and strategies. Beyond that, an "operations manual" for the system is usually prepared in order to fully describe and document any complex algorithm, set of equations or methods of control. It is obvious that a full optimization of a unit operation cannot be described by flowcharts, logic diagrams or other symbols, and must require a description of the mathematics used along with the solutions. However, the PCD is a key document used in all control systems designs in order to document, in one place, all key control strategies. It is used routinely in all control systems designs (for straight analog instrumentation as well as for computer-driven systems) and serves as one of the first "production" documents for a plantwide control system in the C.P.I. Many downstream engineering and design production efforts require this document as a source. For example, the P. and I.D.'s (piping and instrumentation flowsheet) are based on this document.

In the case of a computer project, the design process should include the following three phases.

1. *The conceptual phase.* Systems engineering analysis; control diagrams production; etc.
2. *The hardware design phase.* Configuration and specification of the computer; man-machine interface designs; instrumentation interface design; etc.
3. *The software design phase.* Process control language development; data base building; I/O list production; report and log programming; applications programming; etc.

The Conceptual Phase

The Socratic question and answer method for goal definition also is useful in the development of a detailed scope for the design phase of the project. Questions that might be posed and answered in order to arrive at a set of concise statements concerning the particular control systems design(s) for the project are discussed below. The first question to be considered is as follows:

- Is the computer system to operate in a *closed-loop* or *open-loop* control

*The PCD is used by Stone & Webster Corporation and others.

mode? That is, is it to be an *on-line* or *off-line* system, operating in so-called *real time*?

The terms *closed-loop, on-line, off-line* and *real time* in the corollary question are defined in the Glossary. However, in order to offer further clarification at this point, let us define the terms *closed-loop* control and *open-loop* control.

A *closed-loop* system is one in which the computer is wired to (and receives process variable electronic signals from) primary measuring instrumentation hardware such as pressure, temperature, flow rate, continuous product composition analyzers, etc. This is done via primary sensor transmitter instrumentation, which converts the process variables to a standard 4 to 20 milliamperes dc current (see Figure 2-13) or other standard signal for transmission to the computer. Process variable characteristics, then, such as psi (pressure), °F (temperature) or gpm (gallons per minute) flow rate, are converted to the standard pneumatic (3 to 15 psi) signal for transmission or to the electronic signal previously mentioned. Figure 2-14 illustrates the conversion of a process variable flow rate in a pipeline to the standard transmission signal of 4 to 20 milliamperes dc used in electronic instrument systems.

The digital computer further processes this converted analog signal within its *"front end"* circuitry into the electronic pulse code signals acceptable to its internals. Beyond this, *closed-loop* requires that the computer drive the final control element (usually a control valve) in the loop in one fashion or another (either with a velocity or displacement algorithm) or that it drive the *set point* of a conventional analog instrument which, in turn, will control the valve. The former action is referred to as direct digital control (DDC) and the latter

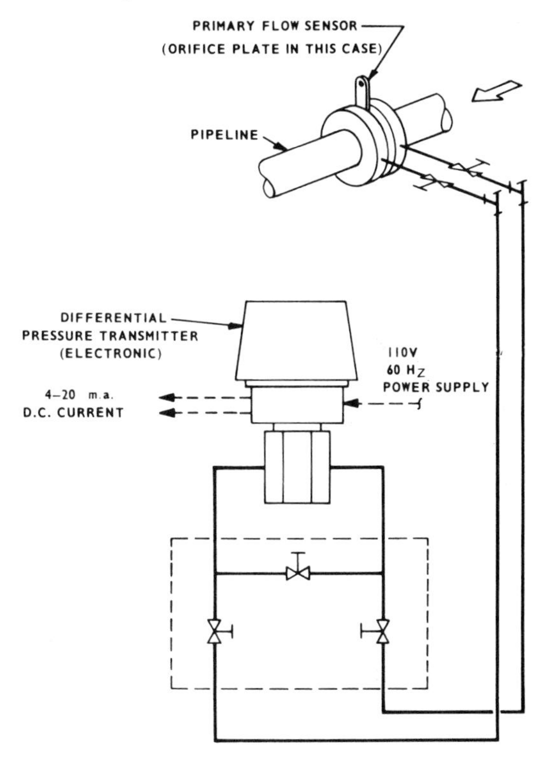

Figure 2-14. Electronic analog flow rate transmitter.

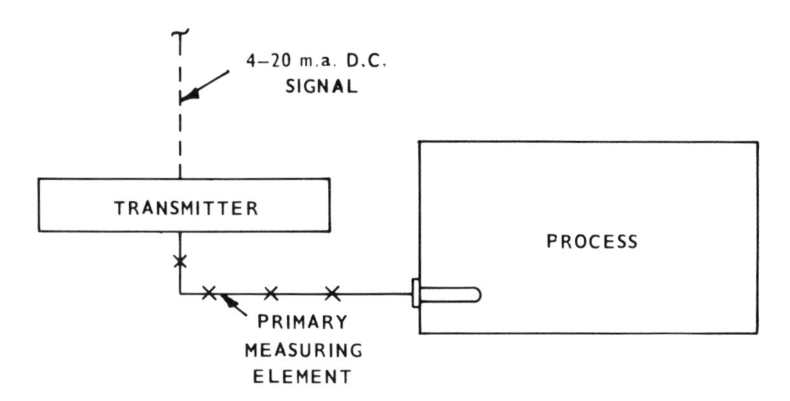

Figure 2-13. Temperature transmitter.

as set point control (SPC). Figure 2-15 illustrates the SPC mode of computer control.

By definition, then, a *closed-loop* system is one in which the computer must affect the manipulated variable (see Figures 2-15 and 2-16) in order to produce a desired change in the primary process variable being measured and controlled. In Figure 2-15, the primary process variable is temperature, illustrating typical elements in such a control loop. In contrast, an *open-loop* system is one in which the computer gathers process measurement information

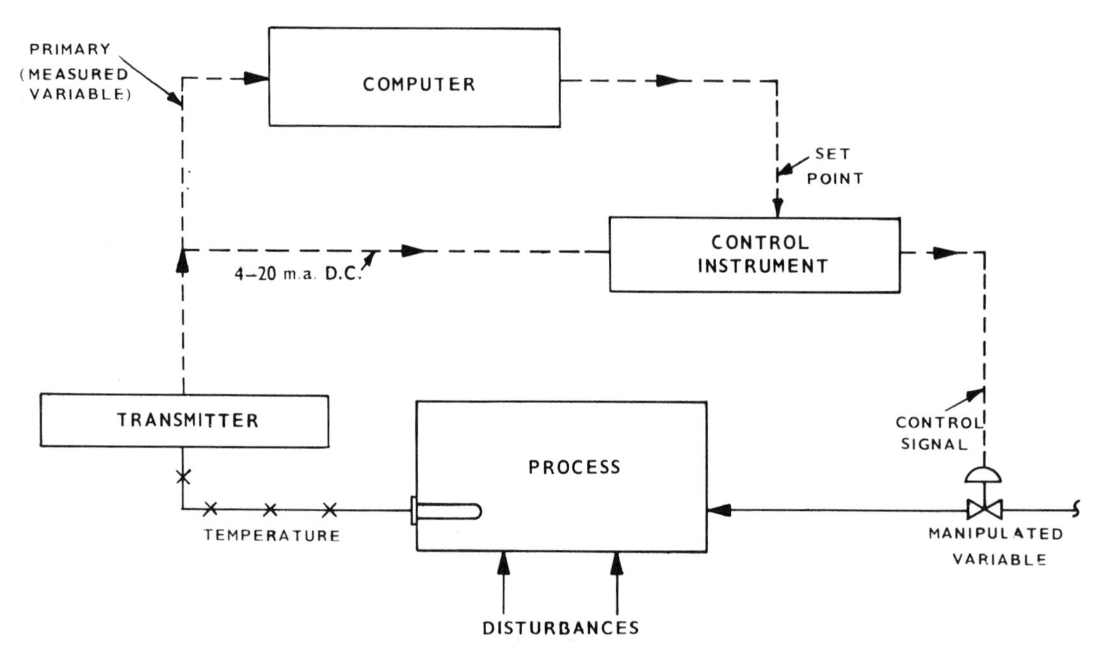

Figure 2-15. A simple feedback loop illustrating computer manipulation of an analog instrument set-point.

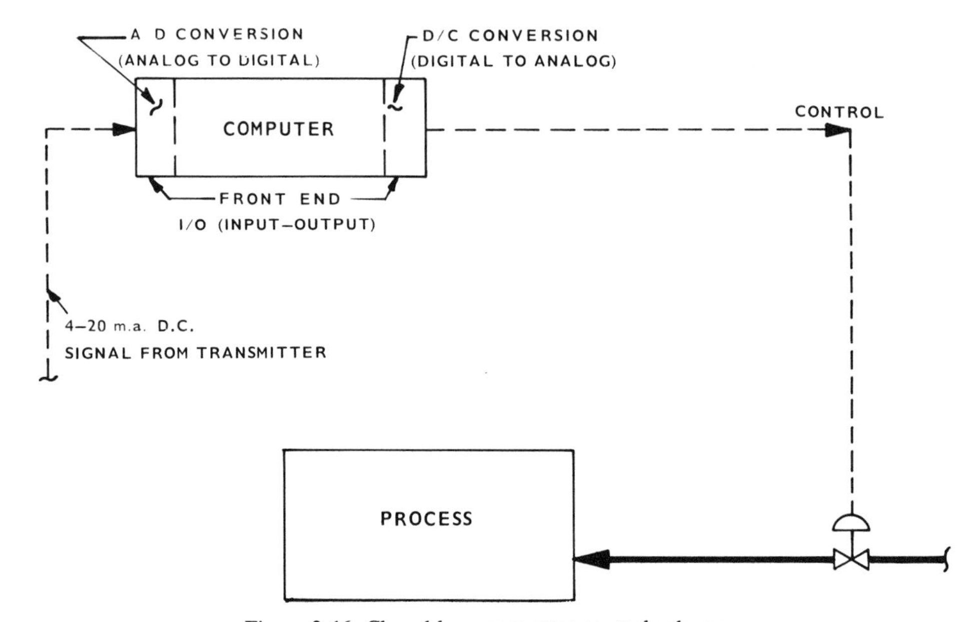

Figure 2-16. Closed-loop computer control scheme.

(as before, via the transmitter) but does not perform any control action. The computer may display the process variable data in a number of ways—as a trend record, as daily or hourly reports (see Figures 2-6, 2-7, 2-8 and 2-9) or as alarm summaries, for example—but no final action is taken. The system, in this case, serves as a *data-gathering* and *logging* system. As such, it may alert an operator that a dangerous situation exists, and may even indicate that he move the set point of a particular analog control instrument in a certain manner to correct the situation, but it is not designed to move a final control element. Figure 2-17 illustrates such an *open-loop* mode of control.

The terms *on-line* and *off-line* versus *real time* can also be defined in a similar fashion. An *on-line* system, then, is one that is wired to primary variable transmitter instrumentation. Therefore, it must operate in *"real time,"* since it is tied to the process by physical hardware. The computer CPU is thus constrained to operate within a time frame set by outside switch closure, analog signal generation or some other such action. An *off-line* system, on the other hand, is one that is physically free of the process for which it is performing a calculation or suggesting an action. That is, any data required for calculation is manually fed into the computer. Any reports generated by the computer for action are also manually dealt with. In other words, there may be no "real time" time frame within which an action *must* take place in order for the plant to continue with its production schedule.

Another question that should be posed is the following.

- What are the control strategies for each unit operation on the project that may, reasonably and economically, be justified for the process in question? For example, shall feedforward/feedback-corrected algorithms be used for column control—or shall optimization-level control be utilized on the cracking furnaces (and to what end)? Are simulation and modeling required because of the complexity of the process—or for later training of operators?

The conceptual design phase involves process analysis and decision-making based on that analysis, concerning the particular control modes and strategies on all units of the plant, and assuming the basic plant unit operations have been designed (and have been decided upon) by the process engineering function (either in the client organization or jointly with the contractor,

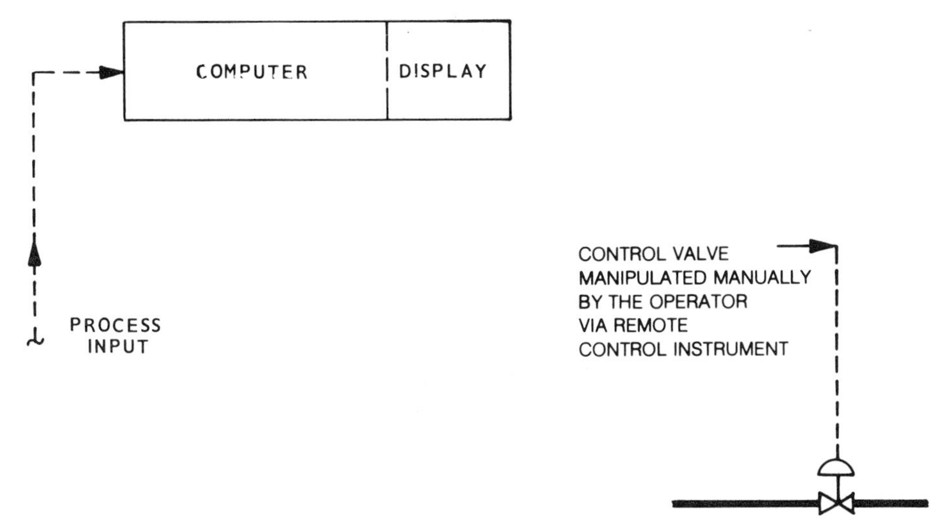

Figure 2-17. Open-loop computer schematic with display of process response to an external correction of the process variable.

or by the contractor on the particular project).

Column control strategies (see Figure 2-18) are fully discussed in Chapter 5, while optimization-level control is partially examined in Chapter 7. The purpose of this chapter is only to address technique and not to solve individual unit operation control problems.

It is assumed that a systems engineering analysis technique is used to promulgate, evaluate, model (if necessary), off-line computer test and simulate each hypothesis presented regarding a particular control strategy. It is, however, beyond the scope of this text to present the principles of systems engineering. There is a large library of textbooks devoted solely to this subject. The reader is referred to D. P. Campbell,[2] G. D.Shilling,[3] Del Toro and Parker,[4] Coughanow and Koeppel,[5] J. G. Rau,[6] and other works covering a wide range of material, from control systems engineering principles to optimization and operations research.

The Hardware Design Phase

Control systems panel designs have traditionally been based on large, vertical panel concepts in which clusters of analog instruments are arranged in sections according to plant operating functions. For example, a large ethylene complex might have over one hundred and fifty lineal feet of such panels sections, each approximately 8 ft high. Figure 2-19 illustrates a plan view of such an arrangement. A typical vertical elevation view of one such section is shown in Figure 2-20. These designs are based on the traditional so-called "high-density" packages of analog instruments and, as can be seen, they still tend to require all peripheral space in the control room. One control room operator can only monitor

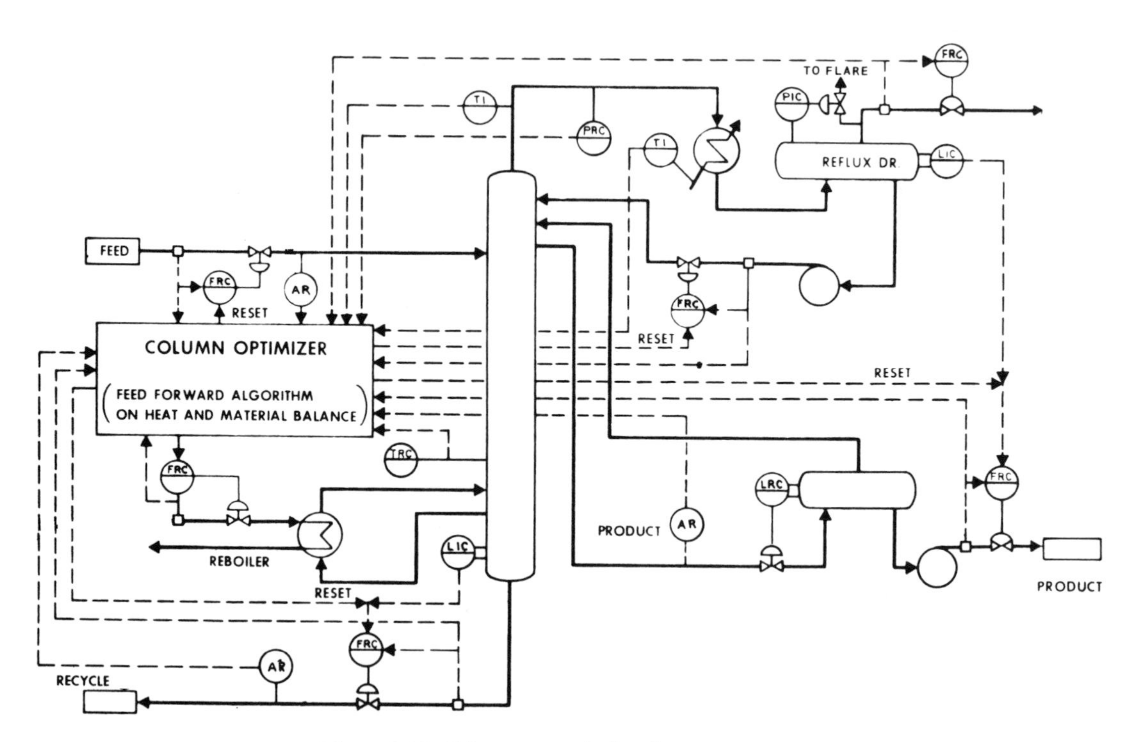

Figure 2-18. Column control via microprocessor.

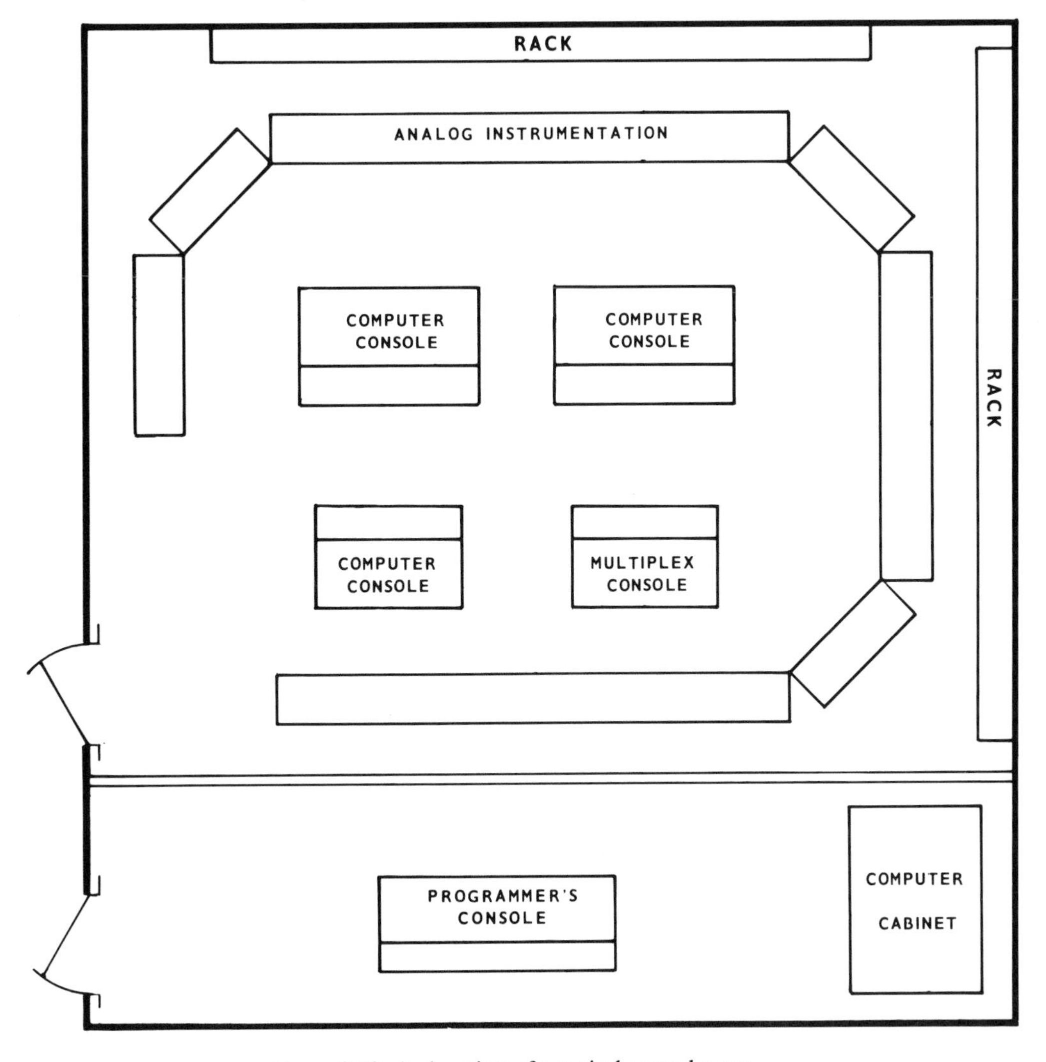

Figure 2-19. A plan view of a typical control room.

one or two major units in the plant. He might be able to scan, for example, all of the furnaces in an ethylene complex, the primary fractionation section or the compressor section. Any of these is an accomplishment, when compared to older control panel designs, but still does not allow the operator a clear and quick view of possible interactivity between sections of the plant. In other words, he cannot readily obtain an "overview" of large sections of the plant in order to prevent or predict upsets or to drastically improve production rate.

The development of a historical perspective on analog instrument and control panel design, at this point, is necessary to a full understanding of what the newer distributed systems of digital hardware can provide.

History of Control Panel and Instrument Development

In the late 1940's and early 1950's, analog instrumentation hardware was generally based on pneumatic, large-case (approximately 18×18 in.) concepts. Each instrument was direct-connected to a process sensing point and usually located near that

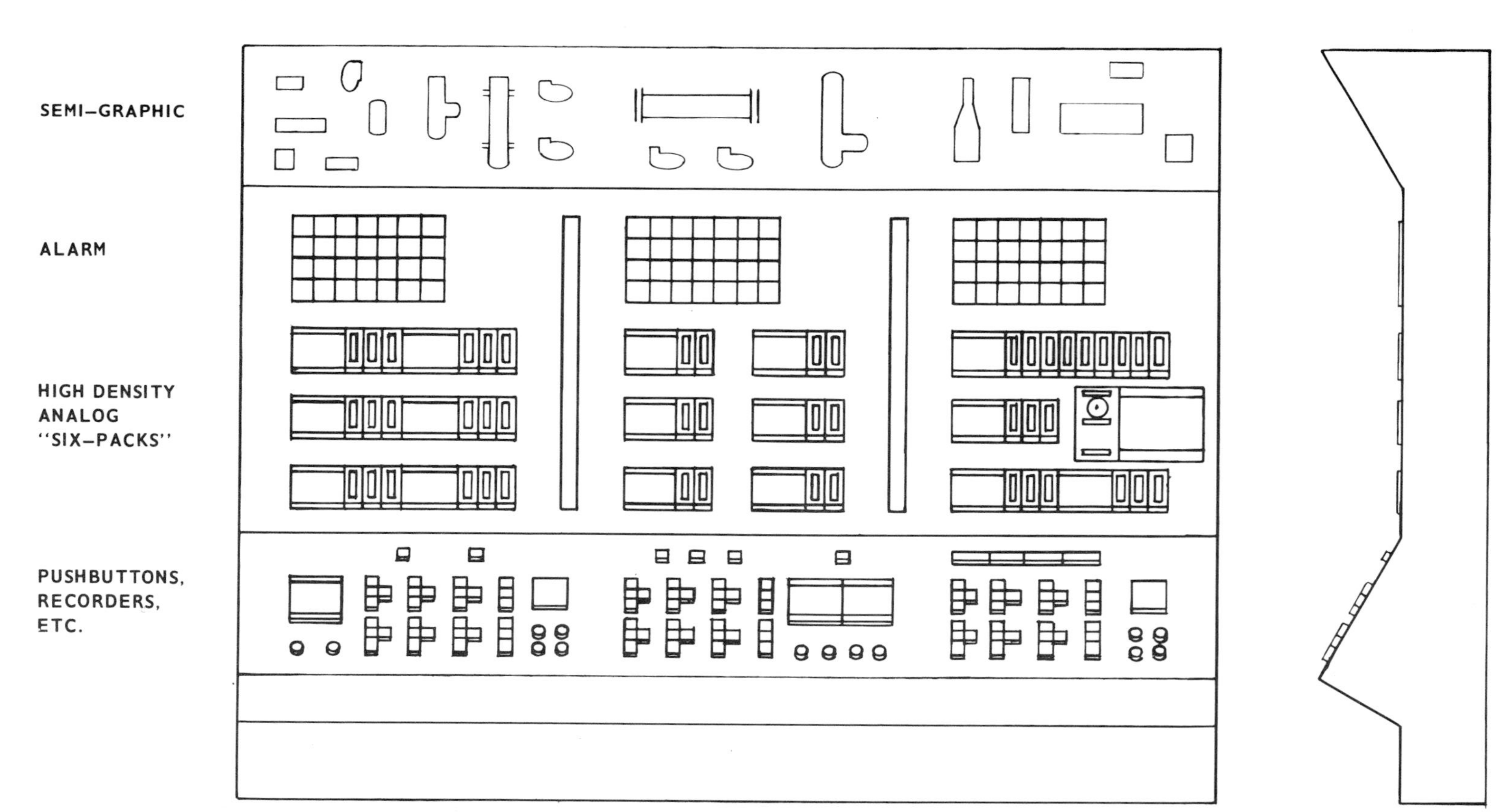

Figure 2-20. An elevation of typical control panel sections.

point. As a result, process control was largely decentralized and the operator could only view one section of *one* unit operation; e.g., one kettle, one condenser, the overheads of one column. With the development of pneumatic transmission techniques, centralized control became possible, gradually permitting more control hardware to be placed in one section of a control panel. However, the receiving instrumentation was still fairly large and cumbersome and was usually dedicated to the control of one process variable.

In the late 1950's, the picture improved somewhat with the miniaturization of the receiver instrumentation. Although still largely pneumatic, the case size had decreased to 6×6 in. and eventually to 3×6 in. and 2×6 in. standards. About this time, electronic instrumentation hardware was inaugurated, based on transistor technology. This allowed for electronic transmission development and a consequent further centralization of instrumentation on one control panel.

During the early 1960's, the digital computer was introduced to process control, adding peripheral hardware to the control room. New interface hardware, such as printers, typers, CRT screens and keyboards, were now introduced to the operator, making the control room scene more complex, as all of the new hardware was still backed by the conventional analog instrument panel. Thus, the operator had to learn new techniques while recalling old ones in an emergency. This was the current state-of-the-art of control panel design until very recently.

About five years ago, a revolution in man-machine interface design philosophy began, with the introduction of a distributed architecture based on microprocessor hardware. This new hardware digitized the usual analog hardware and made applicable new control modes. It also introduced the communications network into the conventional analog loop and enabled the return of some decentralization of control to the field, while at the same time centralizing more information at the main control console(s). The distributed systems now make it possible to place all relevant process information on these control consoles within easy reach of a seated operator. That, essentially, is the revolution. Figure 2-21 illustrates the evolution in control panel design from the 1950's to the current centralized overview CRT consoles of the 1970's and 1980's.

These distributed systems have been introduced by most of the major instrument manufacturers, namely, Honeywell, Inc., Foxboro Corporation, Taylor Instrument Company, the Bristol Company, Fisher Controls Corporation, EMC Corporation and some others. Honeywell, Inc.'s "TDC-2000" was one of the first introduced ("TDC" stands for totally distributed control). The system is based on microprocessor hardware configured into a "data highway" network and connected via a 50-conductor cable. Traffic on the highway is monitored and regulated by a "hiway traffic director," which controls what data is transmitted and performs various security checks on the data to ensure that it is correct. Figure 2-22 illustrates the TDC-2000 "operating center" CRT consoles, recorders and printers.

Foxboro Corporation's "VIDEOSPEC" hardware is pictured in Figures 2-23 and 2-24, illustrating the reduction in panel space possible with such systems. Figure 2-24 shows various CRT screen displays, the equivalent of a hierarchy of high-density analog instrument panel arrangements (the first being a plantwide overview and the last being the equivalent of a single-loop analog controller).

Taylor Instrument Company's "MOD III Command Console" is illustrated in Figure 2-25. It should be noted that one operator can easily monitor three CRT screens in such systems. One screen (Figure 2-26) provides a flowchart representation of the process unit operation, giving pertinent process variable information. Another

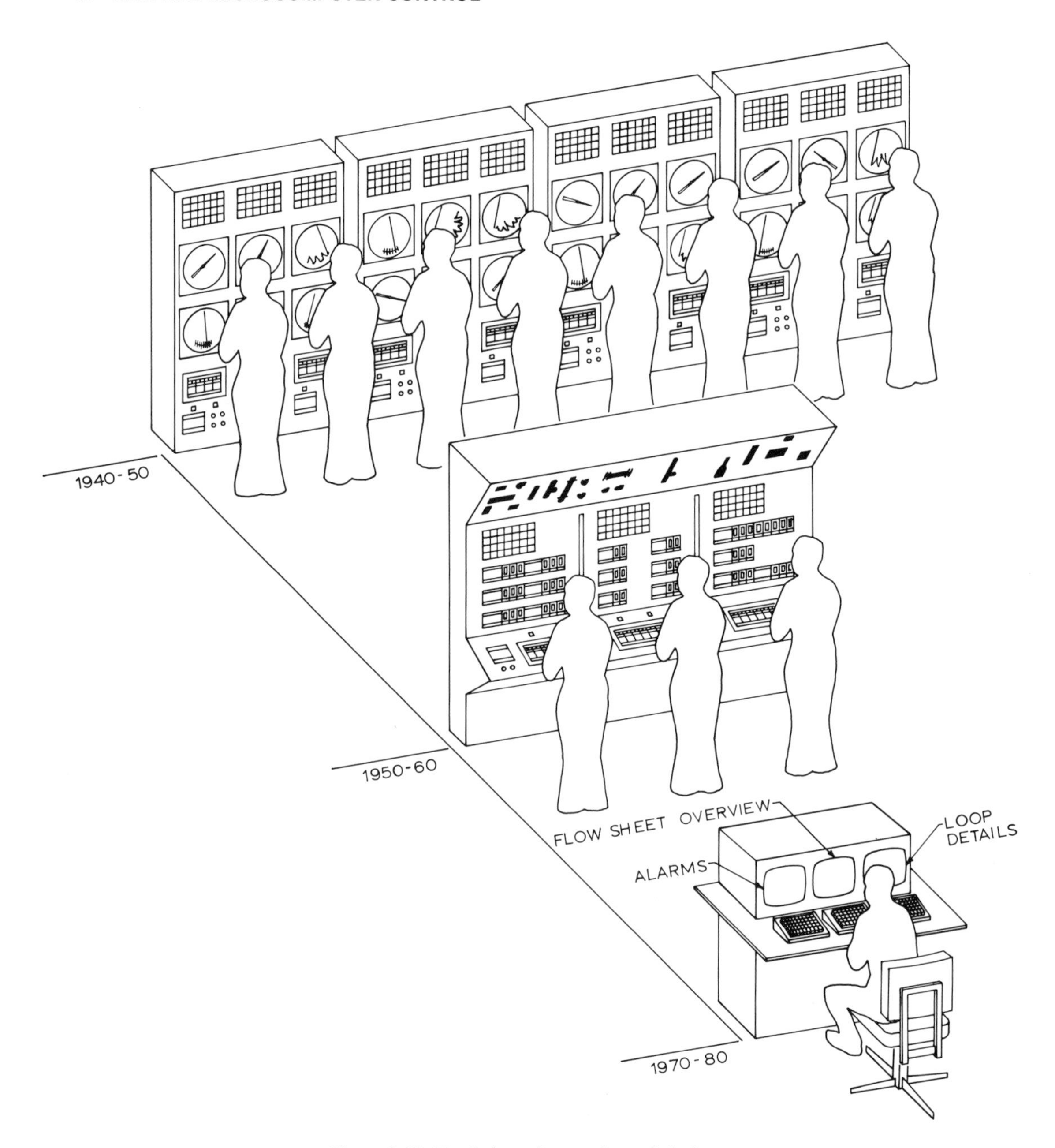

Figure 2-21. Evolution of control panel design.

Figure 2-22. Operating center—"TDC-2000." (*Courtesy of Honeywell, Inc.*)

screen (Figure 2-27) can provide a trend plot for three variables. The third screen (Figure 2-28) is a status display.

The Bristol Company's "UCS 3000" console is another version of this hardware and panel board concept for reduction of all conventional analog instrumentation into one central console. (The UCS-3000 console is illustrated in Figure 2-29.)

It is important to note that although most of these distributed systems use microprocessor hardware, they are not computer systems. They are but a replacement for, and a redesign of, conventional analog con-

Figure 2-23. "VIDEOSPEC" display console. (*Courtesy of The Foxboro Corporation. VIDEOSPEC and FOX are trademarks of The Foxboro Corporation.*)

trol instrumentation, using microelectronic digital hardware. They are, of course, much more sophisticated in many ways, providing such features as self-checking diagnostic routines for preventative maintenance (Honeywell, Inc.'s TDC-2000) and elaborate trend patterns and displays. On control algorithms, the sophistication is not much better than that possible with conventional analog instrumentation (namely, the usual P.I.D., Ratio, Bias, Lead-Lag, Square Root, Hi-Lo Select, etc., functions). To do more in control necessitates the installation of a computer.

Man-Machine Interface Considerations

The Instrument Society of America (I.S.A.) proposed draft-standard (ISA-RP60.3-1977) entitled "Human Engineering for Control Centers,"[8] Dallimonti[17], and others, have presented cogent arguments for adequate study and evaluation of the human engineering aspects of information gathering. Such studies should be undertaken prior to the design of a control panel, console or center. However, this is rarely done in practice. The styles outlined above were designed during each of the decades, beginning with the 1940's, and the method of design has not altered radically over the years. The current microelectronic hardware revolution may change all that. The systems described in the previous section, currently being offered by the major instrument manufacturers, will probably escalate the application of such systems within the next few years. The trend is already apparent. Most human engineering studies conclude that *both* man

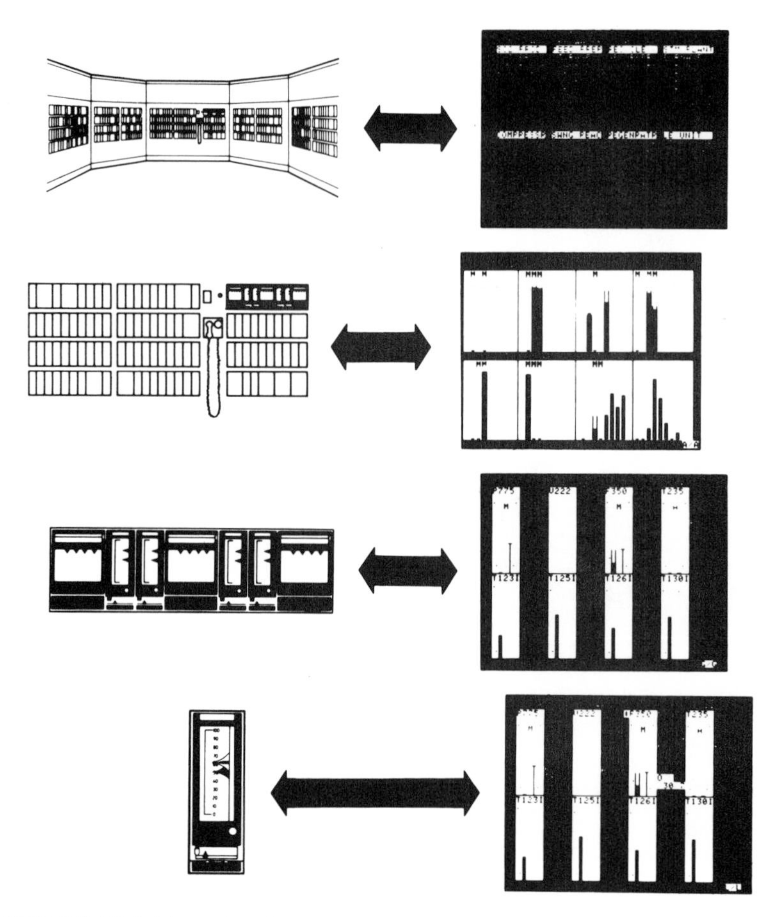

Figure 2-24. "VIDEOSPEC" hierarchy of reports (Series A-E). (*Courtesy of The Foxboro Corporation. VIDEOSPEC and FOX are trademarks of The Foxboro Corporation.*)

Figure 2-25. Taylor ''MOD III'' command console. (*Used with permission of The Taylor Instrument Company, Division of Sybron Corporation.*)

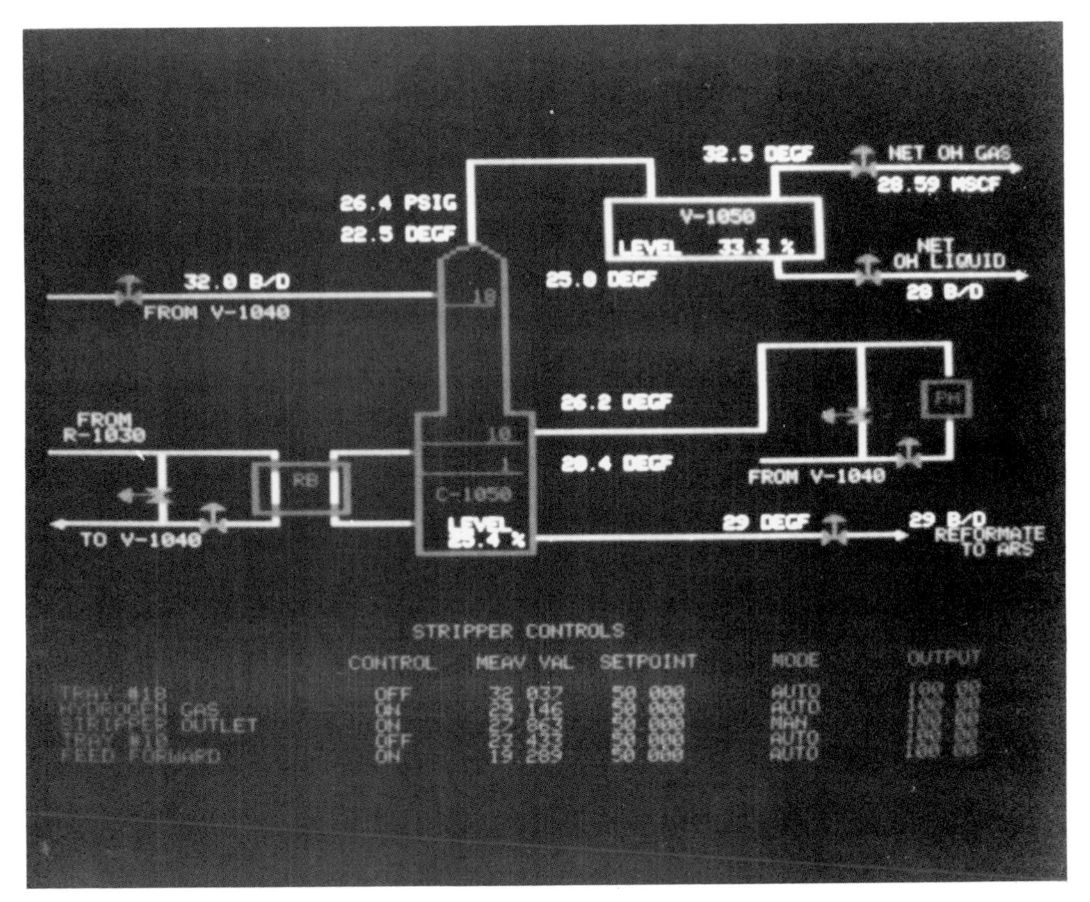

Figure 2-26. ''MOD III'' flowchart screen display. (*Used with permission of The Taylor Instrument Company, Division of Sybron Corporation.*)

and the computer have limited ability to process masses of information submitted simultaneously. Man is similar to a computer in that he can only see and digest one piece of information at any given moment. Some men, like computers, can handle more information faster and with more accuracy than can others, but generally, if a man is flooded with auditory and/or visual inputs in large quantities, he does reach a point of no response and consequent inaction to all of the inputs he is receiving. He tends to disregard most of the inputs in those situations, while settling on only a few that he feels are familiar and useful to him.

Only two of man's four basic senses are really used to any extent in the control room: the visual and auditory senses. The other two senses—cutaneous (touch, heat, cold and pain) and kinesthetic (body position awareness)—are little used in the context of this chapter. Of all the information gathered by man in the control room, 80 to 90% is by visual input. The remaining 10 to 20% is received by his auditory senses. Therefore, control instrumentation should be located and grouped in the following manner, according to the I.S.A.[8]

- Locate all devices in the same sequence that the operator will use during the operation of the unit or subsystem.
- Group together those display devices which are inter-related or are used to provide status information on the system.
- Leave space between adjacent groupings; use a different color or outline a functional grouping.

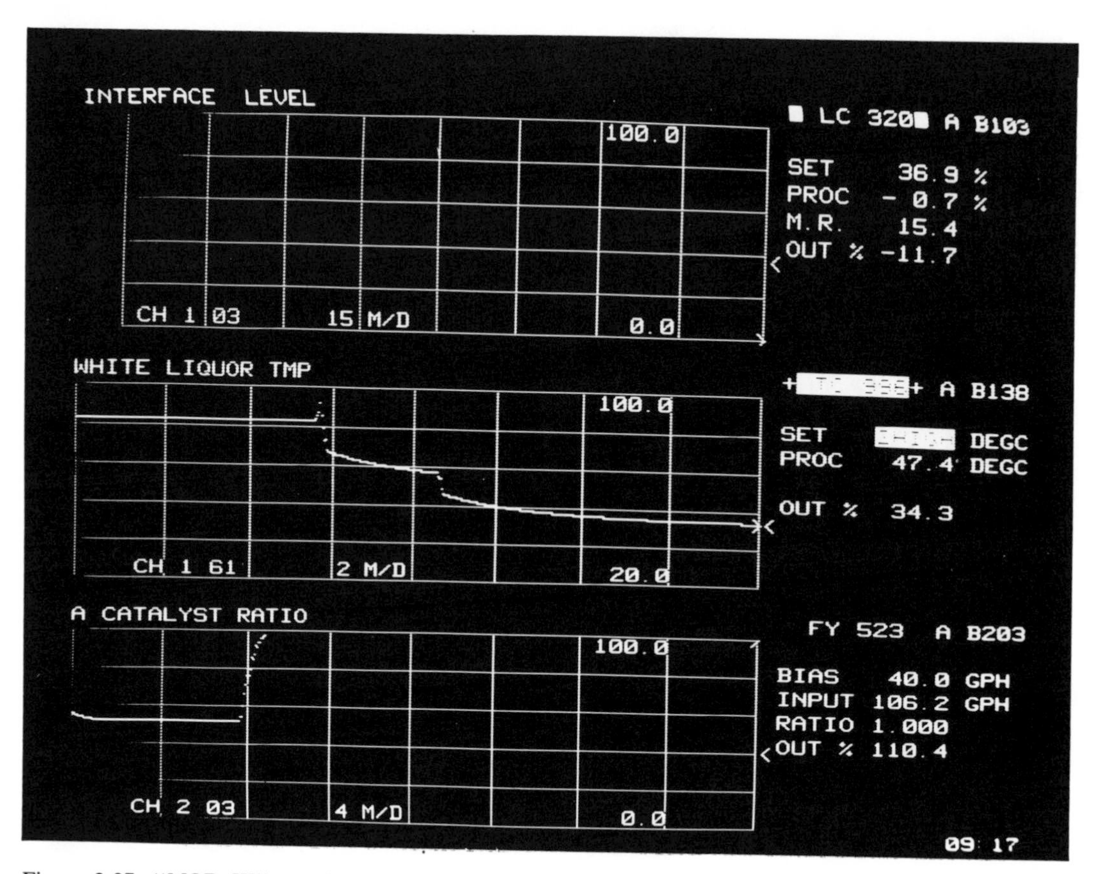

Figure 2-27. "MOD III" trend plot display. (*Used with permission of The Taylor Instrument Company, Division of Sybron Corporation.*)

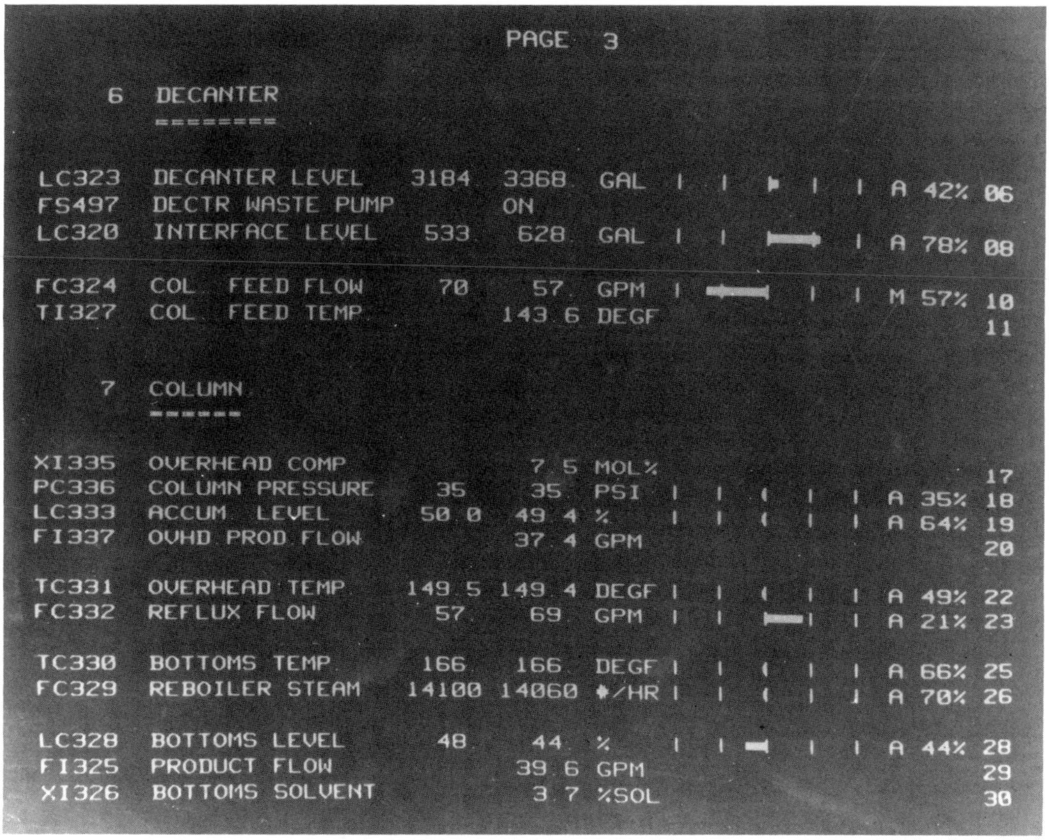

```
                              PAGE 3

      6  DECANTER
         ========

LC323   DECANTER LEVEL      3184   3368  GAL  I  I  ▶  I  I  A 42% 06
FS497   DECTR WASTE PUMP            ON
LC320   INTERFACE LEVEL      533    628  GAL  I  I  ━━━▶  I  A 78% 08

FC324   COL  FEED FLOW        70     57  GPM  I ◀━━━━  I  I  M 57% 10
TI327   COL  FEED TEMP            143 6  DEGF                     11

      7  COLUMN
         ======

XI335   OVERHEAD COMP               7 5  MOL%                     17
PC336   COLUMN PRESSURE       35     35  PSI  I  I  (  I  I  A 35% 18
LC333   ACCUM  LEVEL        50 0   49 4  %       I  (  I  I  A 64% 19
FI337   OVHD PROD FLOW             37 4  GPM                      20

TC331   OVERHEAD TEMP      149 5  149 4  DEGF I  I  (  I  I  A 49% 22
FC332   REFLUX FLOW           57     69  GPM  I  I  ━━━I  I  A 21% 23

TC330   BOTTOMS TEMP         166    166  DEGF I  I  (  I  I  A 66% 25
FC329   REBOILER STEAM     14100  14060  #/HR I  I  (  I  I  A 70% 26

LC328   BOTTOMS LEVEL         48     44  %       I  I ━  I  I  A 44% 28
FI325   PRODUCT FLOW               39 6  GPM                      29
XI326   BOTTOMS SOLVENT             3 7  %SOL                     30
```

Figure 2-28. "MOD III" status report display. (*Used with permission of The Taylor Instrument Company, Division of Sybron Corporation.*)

Figure 2-29. "USC-3000" console. (*Courtesy of Bristol Division of Acco.*)

- Use consistent criteria to delineate status display devices where a large such grouping exists.
- Mount control devices within the functional graphic symbol on the control board (to help the operator identify and recall the function).
- Use consistent criteria for the location of nameplates.

There are also questions of accessability of controls raised in these studies (i.e. locate devices within easy reach of the operator in a given group).

The I.S.A. study, and other such studies, dwell on pattern recognition, shape and type of display, visibility and readability, illuminations, color coding and auditory techniques. All such studies conclude that there is a need to rapidly bring control room information to the operator in a more easily understandable manner and within his visual and physical reach. As a result, the "Star-Trek" space age concepts of control console design currently being marketed by the major instrument manufacturers will probably provide a ready solution and possibly set new styles for control room design within the next few years. The microelectronic revolution, thus, may have already reached the industrial process control console, as well as the computer and the back-up instrumentation.

How do mini- and microcomputers fit into this picture? What are some of the possible hardware configurations with a computer for a plantwide control system?

The Hardware Configuration With Computers

Other questions that might be posed are as follows.

- Is the computer system to be based on a single CPU, a dual CPU or a distributed network?
- What are the advantages and disadvantages of each system?
- What back-up configurations should be (or usually are) considered for operation on computer failure?
- What are some of the problems associated with the instrumentation interface to the computer (i.e., electronic signal grounding problems such as "common mode rejection" levels and signal compatability with the computer)?

These and other questions are answered in Chapter 8. However, the questions on back-up configuration and instrumentation interface will be covered here.

Back-up configuration with computers. Given a unit operation such as a fractionating column or a train of such columns, there are four possible schemes that bear discussion.

- *Configuration No. 1.* Figure 2-30 illustrates this configuration. It is useful for the implementation of a local optimizer such as that depicted in Figures 2-1 and earlier in this chapter. The "column optimizer" microcomputer is located in an existing control panel, and inputs and outputs to the micro are taken at the control board. There is no peripheral hardware in this case (such as typers, keyboards, etc.), as this is a situation in which just one unit operation is being placed on computer control in an existing plant.
- *Configuration No. 2.* This is a possible configuration (Figure 2-31) illustrating the next step in computerization of an existing plant—that of the addition of a "MUX" or MULTIPLEXER as a remote unit. This is done routinely in current plant designs, as there may be a considerable savings in signal and/or thermocouple wire realized with the addition of such a multiplexer. These multiplexing systems are easily incorporated into new plant designs, whether or not a plantwide computer system is scheduled for the plant, since they augment analog instrumentations and provide digital (alphanumeric) displays

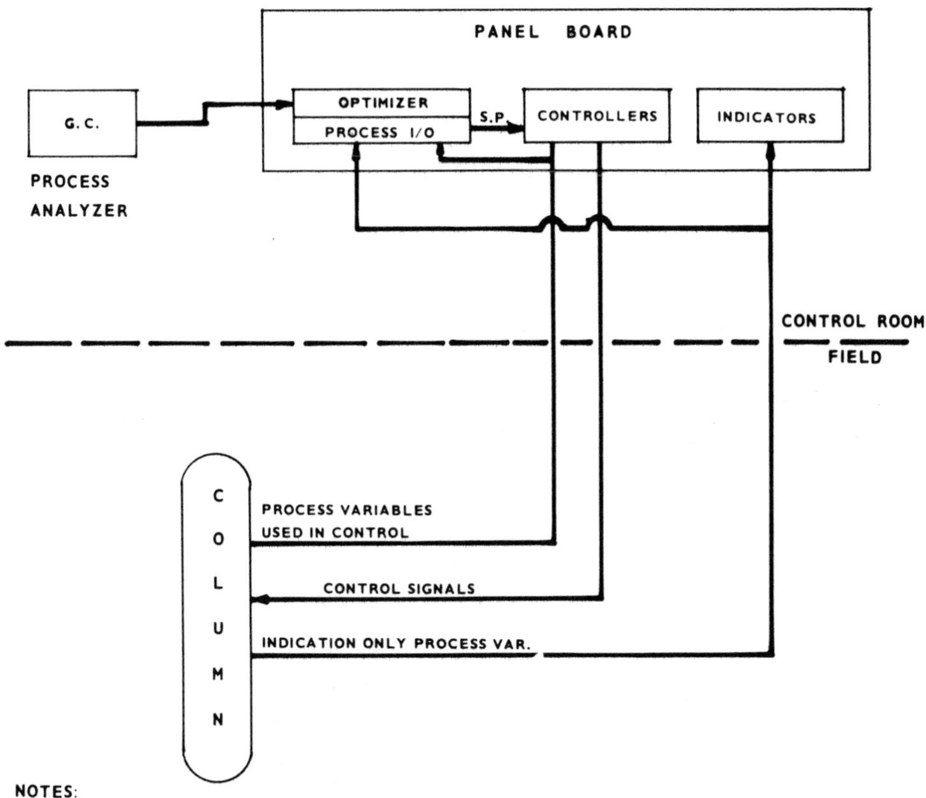

NOTES:

1. THIS APPROACH IS USEFUL FOR ADDITION OF COLUMN OPTIMIZER TO AN EXISTING PLANT.
2. G.C. CAN COMMUNICATE WITH SEVERAL OPTIMIZERS.
3. THE SINGLE COLUMN OPTIMIZER SHOWN IS TYPICAL OF SEVERAL COLUMN OPTIMIZERS.

Figure 2-30. Computer configuration No. 1.

of logs as well as printouts, on demand.

- *Configuration No. 3.* Figure 2-32 illustrates this configuration—one version of distributed control using microprocessors. In Configuration No. 3, the conventional panel with analog instrumentation is completely eliminated in favor of multiple CRT screen displays. The control micro communicates with the display network via a communications micro. It also receives process analyzer information, as does the communications micro.
- *Configuration No. 4.* This (Figure 2-33) illustrates the same network, backed by a conventional panel board of analog (or digital) controllers. (A full discussion of networks in larger systems, with main-frame CPU's, is provided in Chapter 8.)

Papers[12, 13, 14] given at recent A.S.M.E. and I.S.A. conferences on the subject promulgated such designs. In those papers such a special purpose micro was proposed, as indicated in Figures 2-1 and 2-2 earlier in this chapter.

PROPOSAL FOR NEW "COLUMN OPTIMIZER" CHIP

We have proposed an instrument essentially similar in front-face appearance to an analog control station, that would mount in a conventional control board shelf, but one that houses all the leed-lag (dynamic elements) and P.I.D. functions necessary to complete the heat and material balance functions around one distillation column in a feedforward-feedback mode. The control circuitry for such an instrument might be "burned" onto one LSI microcomputer chip. This instrument should sell for between $3,000 and $5,000 per unit, and might enhance the application of DDC, feedforward control and simple optimization in distributed systems.

We envision a drastic reduction in control board space within the next five to ten years as a result of the development of dedicated microprocessor hardware.

A typical billion pound/year ethylene plant can have up to 200 lineal feet of control board, including offsites, for a

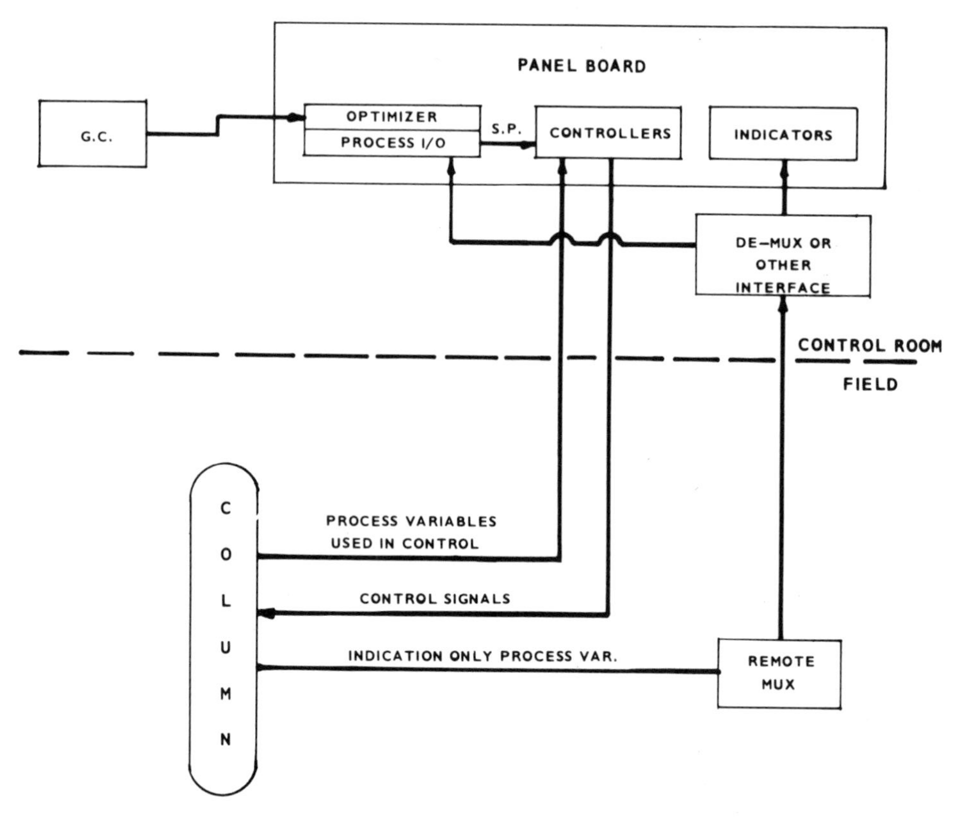

Figure 2-31. Computer configuration No. 2.

grass-roots complex. This might be reducible to approximately 50 to 80 feet, or less. There are approximately 600 loops with control valves, plus approximately 400 interacting loops in such a complex. Most of the valves will probably never be eliminated, but many of the controllers could be combined through the use of dedicated microprocessors. Assuming a reduction of only 10%, a savings of approximately 60 to 100 controllers is possible, if cascade and other such loops are counted. Further, the associated savings in control systems engineering and design man-hours required for specification of hardware, and for drafting of installation details and loop sketches, etc., could amount to approximately 2,500 man-hours.

The next step is for instrument manufacturers to begin the development of the smaller, dedicated microprocessors, in addition to the currently available large multiprocessing systems. Not only will energy savings and design manpower reduction be the result, but operating plants will produce products closer to target, utilizing modern control strategy.

If the larger instrument manufacturers do not produce the "Column Optimizer" type hardware, the gap may be quickly filled by smaller systems houses using available LSI technology and working with LSI manufac-

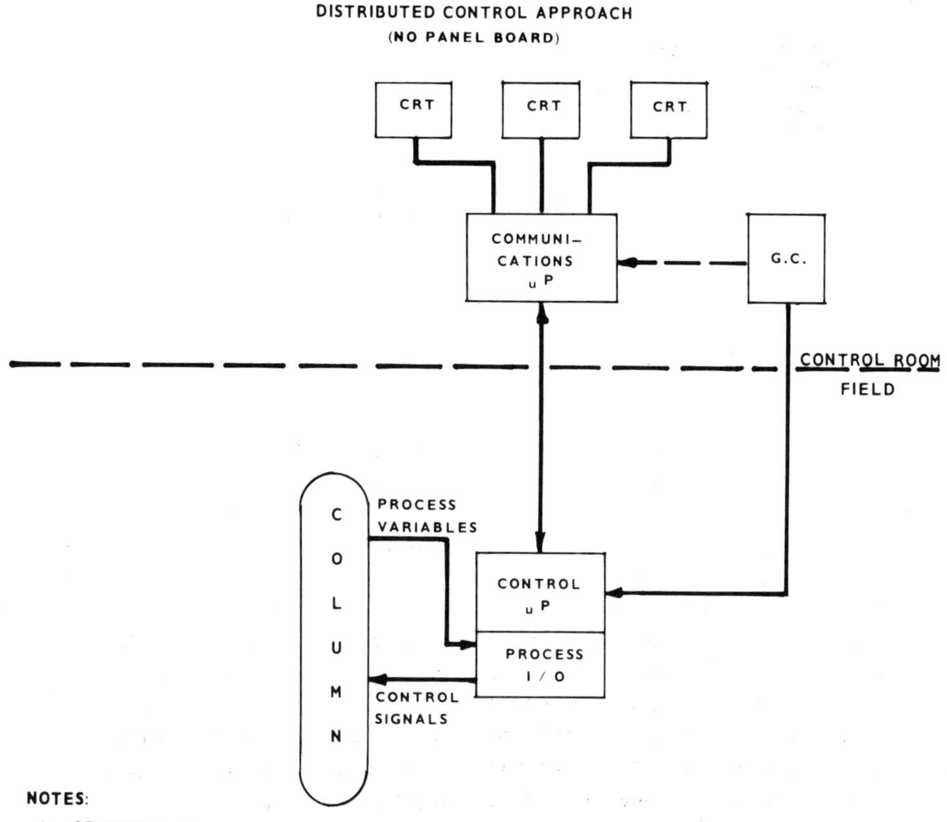

Figure 2-32. Computer configuration No. 3.

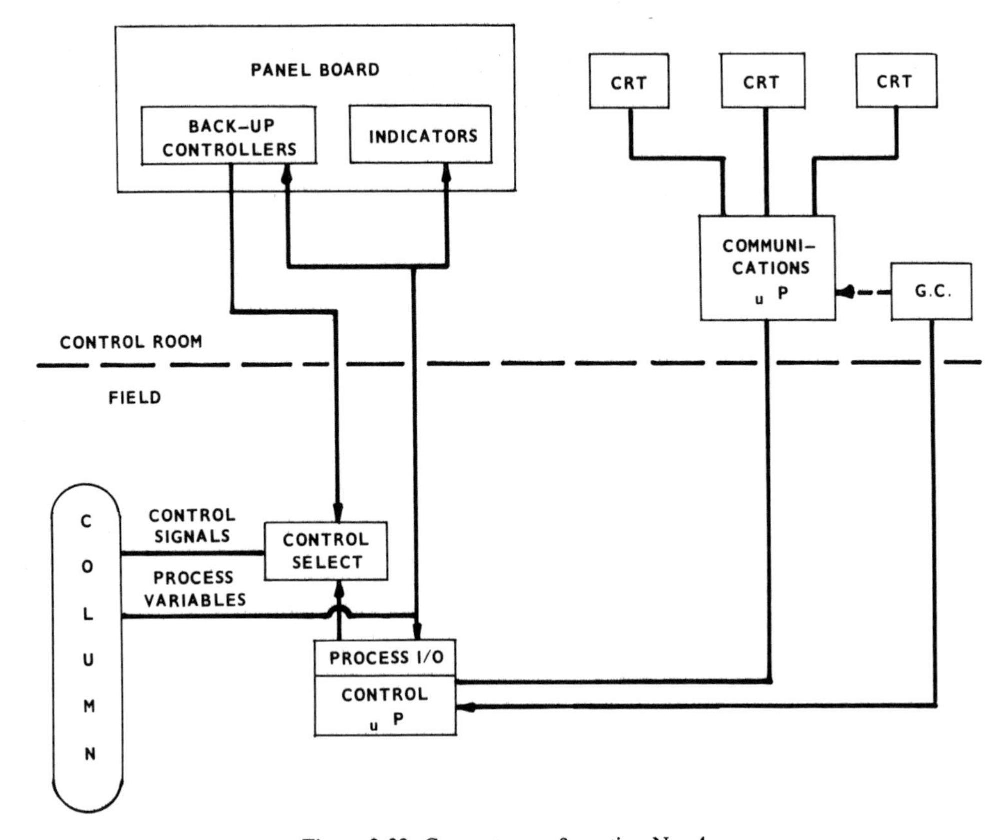

Figure 2-33. Computer configuration No. 4.

turers to produce a specially "masked" chip for petrochemical purposes. Some smaller systems houses are already beginning to fill such a gap.

The implementation of LSI technology for production of "Column Optimizer" type hardware by the larger instrument manufacturers might also help to eliminate many of the problems associated with computer systems and advanced control and, in the process, possibly revolutionize control system application engineering and petrochemical plant operation.

The microprocessor computer should then become just another item of instrument hardware—one for which a standardized specification is available, and one which can be shown as just another special balloon on the Piping and Instrumentation (or Process Control) Flowsheet.

We propose that instrument manufacturers produce standardized digital shelf-size instruments dedicated to the control of single columns, and yet generalized enough in strategy so that a feedforward-feedback algorithm can be masked onto an R.O.M., LSI chip, and housed in approximately a 6″×6″ panel-mounted instrument.

The above, in addition to the distributed hardware that has already been developed (as described earlier: Foxboro Corporation Videospec, Honeywell, Inc.'s TDC-2000, etc.) and which is currently on the market, would be a contribution to control systems design for the reasons cited. The next chapter details the use of such hardware in development systems for hardware and software.

Computer Control Consoles

Figure 2-12 illustrates a typical console design for a process monitoring and control computer. As shown in this design, many of these systems currently house CRT displays that emulate analog control stations, in order to improve the man-machine interface. It is a help to the operator to be able to see a familiar presentation, and his reaction is improved because of such displays. (Such a system is fully described in Chapter 10, and computer systems hardware configurations and networks are detailed in Chapter 8.)

A typical distributed configuration is illustrated in Figure 2-5, wherein satellite microcomputer systems are monitored by the host mini in a redundant communications network, policed by dual communications micros. This system was described earlier in this chapter (Item 5 in the "assumptions" list).

Common Mode Rejection Problems

Signal compatability and the proper grounding and shielding of signal wiring between the primary sensing instrumentation and the computer are important aspects of any sensor based computer system. The acquisition of low-level data by the computer via analog front-end hardware can cause "ground loops" and consequent signal loss, drift or elevation (or depression) of the signal due to inadequate, incorrect, common mode voltage design. In thermocouple installations tied to computers, it is a general practice, therefore, to use ungrounded thermocouple assemblies; e.g., the couple tip does not physically touch (is insulated, thermally and electrically, from physical "ground"). The thermocouple head is then electrically grounded in the field and the receiving instrumentation is left to "float" relative to electrical ground. Figure 2-34 illustrates both the physical arrangement of the thermocouple tip and the electrical grounding of the system.

Common mode rejection (CMR) techniques are design and grounding methods applicable only to analog amplifier instrumentation and are used to prevent the common mode voltage (CMV) from being inadvertently converted into a normal voltage signal which then may appear at the computer terminal as a viable data signal.

Analog voltage measurements on low-level signals emanating from instrumentation tied to computers is largely based on a differential voltage developed between the input wire pair tying the system together. The CMV, then, is that voltage that exists in each of the differential input lead wires, or, as described in the thermocouple example above, it may be the voltage that results from a "ground loop," or the difference in potential to ground between the thermocouple head and the remotely located amplifier-receiver instrument. Various design techniques on CMR problems are amply described in many electronic engineering handbooks, texts and papers (such as Coffee[18]). For the purpose of this chapter, it is sufficient to indicate and illustrate such problems.

The Software Design Phase

Fundamentally, the new distributed computer systems are composed of a network of localized software files in hardware packages, designed so as to allow easy application to process control. As we have seen in distributed hardware systems designs, the distributed processing of software similarly enables systems designers to match the functional organization of the unit operation, process or plant complex being placed under computer control. These new localized software groupings are arranged in a variety of networks or hierarchies according to function (such hierarchies and networks are fully discussed in Chapter 7.) The more important the function of the software package, the higher in

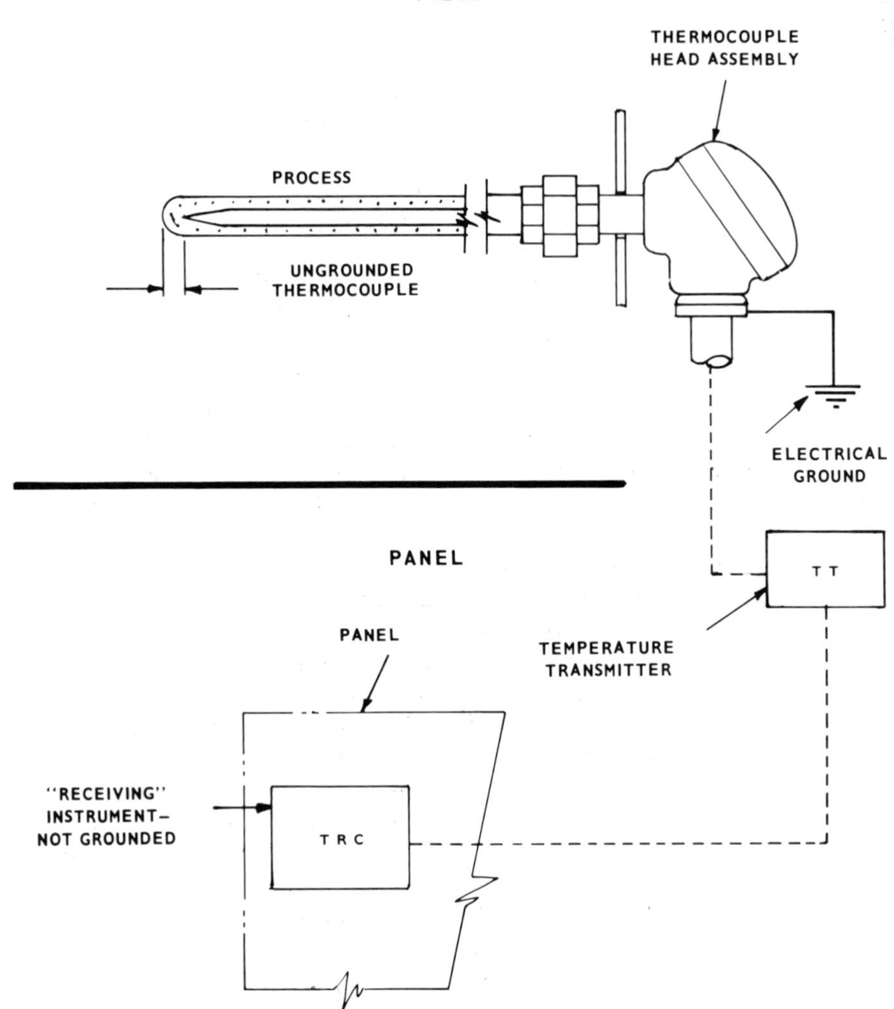

Figure 2-34. Ungrounded thermocouple installation.

the hierarchy it is placed. Such functional organization facilitates systems operation and maintenance, as the causes of software "bugs" or problems can be more readily identified. Since the software is designed along application lines, an application software error can more easily be seen as a true error in application, rather than be assumed an error of calculation or syntax.

Distributed processing of software also minimizes the total traffic flow of information within the network, since the "massaging" of data and its control and reporting requirements are minimized and localized. Unnecessary processing of data by the host computer is eliminated because of the localized control of priority of message flow. Localized control of priority also minimizes interaction of control errors and contention among large message blocks for priority. The software files handled are thus prescreened locally. Therefore, problems arising in software can be analyzed and corrected more quickly and identified more easily. Furthermore, distributed processing allows for the communication of *editing commands,* rather than the transmittal of complete software files, between nodes in the network. Messages between the nodes can be transmitted in small packages when traffic volume is heavy in the system, thus preventing the "queuing" (or the lining up) of blocks of important data (i.e., a "protocol" is more readily enforced).

Types of Software

There are basically two types of software systems: the operating system "standard" software (the software necessary to drive all elements of the system, such as peripheral equipment—typers, printers card readers, etc.) and the "applications" software (that software package specific to the control of the process, unit operation or plant).

The standard software also includes the programming necessary to produce reports, logs, graphic displays and interface with the unit operator. In other words, that software necessary to process information within the system and to provide a means of interface to the outside world.

The applications software package is the custom software designed specifically for the execution of control algorithms and strategies peculiar to the process, unit operation or plant complex being controlled.

Language Systems

In order to execute either the standard software or the applications software, various language systems must be employed. There are three levels of languages usable in process control computer systems: a *high-level process control* language, a *Fortran* based language and an *assembly* language. The high-level process control languages are designed at conversational level, sometimes in question-and-answer format, and are specifically intended to allow non-programmer-type control systems engineers to build the data base for the computer system. Thus, the control engineer may input desired strategies as answers to questions covering the types of control algorithms desired. He also may build operator interface reports on the CRT screens or on the logging printers (see Figures 2-6, 2-7, 2-8 and 2-9 for examples of such logs and reports).

Manufacturer standard process control languages (PCL's) usually have the following features included in their software systems.

- Routines for standard displays and graphics.
- Standardized alarm message programs.
- Software for hard-copy driving.
- Interface routines for keyboard operation.
- Customized operator entry modes.
- Standardized formats for printed and video displays.
- A library of control algorithms.

Some typical examples of high-level process control language blocks are illustrated in Figures 2-35 and 2-36. Figure 2-35 is a typical block of software for inputting analog instrument loop information into a Foxboro computer system. Note the question-and-answer technique employed for defining and building the parameters in the block. Figure 2-36 illustrates a simple program for the implementation of one task on a "Fox 3" Foxboro computer. Here, the level and temperature parameters are described by the system software, as indicated.

The next chapter outlines in some detail the development of languages at various levels. Dr. Toong discusses the development at five levels, beginning with the bare instruction set at the binary machine code level, and progressing through assembly, simple higher-level languages, Fortran based languages and, at the top, complete distributed systems software modules for file management and application.

APPLICATIONS

Digital computer applications have been made in many industrial processes since the late 1950's. Initially, however, all were primarily monitoring systems (so-called "data logging" applications). As discussed in Chapter 1, some well-publicized com-

One typical block (Analog Input, AIN) demonstrates the simple question-and-answer procedures employed in defining and entering the block's parameters.

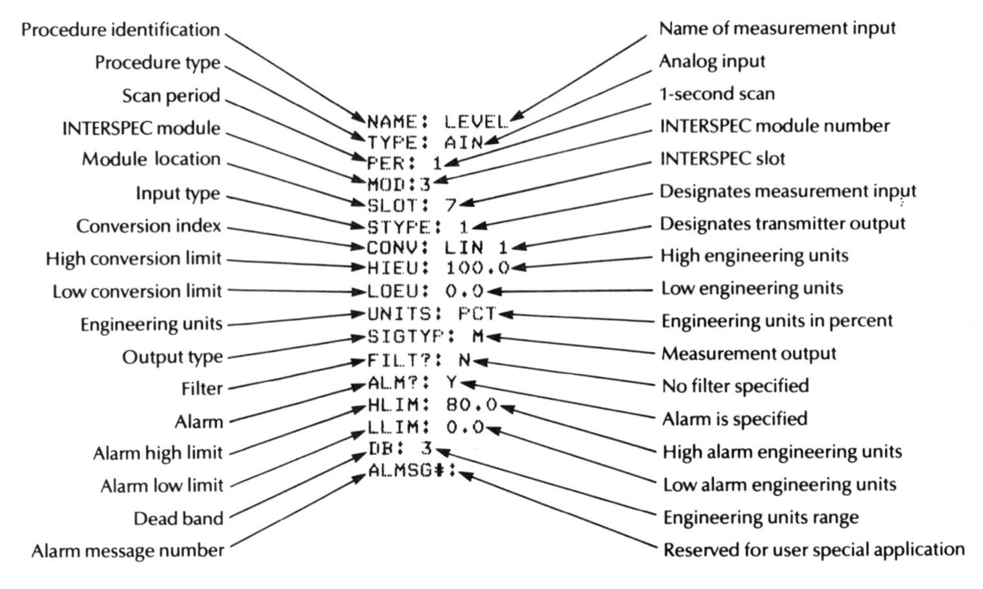

Figure 2-35. Typical "fill-in-the-blanks" language. (*Courtesy of The Foxboro Corporation.*)

puter control applications were also attempted in the paper, steel, chemical and petrochemical industries. Many of these early installations did not produce all of the benefits expected (or projected), for the reasons cited. One such reason was the lack of a systems engineering analysis prior to implementation of detail design. Another may have been due to the massive hardware that was available at the time, which may have helped create the "all eggs in one basket" syndrome regarding the use of one large CPU monitoring many aspects of the process. Still another was the lack of an adequate hardware interface design for the sensing instrument-computer interface. Finally, inadequate operator training on the systems did not help the situation. Whatever the cause, many of these problems are being addressed in the 1970's. One result seems to be the advent of distributed designs of computer hardware and software blocks in various network arrangements. Although distributed control concepts may not be the panacea for computer application to process control, and may be, as some say, just a passing phase (Harrison,[11] in his chapter on minis and micros, indicates that distributed control may be just a passing style), there is beginning to appear a rather strong proliferation of distributed network designs for process control. If strong enough, this trend may indeed set the style for control systems designs of the 1980's, both in analog and digital systems. The usefulness of such designs is just beginning to be proven, particularly in large plant complexes where many unit operations are spread over wide plot-plans and where interaction between key units becomes important. One such application discussed throughout this text is the ethylene process.

As indicated earlier, the ethylene complex has been chosen as a basis of discussion because it embodies almost all unit operations of other petrochemical processes (e.g., cracking furnaces, fractionating towers, compressors) and problems solved for ethylene are applicable to most other petrochemical processes.

SUMMARY

This chapter has provided an overview on the planning and execution aspects of a computer project in the C.P.I. and has detailed highlights of various systems design aspects as an introduction to later chapters.

Given the assumptions raised early in this chapter, concerning the type of project, its scope size and process details, major points on each subject will serve as a base of expansion for later chapters. Thus, the critical path chart (CPM), typical scope document, man-machine interface, common mode rejection considerations, systems design and hardware and software design aspects can be used as a guide to details given later in the text. For example, Chapter 3, on microprocessor hardware and software development systems design, illustrates a number of development techniques for the use of this new hardware and associated software. Chapter 4 gives a lengthy treatment on fractionating column control strategies with computers, and Chapter 5 describes applications of sequence control systems. Chapter 6 treats an important sub-system of any process control computer project for the C.P.I.—product and feedstock composition sensing, analysis and control. This analyzer-instrumentation interface is vital for closed-loop control and implementation of even the simplest of strategies in the petrochemical industries. Chapter 7 is a guide to the writing of a good specification for a com-

Example below illustrates implementation of a task for a simple batch type process. Process input/output blocks are defined in a manner similar to that on page 9. The task, called CHARG, involves the following steps: filling the tank to a level of 50 percent with material "A", then to a level of 75 percent with material "B", starting an agitator, heating the contents to 95° C, holding at that temperature for 20 minutes, stopping the agitator, starting the discharge pump, waiting until the level is less than 5 percent, stopping the pump, and deactivating the task.

```
TASK CHARG
LET FILL A = ON
WAIT UNTIL LEVEL >=50.0
LET FILL A = OFF
LET FILL B = ON
WAIT UNTIL LEVEL >=75.0
LET FILL B = OFF
LET AGIT = ON
CALL SET (TEMP,'SP',95.0)<---- CALL SET means change the set point
WAIT UNTIL TEMP >94.5
WAIT 1200<---- 1200 seconds = 20 minutes
CALL SET (TEMP,'SP',0.0)
LET AGIT = OFF
LET PUMP = ON
WAIT UNTIL LEVEL <5.0
LET PUMP = OFF
DEACT CHARG
END
```

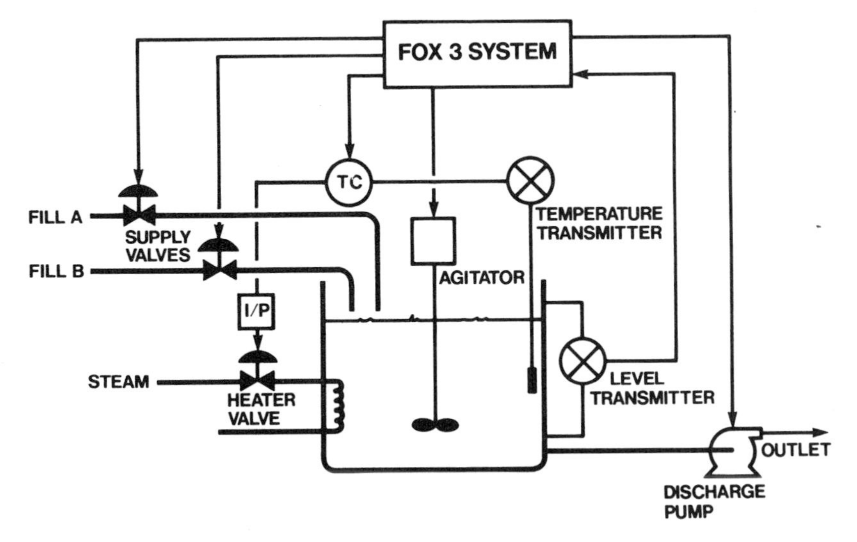

Figure 2-36. Typical batch language format. (*Courtesy of The Foxboro Corporation.*)

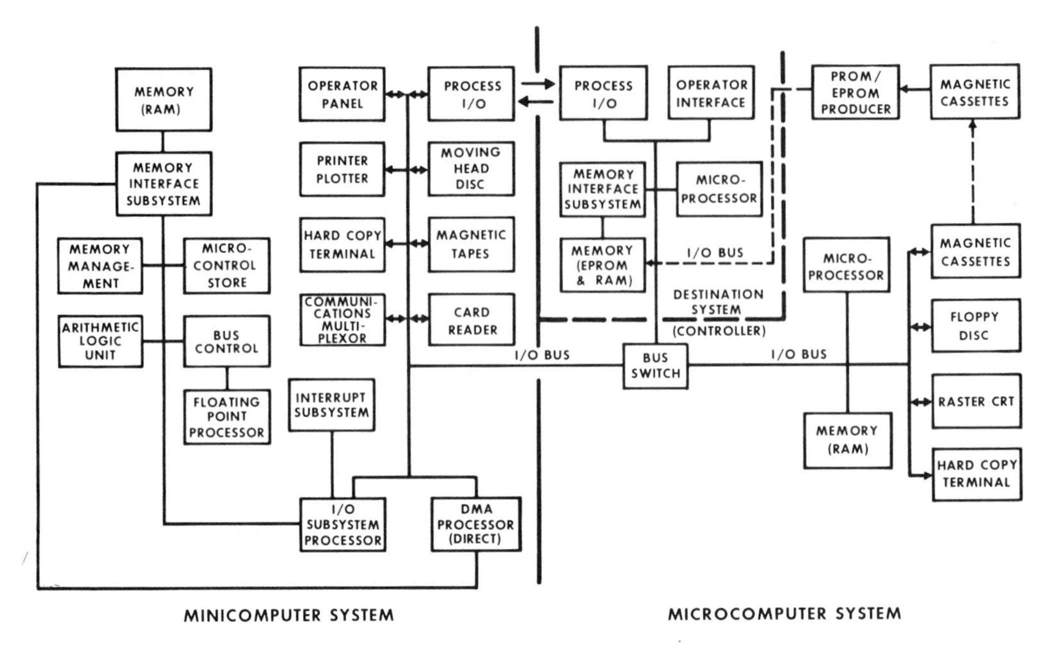

Figure 2-37. A typical development system for process control.

puter system in a closed-loop control scheme for a world-sized plant, and Chapter 8 describes some typical hardware configurations and networks for such projects.

Chapter 9 on special control systems, covers only compressor surge control. (Volume II of this handbook series will cover boiler control, batch processes, oil movement control systems in petroleum refineries, C_3 actelyene hydrogenation package controls, de-mineralizer and other utility package systems, energy savings systems, solids handling and weigh-belt-type process applications using computers.)

We conclude with Chapter 10, which discusses a systems manufacturer's implementation of a process control computer system. A typical distributed system is examined and is carried through the manufacturing process, from the specification generation stage to the building and implementation of such systems with the client or prime contractor.

APPENDIX

The following is a listing, by category, of some major U.S. Corporations cited in this chapter.

Chemical and Petroleum Corporations

- EXXON Engineering and Research Company
  Florham Park, New Jersey
- The TEXACO CORPORATION
  Houston, Texas
- MOBIL CHEMICAL CORPORATION
  Research and Engineering Center
  Princeton, New Jersey
- MONSANTO COMPANY
  St. Louis, Missouri
- E.I. DUPONT DE NEMOURS & COMPANY
  Wilmington, Delaware

- UNION CARBIDE CORPORATION
  South Charleston, West Virginia
- AMERICAN CYANAMID COMPANY
  Wayne, New Jersey

Engineering Contracting Organizations

- STONE & WEBSTER ENGINEERING CORPORATION
  One Penn Plaza
  New York, New York
- The LUMMUS CORPORATION
  Livingston, New Jersey
- The BECHTEL CORPORATION
  San Francisco, California
- The FLUOR CORPORATION
  Los Angeles, California
- M. W. KELLOGG COMPANY
  Division PULLMAN COMPANY
  Houston, Texas

Instrument and Process Computer Systems Manufacturers

- The FOXBORO CORPORATION
  Foxboro, Massachusetts
- TAYLOR INSTRUMENT COMPANY
  Rochester, New York
- HONEYWELL, INC.
  Fort Washington, Pa.
- The BRISTOL COMPANY
  Waterbury, Connecticut
- FISHER CONTROLS, INC.
  Marshalltown, Iowa
- FISCHER & PORTER COMPANY
  Warminster, Pennsylvania

Computer Manufacturers

- IBM (International Business Machines Corporation)
  White Plains, New York
- (D.E.C.) (Digital Equipment Corporation)
  Maynard, Massachusetts
- EMC CORPORATION
  Timonium, Maryland
- HEWLETT-PACKARD CORPORATION
  Palo Alto, California
- GENERAL AUTOMATION CORPORATION
  Anaheim, California
- MODCOMP, INC.
  Ft. Lauderdale, Florida

Computer Systems Organizations Knowledgeable in the C.P.I.

- BILES AND ASSOCIATES
  Houston, Texas
- C.A.T.C.O. (Control Automation and Technology, Inc.)
  Houston, Texas
- EMC CONTROLS, INC.
  Cockeysville, Maryland
- METROMATION
  Applied Technology, Inc.
  Houston, Texas

REFERENCES

1. Kepner, C.H. and Tregoe, B.B., *The Rational Manager: A Systematic Approach to Problem Solving and Decision Making*, McGraw-Hill, New York, 1965.
2. Campbell, D.P., *Process Dynamics,* John Wiley & Sons, New York, 1958.
3. Shilling, G.D., *Process Dynamics and Control,* Holt, Rinehart and Winston, New York, 1963.
4. Deltoro, V. and Parker, S.R., *Principles of Control Systems Engineering,* McGraw-Hill, New York, 1960.
5. Coughanowr, D.R. and Koppel, L.R., *Process Systems Analysis and Control,* McGraw-Hill, New York, 1965.
6. Rau, J.G., *Optimization and Probability in Systems Engineering,* Van Nostrand Reinhold, New York, 1970.

7. Hillier, F.S. and Lieberman, G.J., *Operations Research,* 2nd Edition, Holden, Day, San Francisco, 1974.
8. Instrument Society of America Draft Standard (ISA-RP60.3), *Human Engineering for Control Centers,* Pittsburgh, 1977.
9. Purdue Workshop Committee—Man Machine Interface, *Guidelines for the Design of Man/Machine Interfaces For Process Control,* Purdue University, Indiana and I.S.A., 1976.
10. Bristol, E.H., The Foxboro Company, ''Organization and Discipline For Distributed Process Control,'' Intech, pp. 41–44, January 1979.
11. Harrison, T.J. (Ed.), *Minicomputers In Industrial Control: An Introduction,* I.S.A., Pittsburgh, 1978.
12. Skrokov, M.R., *Microprocessor Control Benefits in Olefins Plant Design,* American Society of Mechanical Engineers' Conference, Mexico City, September 19–24, 1976.
13. Weiss, M.D. and Skrokov, M.R., *Energy Conservation in Olefins Plants,* Control Engineering's Fourth Annual Advanced Control Conference, Chicago, April, 1977.
14. Thor, M.G., Skrokov, M.R., Weiss, M.D., *Energy Conservation Control in Olefin Plants By Mini and Micro Computers,* Instrument Society of America: National Conference, October, 1978.
15. *Introduction to Minicomputer Networks.* A publication of the Digital Equipment Corporation, 1974
16. Foster, C.C. *Computer Architecture,* 2nd Edition, Van Nostrand Reinhold, New York, 1976.
17. Dallimonti, R., Honeywell, Inc., *Operator Interfaces Past, Present and Future,* I.S.A. Paper NF-77-571, presented at I.S.A. Conference, October 17–20, 1977.
18. Coffee, M.B., Computer Products, Inc., ''Common Mode Rejection Techniques for Low Level Data Acquisition,'' *Instrumentation Technology Magazine,* pp. 45–49, July 1977.

3.
Microprocessors and Industrial Process Control*

Dr. Hoo-min D. Toong, PhD.

Assistant Professor, Department of Electrical Engineering and Computer Science and the Alfred P. Sloan School of Management.
Massachusetts Institute of Technology
Cambridge, Massachusetts

INTRODUCTION

A microprocessor is the central arithmetic and logic unit of a computer, together with its associated circuitry and scaled down so it fits on a single silicon chip (sometimes several chips) holding tens of thousands of transistors, resistors and similar circuit elements. It is a member of a family of large scale integrated circuits that reflect the present state of evolution of a miniaturization process which began with the development of the transistor in the late 1940's. A typical microprocessor chip measures half a centimeter on a side. By adding anywhere from 10 to 80 chips to provide timing, program memory, random-access memory, interfaces for input and output signals and other ancillary functions, one can assemble a complete computer system on a board whose area does not much exceed the size of this page. Such an assembly is a microcomputer, in which the microprocessor serves as the master component. About 20 U.S. companies, and the same number of foreign firms, are now manufacturing some 40 different designs of microprocessor chips, ranging in price from $10 to $300. More than 120 companies are incorporating these chips in microcomputer systems selling for $100 and up. The number of applications for microprocessors is proliferating daily in areas such as industrial process control, banking, power generation and distribution, telecommunications and scores of consumer products ranging from automobiles to electronic games.

The potential use of microprocessors for monitoring and control functions in the process control environment is enormous. All levels of the control hierarchy can benefit, including simple closed-loop feedback control systems as well as complex distributed plantwide data-gathering and communication systems. However, a major roadblock to their use at the local level in industrial settings has been the lack of development of reliable and economic sensor/actuator devices that maintain compatibility with the microprocessor technology over extended periods of time.

THE MICROPROCESSOR

As in the central processing unit (CPU) of a larger computer, the task of the microprocessor is to receive data in the form of strings of binary digits (0's and 1's), to store the data for later processing, to perform arithmetic and logic operations on the data in accordance with previously stored instructions and to deliver the results to the user through an output mechanism such as

*This chapter is based in part on an article entitled "Microprocessors," which appeared in "*Scientific American*, September, 1977.

an electric typewriter, a cathode ray-tube display or a two-dimensional plotter. As shown in Figure 3-1, the block diagram of a typical microprocessor has the following units: a decode and control unit, to interpret instructions from the stored program; an arithmetic and logic unit (ALU), to perform arithmetic and logic operations; registers, which serve as an easily accessible memory for data frequently manipulated; an accumulator (a special register closely associated with the ALU); address buffers that supply the control memory with the address from which to fetch the next instruction; and input/output (I/0) buffers, to read instructions or data into the microprocessor or to send them out.

Present microprocessors vary in their detailed architecture, depending on their manufacture and, in some cases, on the particular semiconductor technology adopted. One of the major distinctions is whether all the elements of the microprocessor are embodied in one chip or are divided among several identical modular chips that can be linked in parallel, the total number of chips depending on the length of the "word" the user wants to process: 4 bits (binary digits), 8 bits, 16 bits or more. Such a multi-chip arrangement is known as a bit-sliced organization. A feature of bit-sliced chips made by the bipolar technology is that they are "microprogrammable": they allow the user to create specific sets of instructions, a definite advantage for many applications. However, the majority of microprocessor applications make use of the one chip CPU architecture.

Another major organizational distinction arises from those single chip microprocessors that also incorporate a limited amount of program/data memory and I/0 interfaces

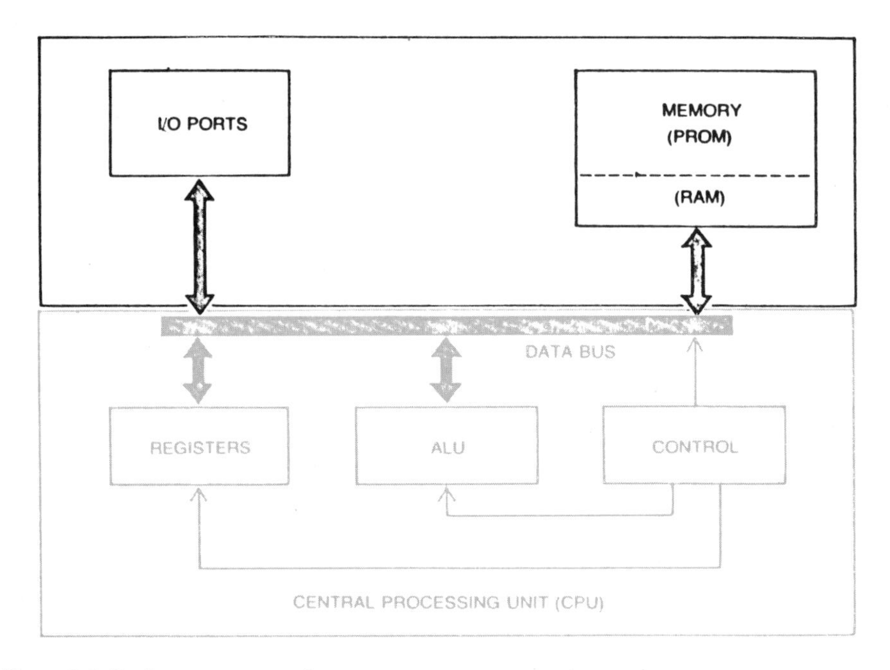

Figure 3-1. Basic components of computer system can now be compressed onto a single chip, as in the Intel 8748. In this block diagram "control" includes control logic and instructions for decoding and executing the program stored in "memory." "Registers" provide control with temporary storage in the form of random-access memories (RAMs) and their associated functions. "ALU" (for arithmetic and logic unit) carries out arithmetic and logic operations under supervision of control. "I/O ports" provide access to peripheral devices such as a keyboard, a cathode-ray tube display terminal, "floppy disk" information storage and a line printer. The functions that are in black convert a microprocessor (*shaded*) into a complete microcomputer.

(e.g., A/D and D/A; digital I/0) on the same chip to form a complete computer. These devices are called single chip microcomputers (versus microprocessors). Their computational power is generally less than their microprocessor counterpart, because of the space needed for the on-chip memory. They are increasingly found as remote devices for communications or for control loop computations in distributed process control systems (See Figures 3-2, 3-3 and 3-4.)

THE MICRO HIERARCHY

The flood of microprocessors and microcomputers reaching the market, combined with the rapid rate of innovation, guarantees that any attempt to catalogue them will be instantly obsolete. A more fruitful intro-

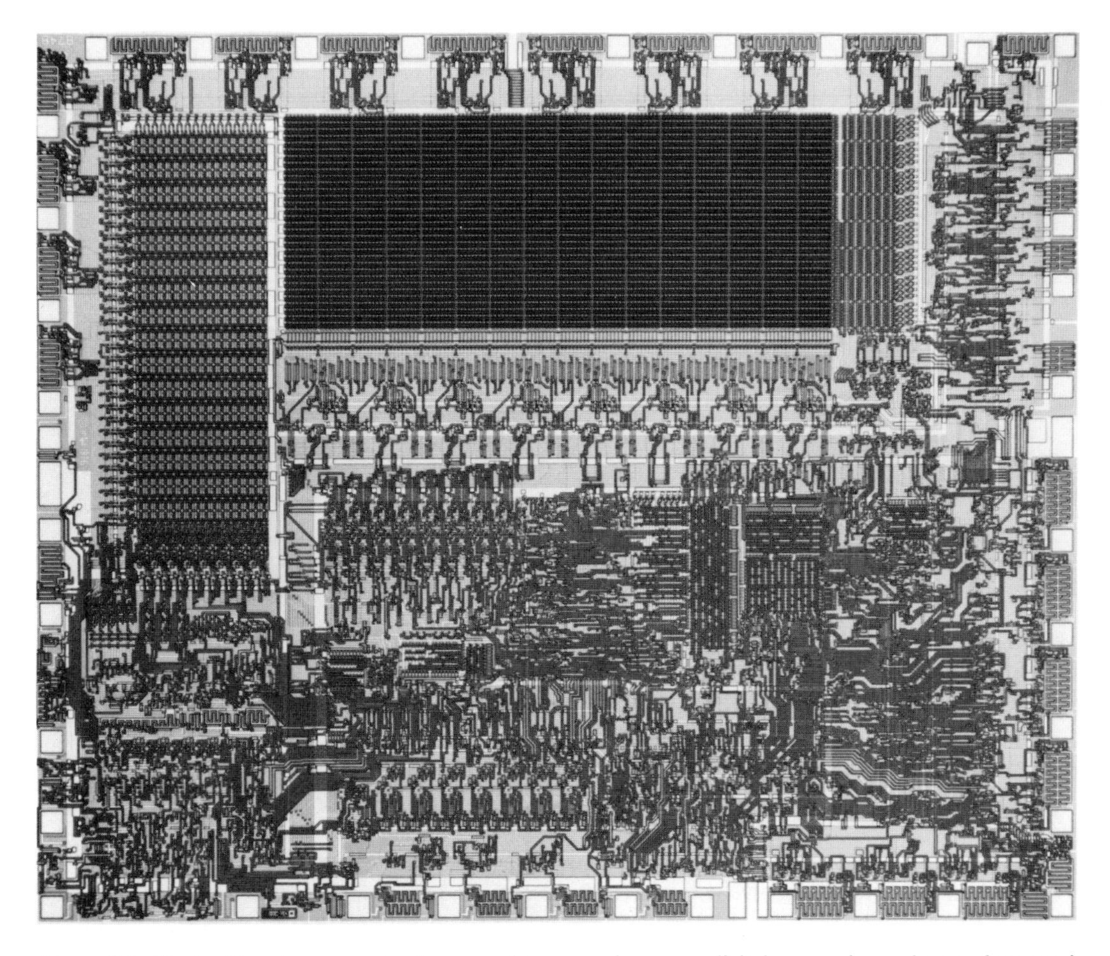

Figure 3-2. Single-chip microcomputer is a complete general-purpose digital processing and control system in one large-scale integrated circuit. The device combines a microprocessor, which would ordinarily occupy an entire chip, with a variety of supplementary functions such as program memory, data memory, multiple input-output (I/O) interfaces and timing circuits. The device shown, the 8748 made by the Intel Corporation, measures 5.6 by 6.6 millimeters. The program is stored in an erasable and reprogrammable read-only memory (EPROM), which has a capacity of one kilobyte, or 8,192 bits (binary digits). The program is erased by exposing the circuit to ultraviolet radiation, which causes the electric charges stored in the EPROM to leak away, after which a new program can be entered electrically. In volumes of 25 or more a packaged 8748 sells for $250.

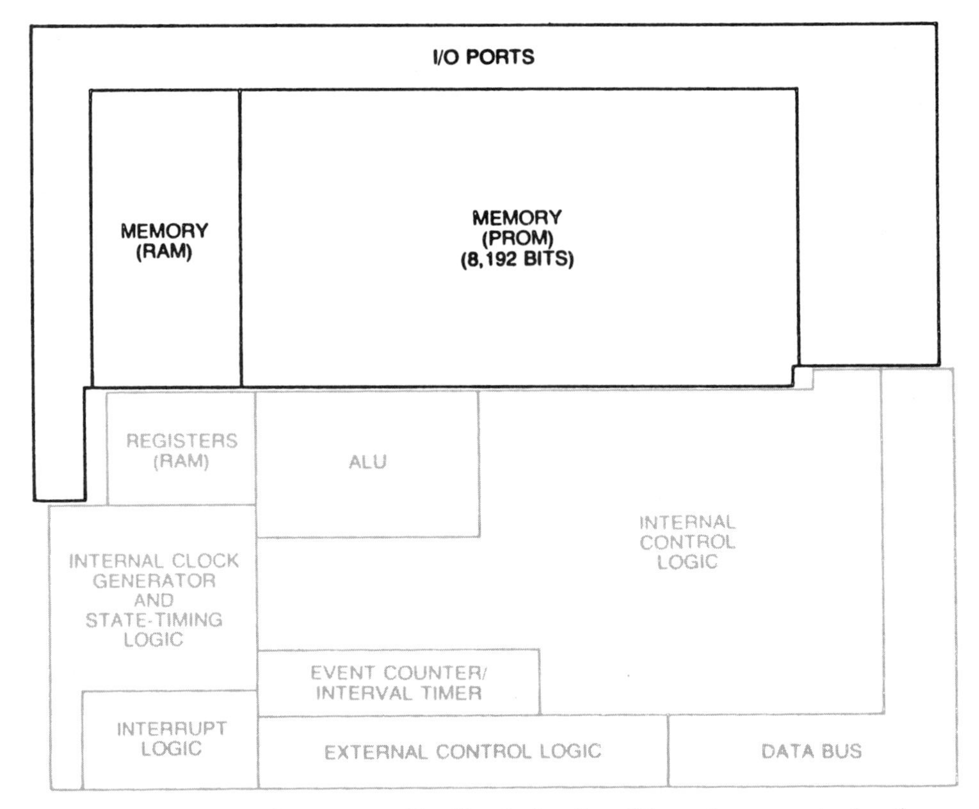

Figure 3-3. Map of 8748 microcomputer identifies the location of the various computer functions. The color scheme used in Figure 3-1 is repeated here. Each function can be assigned to one of the five basic functional blocks; control, memory, registers, ALU and I/O ports. The portions of the chip outlined in black represent the functions that transform the 8748 from a simple microprocessor into a microcomputer. Device holds some 20,000 transistors fabricated by *n*-channel silicon-gate metal-oxide-semiconductor (*n*MOS) technology. Eight-bit central processor responds to 96 instructions in average time of 2.5 microseconds.

duction to the "micro" marketplace is to classify systems hierarchically, according to their capability and function. Along these two dimensions, there is a well defined upward progression in both hardware and software. In hardware, the levels are chips, modules, "bread-board" systems, small computer systems, full development systems and multi-processor systems (see Figure 3-5).

This hierarchy is not absolute because the evolving technology creates ever more powerful chips, some of which can bridge two or three hierarchic levels. Chips are used to construct a module, modules to construct a small computer system (SCS) and small computers to construct a full development system (FDS). Multi-processor systems can incorporate modules, SCSs or FDSs, depending on the application and complexity.

Microprocessor Chips

At the first level of the hierarchy are the microprocessor chips, representing the large scale integration of tens of thousands of individual electronic devices: transistors, diodes, resistors and capacitors. At this level, there are also more specialized chips: random-access memories (RAM), read-only memories (ROM), programmable read-only memories (PROM), I/0 interfaces and others. The cutting edge of the technology works most directly at the chip level, providing, for example, RAM of ever higher storage capacity (for instance, currently 65,536 and soon 262,144 bits per chip).

Generally, the various kinds of chips are grouped into families that are compatible with particular microprocessors. The families will include a series of RAM, ROM and PROM chips to create a memory system, a series of interface chips capable of handling both parallel and serial I/0 functions and miscellaneous chips to enhance system capabilities, such as high-speed arithmetic operations. Master control chips are needed to establish priorities and to keep signals flowing smoothly through the complex maze of interconnections. The compatibility of chips and chip families made by different manufacturers varies widely. For example, the microprocessors of different builders are generally not physically interchangeable, whereas several types of memory chips often are.

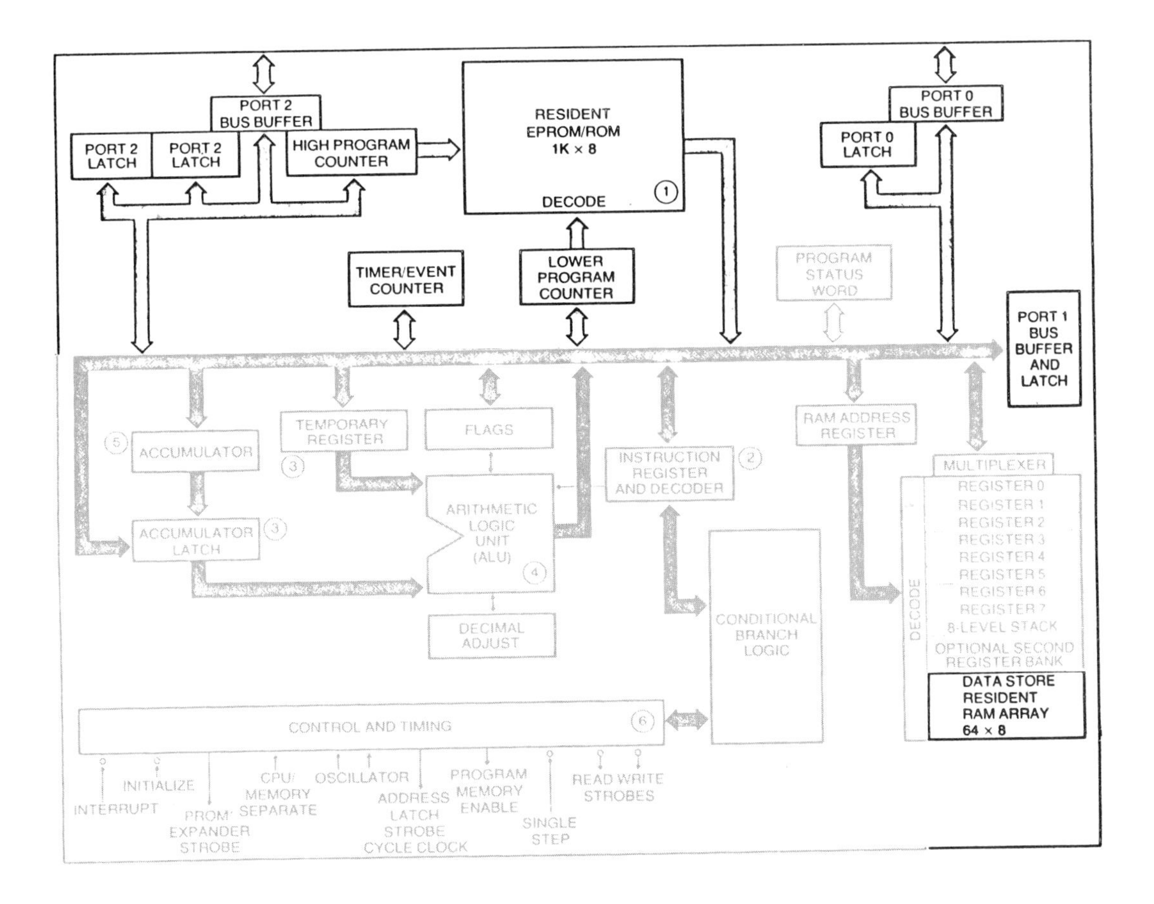

Figure 3-4. Functional block diagram of the 8748 microcomputer can be used to follow the sequence of steps involved in a simple operation, for example the addition of the contents of two registers, *A* and *R*, where *A* is the accumulator and *R* is any of the registers in the array at the lower right. The computer's first step (*1*) is to fetch the instruction from memory: "ADD A, R." The next step is to place the instruction in the instruction register and decoder (*2*), where the decoder finds the instruction to add *R* to *A* and to leave the result in *A*. In the next step the contents of register *R* are sent to the temporary register (*3*) and the contents of the accumulator to the accumulator latch (*3*). The ALU (*4*) then adds the contents of the two registers and the result is returned to the accumulator (*5*). Instruction ends and a signal is generated (*6*) to fetch next instruction. The 8748 microcomputer is capable of performing some 400,000 such additions per second.

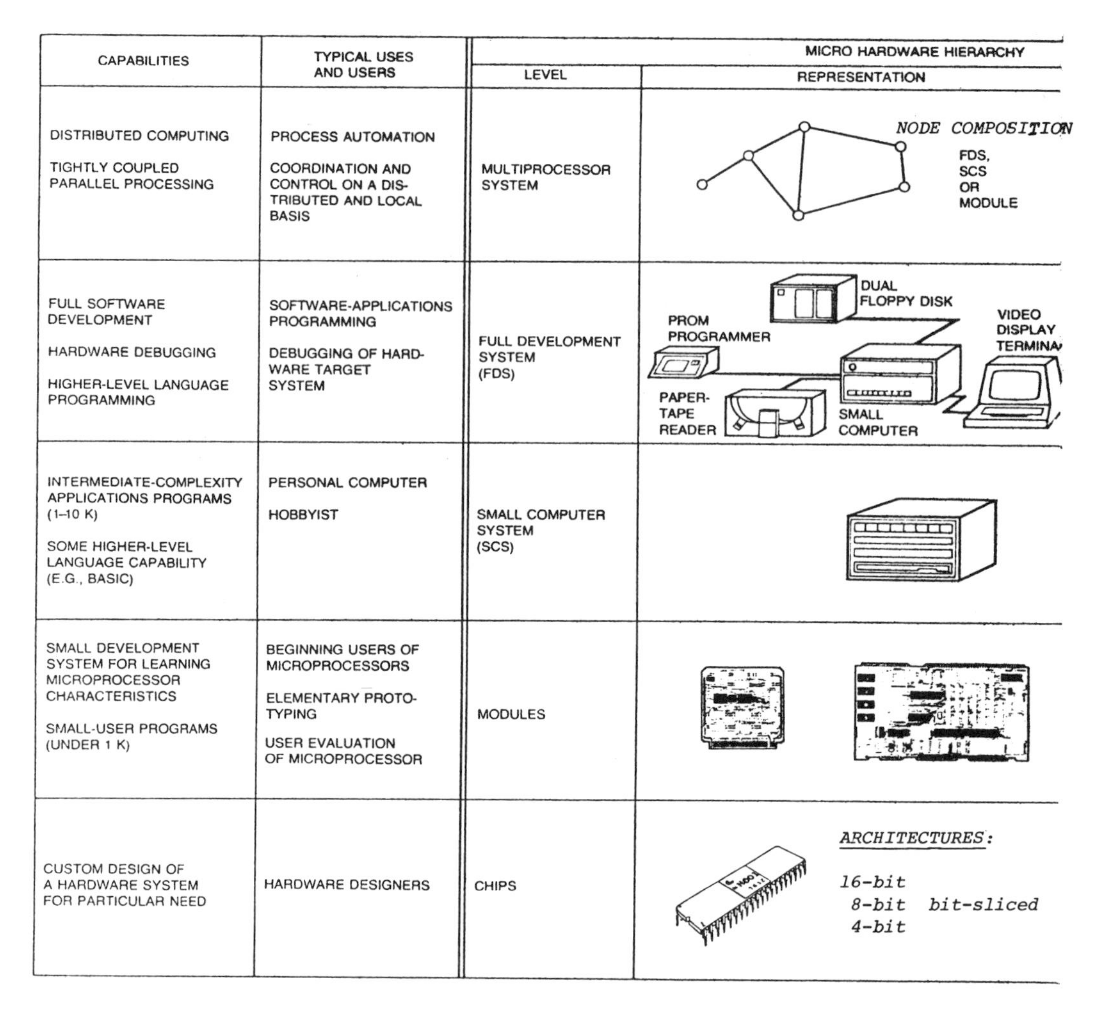

CAPABILITIES	TYPICAL USES AND USERS	MICRO HARDWARE HIERARCHY LEVEL	REPRESENTATION
DISTRIBUTED COMPUTING TIGHTLY COUPLED PARALLEL PROCESSING	PROCESS AUTOMATION COORDINATION AND CONTROL ON A DISTRIBUTED AND LOCAL BASIS	MULTIPROCESSOR SYSTEM	NODE COMPOSITION FDS, SCS OR MODULE
FULL SOFTWARE DEVELOPMENT HARDWARE DEBUGGING HIGHER-LEVEL LANGUAGE PROGRAMMING	SOFTWARE-APPLICATIONS PROGRAMMING DEBUGGING OF HARDWARE TARGET SYSTEM	FULL DEVELOPMENT SYSTEM (FDS)	DUAL FLOPPY DISK PROM PROGRAMMER VIDEO DISPLAY TERMINAL PAPER-TAPE READER SMALL COMPUTER
INTERMEDIATE-COMPLEXITY APPLICATIONS PROGRAMS (1–10 K) SOME HIGHER-LEVEL LANGUAGE CAPABILITY (E.G., BASIC)	PERSONAL COMPUTER HOBBYIST	SMALL COMPUTER SYSTEM (SCS)	
SMALL DEVELOPMENT SYSTEM FOR LEARNING MICROPROCESSOR CHARACTERISTICS SMALL-USER PROGRAMS (UNDER 1 K)	BEGINNING USERS OF MICROPROCESSORS ELEMENTARY PROTOTYPING USER EVALUATION OF MICROPROCESSOR	MODULES	
CUSTOM DESIGN OF A HARDWARE SYSTEM FOR PARTICULAR NEED	HARDWARE DESIGNERS	CHIPS	ARCHITECTURES: 16-bit 8-bit bit-sliced 4-bit

Figure 3-5. Microprocessor systems can be arranged in an ascending hierarchy of hardware and software in which smaller components are assembled into successively larger systems with more powerful capabilities. The building blocks are the families of chips designed for various functions. To solve an application problem, for example the control system of an airplane, designers usually assemble modules or small computer systems and provide them with a suitable program for the task. Chips and modules have become so cheap (less than $30 for a microprocessor and less than $300 for a single-board module) that a major cost in engineering an application

Bread-boards: Single Board Computers

The second level of the hierarchy, the module and bread-board systems, represents the simplest true computer systems. They can be created by combining a microprocessor with a limited array of memory chips (RAM and ROM) and I/0 chips. In order to communicate with such a minimal system the user will also need a simple device such as a numeric keyboard, as well as a device capable of displaying or recording the computer output. Such single board systems are useful for introductory teaching purposes or as bread-board prototypes for more sophisticated systems. For a modest investment (usually under $300), a beginner can learn the fundamentals of microprocessor programming. However, because of the system's limited memory, its lack of software development tools and its crude in-

	MICRO SOFTWARE HIERARCHY		
COMPONENTS	LEVEL	REPRESENTATION	COMPONENTS
MICROCOMPUTER SYSTEMS COMMUNICATION SUB-SYSTEMS REAL-TIME CONTROL INTERFACES WITH SENSORS AND ACTUATORS	DISTRIBUTED-SYSTEMS SOFTWARE	DISTRIBUTED NETWORK MODULES FOR FILE MANAGEMENT, DEVICE CONTROL, COMMUNICATION APPLICATIONS, ETC.	DISTRIBUTED OPERATING SYSTEM
MICROCOMPUTER, VIDEO DISPLAY TERMINAL, FLOPPY DISK, PROM PROGRAMMER, ETC. OR MINICOMPUTER-BASED CROSS-SOFTWARE SYSTEM	DEVELOPMENTAL SOFTWARE	DISK-BASED OPERATING SYSTEM HIGHER-LEVEL LANGUAGES (E.G., FORTRAN, BASIC, PL/M) EXAMPLE X X - 1 WHERE X IS AN 8-BIT NUMBER AND Y IS ITS ADDRESS	FLOPPY-DISK/CASSETTE OPERATING SYSTEM COMPILERS IN-CIRCUIT DEBUGGERS

PACKAGED VERSION OF MODULES WITH EXPANSION CAPABILITY FOR MEMORY AND INTERFACES	STAND-ALONE SOFTWARE (NO PERIPHERALS)	SIMPLE HIGHER-LEVEL LANGUAGES (E.G., BASIC)
		ASSEMBLY LANGUAGE (E.G., 8080)
		LXI H,Y
		MOV A,M
		ADI 1
		MOV M,A
			STAND-ALONE MONITOR
			ASSEMBLER
			EDITOR

BARE PLUG-IN CIRCUIT BOARD INCORPORATING CHIP FAMILY WITH EXPANSION CAPABILITY FOR MULTIPLE BOARDS	ELEMENTARY SOFTWARE	ASSEMBLY LANGUAGES (E.G., HEXADECIMAL) 21 Y 7E C6 01 77	SIMPLE MONITOR LIMITED DEBUGGER
CENTRAL PROCESSING UNITS (CPU'S) RANDOM-ACCESS MEMORIES (RAM'S) READ-ONLY MEMORIES (ROM'S) PROGRAMMABLE ROM'S (PROM'S) INPUT-OUTPUT (I/O) INTERFACES OTHER SPECIAL-PURPOSE CHIPS	BARE INSTRUCTION SET	BINARY MACHINE CODE 0010 0001 Y 0111 1110 1100 0110 0000 0001 0111 0111	OPERATIONS HARD-WIRED OR MICROPROGRAMMED AT CHIP LEVEL

is the cost of developing the software to create the final program for the "target" system. Improvements in semiconductor technology are steadily making it possible for systems at each level to include more of capabilities once assigned to level above. Symbols and characters in color show how the same instruction looks when it is written in different languages of an increasingly higher level, beginning with binary machine code.

terface with the user, even a novice is apt to outgrow a single board system quickly.

Small Computer Systems

At the next level in the hierarchy of capability and function are the small computer systems that are prepackaged as stand-alone units. Unlike the single board modules, they have a self-contained power supply, the capability for memory expansion and room for a series of plug-in interface modules. Some of the more powerful single board development modules can be expanded with the appropriate hardware to create such a single box computer system. All the small computer systems have software capabilities that approach in sophistication those found in much larger conventional systems. They also provide an interface for a cathode-ray tube (CRT) or keyboard display console. In addition, many of the small systems can be interfaced to such peripheral devices as "floppy disk" memories, tape cassettes, paper tapes and line printers. With such enhancements, a small computer system could serve as a full development system. For the most part, however, the single box computers find a major market today among computer hobbyists—who employ them for small programming tasks, word processing, general computation and game playing.

Full Development System

At the next level in the hierarchy we come to the FDS, or full development system. Perhaps its most important role is to provide a quick and efficient means for developing a low-cost microprocessor module that will later be manufactured in volume to solve a manufacturing, telecommunications or business problem. In other words, the FDS is a full capability microelectronic system that can serve as a vehicle for helping to develop a smaller target system.

Whereas the FDS may represent an investment of some $15,000, counting both hardware and software, the target system will be a microprocessor or microcomputer costing perhaps $500, or even less when it is manufactured in volume. For example, if one wanted to develop a microprocessor to optimize the performance of an automobile engine (by continuously adjusting the amount of fuel, the ignition timing and the fuel mixture), the final unit might be a small integrated circuit module affixed to each engine and costing well under $100. One would use an FDS to develop the programs that would ultimately be placed in the ROM and PROM memories of the target system, a dual floppy disk computer system, a dual floppy disk drive with a controller, a line printer, a CRT display terminal or teleprinter console, a ROM or PROM programmer and possibly a few other specialized pieces of hardware.

Multi-Processors

The final level of microcomputer usage is the multi-processor system. The microprocessor represents truly low-cost computing. Its economics are so compelling that microcomputers are serving not only in many applications where computing power was previously too costly, but also in applications where several dozen dedicated microprocessor modules can now be teamed to monitor and control parts of existing industrial or commercial systems where computer control was formerly unthinkable. Such an assembly of microprocessors or microcomputers can be organized in two functionally distinct ways.

In the first type of organization, a tightly coupled group of microprocessors is designed to exchange data at high rates over short distances, with a high degree of parallelism, to achieve a maximum of computational power. Such a system could be used to emulate a large computer, to provide high reliability or to handle a specific problem that can take advantage of several processors operating in parallel.

The second organization, with by far the greatest application potential, is a loosely coupled system in which several widely distributed microcomputers communicate at low data rates with little or no parallelism. Examples of such distributed systems would be applications to factory automation, the control of oil refineries and chemical plants and the control of electrical devices in a home. Distributed systems are approaching reality in sizable numbers.

As can be imagined, the design and development tools for multi-processor systems are much more primitive than those for single processor systems. Software problems that are difficult enough to solve for one microcomputer expand almost geometrically in complexity as more units are added to the system. The problems include the organization of distributed files, or information storage systems; process scheduling; and achieving what programmers like to call "graceful" degradation, which means that the system should not merely "fail safe" but fail in gentle stages—gracefully. On the hardware side of the problem, manufacturers have so far given scant attention to the configuration of microcomputers that would lend themselves efficiently to distributed installations.

SOFTWARE

So far, we have used the term software without being very specific about its meaning. Since an understanding and a manipulation of software are fundamental to all computer usage, we shall now be somewhat more explicit. Like the hardware described above, software has its hierarchies. In the broadest sense, software provides the means for telling a computer explicitly what to do through a step-by-step sequence of instructions that form a program. Each computer is provided with an "instruction set": a list of all the basic operations the computer is capable of performing. Each instruction is written in binary machine code: a sequence of 0's and 1's, typically 8 or 16 bits long. Although a complete program could be written in this low-level language, the task is so tedious that an intermediate representation known as *assembly language* was developed and is currently the most common language employed for programming microprocessors. Usually, each symbolic instruction written in assembly language represents a single instruction in machine language. The translation is done by the computer itself with an "assembler" program.

To make programming still easier, "higher level" languages were developed in which the instructions more nearly approximate ordinary English and the notations of mathematics. Examples are FORTRAN, ALGOL, COBOL AND PL/1. One statement in such languages usually corresponds to many statements in machine language. The translation is done by the computer, with the aid of a program called a compiler.

Even with such simplifications, writing a program is arduous. Discovering the inevitable mistakes and then correcting them is known as debugging. To simplify making changes in a program, the user can employ a special editing program, which facilitates the changing of individual instructions. Once the program is debugged, it is usually stored in some non-volatile memory device, such as a magnetic tape or disk. When the program is ready for the computer, it is rapidly transferred from the tape or disk into the computer's high-speed random-access memory. As we have seen, however, the usual goal of a full development system is to create a program that can be stored in the permanent (ROM or PROM) memory of a microcomputer targeted to solve a specific problem repetitively.

Over the years, there has been a proliferation of symbolic and higher-level languages for special purposes, each with its own assemblers or compilers for making it intelligible to particular models of computers. As a result, "cross-software" systems have been developed to facilitate communication between computers. Thus users of large computer systems and time-sharing services have access to cross-software assemblers, compilers and simulators (programs that enable a computer of one make or model to duplicate the actions of another). At present, this represents an expensive alternative to a full development system for creating microprocessor software. In addition, software simulators do not readily duplicate real-time I/O and execution speeds of the target microprocessor.

MICROPROCESSOR PROJECT MANAGEMENT

The successful execution of a microprocessor application requires a top-down approach to the management of such a project. Upon the initiation of a project, too many designers immediately commence with a bottom-up approach to the application process. This bottom-up synthesis approach may be further encouraged by such factors as competitive deadlines, ready availability of chips and high-technology devices, the fear of technology obsolescence and the desire to show immediate, tangible results. A certain amount of the synthesis approach is necessary, but major emphasis must be placed on a top-down applications-driven approach for the successful execution of the microprocessor application.

There are three distinct sequential phases that form the top-down operational framework for a microprocessor project. These phases are a design phase, an implementation phase and an operations/update phase, as shown in Figure 3-6.

In the design phase, it is important to establish the user needs for the system under consideration. The objectives of the particular proposed system have to be established, and the relationship of the needs for such a system must be considered in light of corporate goals and long-term product planning. Once these needs have been established, a system requirements specification can be generated. The requirements specification differs from the top-level design in that it outlines the functional scope and actions of the system as the user would view them. This specification is then used to generate a preliminary architectural design. These three steps within the design phase are iterated several times before the second phase, implementation, is initiated.

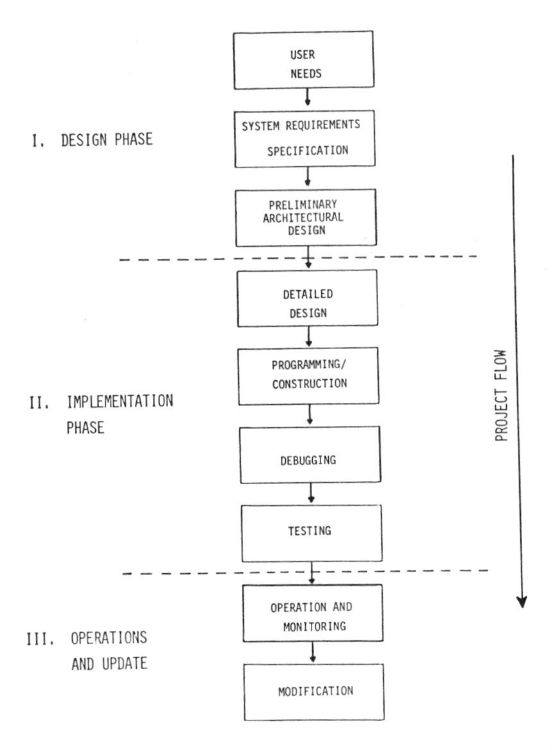

Figure 3-6. Operational framework for software/hardware microprocessor project.

Too many designers *begin* their microprocessor application project at the second phase. They often have a preliminary architectural design in mind which has no relationship to user needs or a requirements specification. If the design phase is not done correctly (or is not done at all), then the entire project runs the risk of developing a product which satisfies no real user needs.

The key factor to the successful completion of the implementation phase is the integration of software and hardware skills in the debugging and testing of the engineering prototype. Software and hardware personnel have traditionally viewed each other's areas of expertise as being separate and worlds unto themselves. However, in a microprocessor environment, the software and hardware are intimately related and intertwined at all levels of design and execution. Minor changes in the hardware design could radically affect the ability of the software to execute control or computation algorithms. Conversely, software evolution and the need for upward compatibility in programs will often restrict hardware architectural changes or modifications. (The implementation task is more fully discussed below, under the heading "Implementation Phase.")

The implementation of a working prototype of a microprocessor application still requires a third phase for the successful completion of the project, operations/update. It is crucial to operate the prototype in the actual industrial environment for which it was designed. This is necessary to evaluate the degree to which a product satisfies the user needs as given in the system requirements specification. This typically involves an operational monitoring and evaluation phase, during which time modifications and updates will be made. The ability to perform maintenance is also key to the long-term survival of the microprocessor application in its industrial setting.

The importance of this top-down phased approach to the successful completion of a microprocessor project cannot be overstressed. The neglect of the design phase and the use of ad hoc assumptions for a quick implementation often results in project trouble signs such as the following.

Project Warning Signs

- Project phases are consistently completed late and cost more than estimated.
- Documentation is inadequate in quantity and quality.
- Managers and users do not know where things stand.
- Systems do not adequately take advantage of new design methodologies or new technologies.
- Users are not satisfied with the final product.

The increasing complexity of distributed monitor/control systems is greatly increasing their cost. As the microprocessor becomes distributed throughout such systems, the costs of software programming and debugging, coupled with system integration and testing, vastly outweigh the hardware costs. Because of the huge manpower investments involved in such tasks, it is difficult to abandon such a project in mid-stream when serious design defects or deficiencies arise. Consequently, much more emphasis must be given to the correct management of these projects and to their adherence to a top-down approach.

Implementation Phase

Given that the design phase has been accomplished and a preliminary architectural design has been generated, how does a manufacturing company go about developing a suitable program? As indicated above, the prospective user does not, as a rule, try to design a special microprocessor chip for his particular task. He starts with one of the chips already on the market and selects from the wide assortment of other available chips: RAM, ROM, PROM, I/0 interfaces or whatever may be needed to construct a module capable of carrying out the task he hopes to perform. In many cases, commercially available general-purpose, single board microcomputer modules will be satisfactory for his task. Microprocessors have become so inexpensive that it is usually cheaper to exploit as little as 10% or even 5% of the computing power of an existing chip or module than to invest in the design and programming of a special unit that would do the job with only the minimum number of electronic components. Given a sufficient volume of production, however, development of a special unit may be justified.

The process of developing a microprocessor application begins with the identification of a need. Often the person in an organization who perceives the need is unfamiliar with the details of the new microelectronic technology. As a result, in most cases, the need is communicated to an engineering project manager, who makes an evaluation to determine whether the use of a microprocessor is justified. Such an evaluation would include an analysis to determine which, if any, of the available microprocessors have suitable capabilities, and to estimate the time and manpower needed to develop the necessary software—by far the most time-consuming and costly part of the job.

If a microprocessor appears to be justified, the task is broken down into two distinct paths: the hardware requirement is turned over to a design engineer and the computation and control requirements are given to a software programmer. In the typical case, hardware and software efforts are carried out in parallel by two separate groups. The key to a successful application is close communication between the two as the system evolves. Unfortunately, there is no standard methodology for achieving a good design. As in all engineering, much depends on intuition, a good working knowledge of available products and past experience.

With "ad hoc" design procedures for both hardware and software, a prototype system is developed. Hardware prototyping mechanisms commonly include wire-wrap bread-board models, plug-board set-ups or printed-circuit prototypes. The corresponding prototyping system in software is a target (or object) program that can be derived from any of several translation mechanisms: hand assembly (by means of a code book), the use of a resident assembler on a full development system or the use of a cross-assembler on a time-sharing system.

When prototypes of both hardware and software are sufficiently advanced, the two must be mated—a task usually performed by a systems engineer. The target program is loaded into the hardware prototype to see if the resulting system meets the original specifications of the program manager. Deficiencies in either hardware or software at this level must be fed back through the hardware and software "loops" in iterative fashion until satisfactory performance is achieved. This process is shown in Figure 3-7.

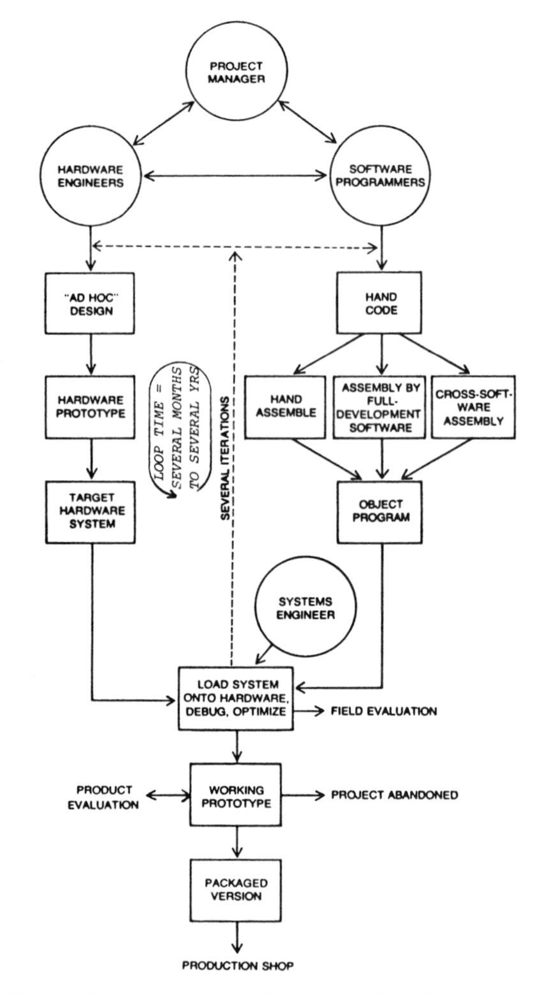

Figure 3-7. Task of developing applications for microprocessors requires close cooperation between hardware engineers, concerned with the selection and interfacing of chips, and software programmers, whose job is to provide a "debugged" program that will meet the project objectives. Because the time needed for programming is often underestimated the original objectives are sometimes obsolete by the time prototype is debugged and ready for production. As a result many projects are abandoned at this stage. It is not unusual for development programs to take two years. Author believes the time can be greatly compressed with Triad.

Now a major decision must be made: whether or not to put the microprocessor module into production. The length of time from the start of development to a successful prototype is often critical. Many prototypes are shelved at this stage because development has taken so long (it can take as long as two years) that the original problem requirements have changed, the perception of the problem has changed or a competitor has reached the market first with an equivalent product, making it necessary to come out with something better. Since time is often critical, software and hardware design groups have an urgent need for development tools that will make it possible to compress the development time of prototype systems into a few weeks.

TOOLS FOR MICROPROCESSOR DEVELOPMENT

In order to effectively execute a microprocessor application, the correct tools, both hardware and software, must be used. In fact, the development of a prototype microprocessor application involves the use of software and hardware tools at all levels of development. In order to understand the problems encountered during this application process, it is best to trace the current use of these tools in the application process.

Generation 0

A large number of people begin their microprocessor application with a small target system, often composed of a single board computer (SBC). This target system has been selected on the basis of a microprocessor CPU evaluation for the particular task and it is chosen as the most inexpensive, cost-effective way to develop the software application program. An example of this technique is shown in Figure 3-8, where the SBC is being used as a set point controller in a simple feedback configuration to monitor and control flow in a pipe. These SBC target systems are offered by the manufacturers of nearly all microprocessor chips. They are designed to interface to a teletype or CRT with an on-board mini-monitor program. This approach is the most primitive in terms of development techniques and is generally abandoned after the first few weeks.

The problems that are encountered with this approach are many fold. Primary among them is that the target system has almost no software support to do program development for the particular application. In most cases, it has no on-board assembler and requires the user to hand-assemble his programs before entering them through the

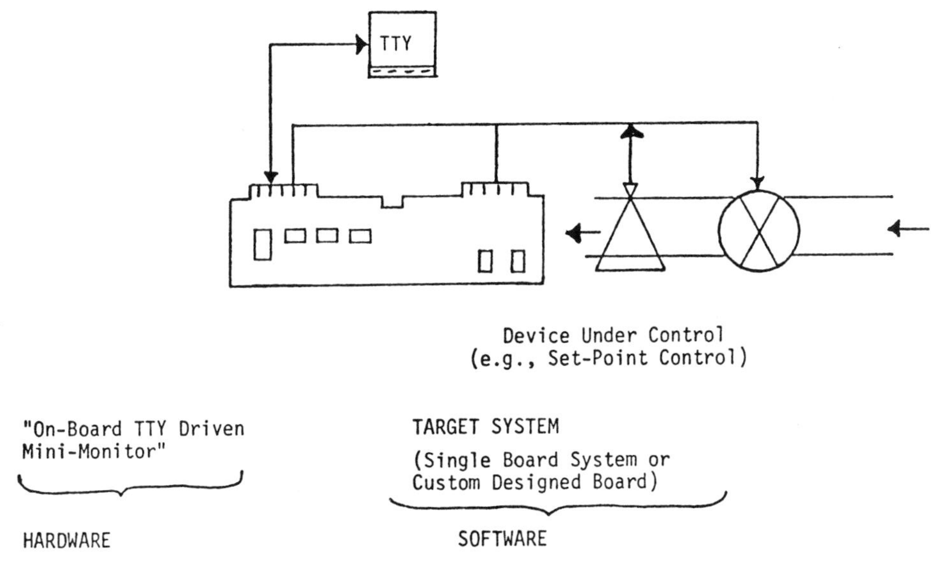

Problems with this approach:
1. Target System too limited for software development
2. Limited Memory Space
3. No permanent storage of programs
4. Limited editing, debugging facilities

Figure 3-8. Most primitive development techniques involving the target system.

teletype. Second, the target system has very limited memory space, confined to the available on-board RAM and PROM memories. Third, the RAM memory, which may be used to store prototype versions of the application program, cannot permanently retain data. Upon a power down condition, the RAM memory loses all data. The final problem of this approach is the limited editing and debugging facilities available through the mini-monitor on such a single board target system.

The principal reason why many individuals choose this approach is that the target system is very close to the microprocessor system that will be installed in the field to accomplish the control and monitoring functions. The size and complexities of the software development effort are nearly always underestimated. For the development of applications programs greater than several hundred lines of code in length, such a target system is severely inadequate.

Generation 1

The development engineer rapidly outgrows the capabilities of such a primitive development system and subsequently takes the step to acquire a full development system, as shown in Figure 3-9. Typical full development systems include dual floppy disk drives, line printers, PROM programmers, in-circuit emulators (ICE), and paper tape equipment. Their software consists of editors to help prepare programs,

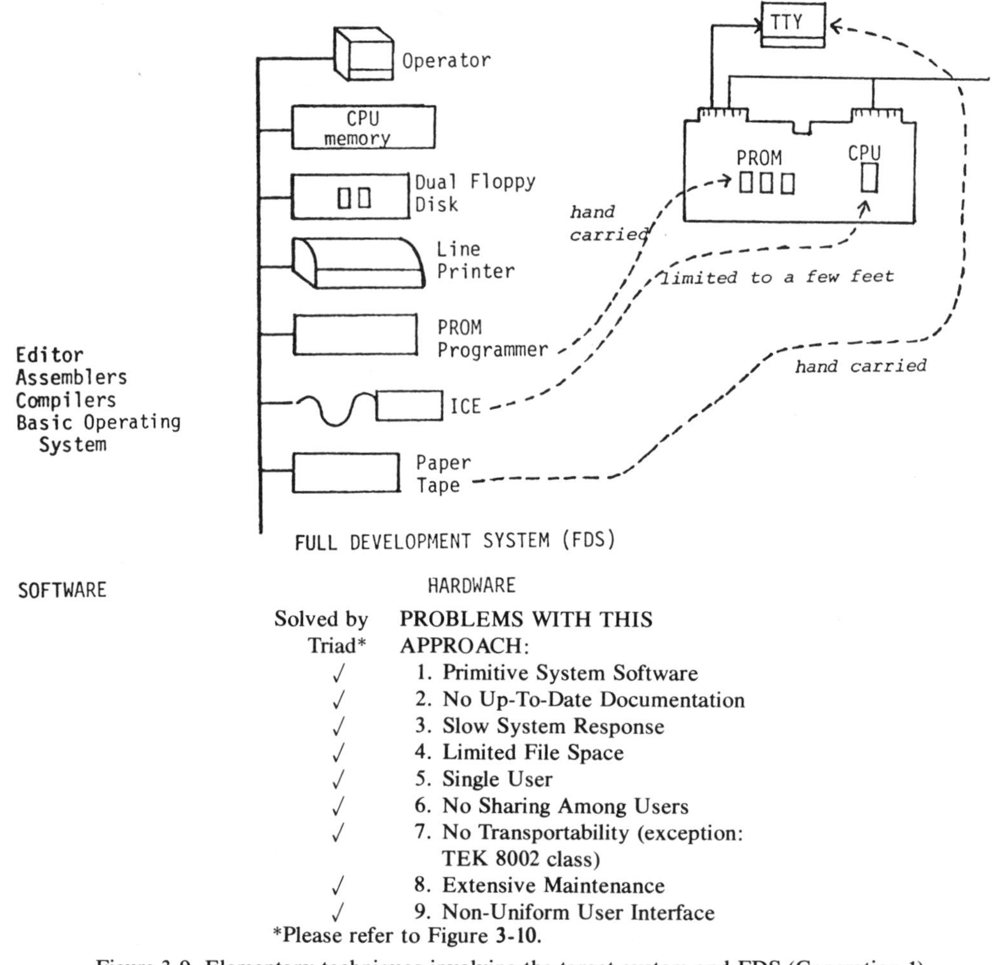

Figure 3-9. Elementary techniques involving the target system and FDS (Generation 1).

assemblers and compilers to translate source into machine code and a basic operating system to manage user files and I/0 devices. The FDS is primarily a software development station used to develop trial programs that can be tested on the target system. The development procedure is as follows.

1. The user employs the editor and floppy disk to create the source level of his application programs.
2. The user invokes the assembler/compiler resident of the FDS to produce the object code.
3. The user now has the option to either execute this program for debugging purposes or to put the object file in a format that is transportable to the target system for testing purposes. These two formats are generally in PROM or in paper tape.
4. The object programs to be tested are hand-carried from the FDS to the TS (in some cases, the ICE facility is used; however, this is limited physically to a distance of a few feet).
5. The user then tries the test program in its target system environment, possibly with a small on-board debugger to highlight program errors at the target system level.
6. Where errors are found, the user returns to Step 1 for a re-editing, reassembling and transport of the next iteration of test programs to the target system for the FDS. Or, more often than not, the user starts to implement patches in his code at the target system level to avoid the delay in Steps 1 through 5. Depending on the utilization of the FDS system by several groups, the typical delay for Steps 1 through 5 can range from an hour to a day. This delay is generally intolerable for development engineers and forces a patching mode at the object code level. This, of course, leads to problems in keeping documentation up to date with the source code level.

The procedure as outlined using these two tools, FDS and TS, has a myriad of problems, as indicated below.

- *Primitive system software*. The systems programs available on the FDS, such as the editor, the file system and the disk operating system, are fairly primitive with respect to current programming techniques. This is an evolutionary phenomenon, but the FDS software has always lacked those capabilities available on larger mini- or midicomputer systems.
- *Difficulty of maintaining up to date documentation*. Because of the long iteration time to develop a new test program, development personnel often patch object programs in the target system. This leads to very severe documentation problems which can cause large amounts of after-project effort to update programs after their development.
- *Slow system response*. Because the FDS is generally floppy disk based, and because the CPU is generally an 8 bit processor, system response has been very slow. This includes file access time as well as searching, sorting and other data base manipulations over user files.
- *Limited file space*. A file space provided by the dual floppy disks is marginally adequate for small development efforts. As the complexity and size of the application efforts grows, as the number of duplicate copies and historical backups kept grows, and as documentation grows, one often finds the file space available per diskette in such a floppy based system to be inadequate. In addition, development personnel frequently like to segment project work per floppy. This cuts down any potential sharing between projects because of the separate diskettes used per project.
- *Single user*. The FDS system is a single user system and consequently will

present a bottleneck problem when several groups must schedule their time on one system. A common approach is to buy several systems, one for each project group. This is an expensive approach as well as one that engenders problems of sharing and maintenance.

- *No sharing among users.* Because each user/project has his own floppy disk for his applications program under development, there are no easy mechanisms for sharing of programs, data, documentation and experience of a project among several users. In other words, it is not easy for one user to have access on a constant and immediate basis to the programs and files of another user's diskette. This problem of sharing is very crucial to prevent duplication of effort among project groups and to allow the experience gained in one project to be used in the next.
- *No transportability.* Generally, FDS systems are dedicated by manufacturer to a specific line of microprocessor products. For instance, a Motorola development system will not support Intel microprocessors. An exception of this is a class of universal FDS systems such as the Tektronix 8002. However, even these systems have the same problems as outlined above; in addition, they offer only limited higher-level software support.
- *Extensive maintenance.* The FDS system essentially represents a scale-down, low-cost minicomputer system. Because its components are not of the same quality as found on larger systems, more maintenance problems are experienced. This particularly applies to the floppy disk system and the line printer. In general, FDS manufacturers do not offer maintenance contracts.
- *Non-uniform user interface.* When several FDS systems are used to accommodate different manufacturers' microprocessors, the user is faced with a variety of operating systems and systems programs that he must be familiar with, depending on the particular microprocessor he is using at that time. In general, the user interface offered by the manufacturers of FDS systems varies from manufacturer to manufacturer. This creates a problem of non-uniform documentation and the need to train personnel on different systems.

It is useful at this juncture to describe a progression of attitudes best categorized as the "seventh heaven" phenomenon. When the traditional hardware/digital engineer first acquires his target system (TS) to do the microprocessor application, he is in "seventh heaven". Having been accustomed to hard-wired digital circuits, he has now under his control a complete computer system that can execute programs and perform small control tasks. However, it takes the engineer only a matter of a few weeks to realize that the target system is woefully inadequate for his development task and he becomes rapidly disillusioned with this level of support. He then purchases an FDS system with its floppy disk, line printer and other peripherals (for about $20K to $25K) and he is again in "seventh heaven." The FDS system has more power than he ever envisioned. It is a matter of time before the problems outlined above appear, but once again he finds the FDS system to be inadequate for his needs, and *again* disillusionment sets in. The time constant for this second disillusionment is generally longer, on the order of a year to a year and a half, depending on his other groups' usage of the FDS. Because there is nothing better that is available commercially, he must often make do with this level of support. Generation 2 is his next "seventh heaven."

Generation 2

The FDS does not provide all the development features that many systems designers would like (or should use), such as large shared file-handling capabilities, quick response to system commands, powerful editing programs and a library of software for the microprocessors of different manufacturers. Designers who require such features are finding it desirable to use a minicomputer (the small but high-performance computing systems that first reached the market in the early 1960's) as a supplementary development tool.

Although some cross-software packages exist for enabling minicomputers to develop microprocessor programs, there are very few systems for closely coupling minicomputers, FDSs and the microcomputer target system (TS). Such coupling is desirable because it enables a development engineer to move easily within this hierarchy as well as to exploit the distinctive features of each system: the minicomputer, to provide efficient editing, mass storage, documentation tools and shared data bases; the FDS, to emulate the real-time performance of the microcomputer as programs are evolving and to debug the hardware; and the microcomputer target system itself, to evaluate the final programs and control routines under the actual environmental and electrical constraints of the application setting.

The cleanest integration of these three systems is the "Triad," a tool that enables the programmer or engineer to work at any level in achieving a desired program (see Figure 3-10). The Triad closely couples three systems: the minicomputer, with its powerful software capabilities; The FDS, with its real-time emulation and hardware debugging facilities; and the target system, with its application-defined construction. The Triad provides fast and direct access to all levels of the hierarchy.

The hardware designer and the software programmer use the Triad as follows. The hardware engineer develops the target system, which might consist of a commercial general-purpose, single board computer module, in conjunction with an additional module, of his own design, that provides an interface to the control system to which the microprocessor is being applied. In order to debug the target system, he will need some test programs. Such programs can be quickly prepared with the editor on the minicomputer, cross-assembled into the machine code of the microprocessor and loaded directly into the target hardware that has not been debugged. As the engineer works out the errors in his hardware, he will continually update his test programs. It takes only a few minutes to reassemble and load each change.

Concurrently with the hardware development, the programmer can be editing, assembling and simulating his programs on the minicomputer and on the FDS. He would maintain all his programs on the central file system of the minicomputer and share a library of applications software with other users of the Triad. System integration can proceed smoothly, because the same facilities the hardware engineer has been using to load test programs into the target system hardware can also be used by the programmer to load the final applications program downward into the hardware. Changes at this stage are readily achieved by a simple process of re-edit, assemble and download. In this way, the Triad system can sharply reduce the time needed for development of a microprocessor application.

Several Triad systems can be supervised simultaneously by a single central minicomputer system. Such an arrangement enables a user to develop application packages utilizing several different microprocessors while he is still maintaining all his programs, documentation and engineering reports within the file system of the central minicomputer. Moreover, other users can share his programs, thus making the development cycle more efficient. As the next step, the minicomputer operating system could be rewritten to allow time-sharing among simultaneous users. This gives rise to the most powerful variation of the Triad concept, as shown in Figure 3-11.

Not only can programmers and engineers work simultaneously on development efforts concerning different microprocessors with the time-shared Triad system, but responsibilities for development and testing

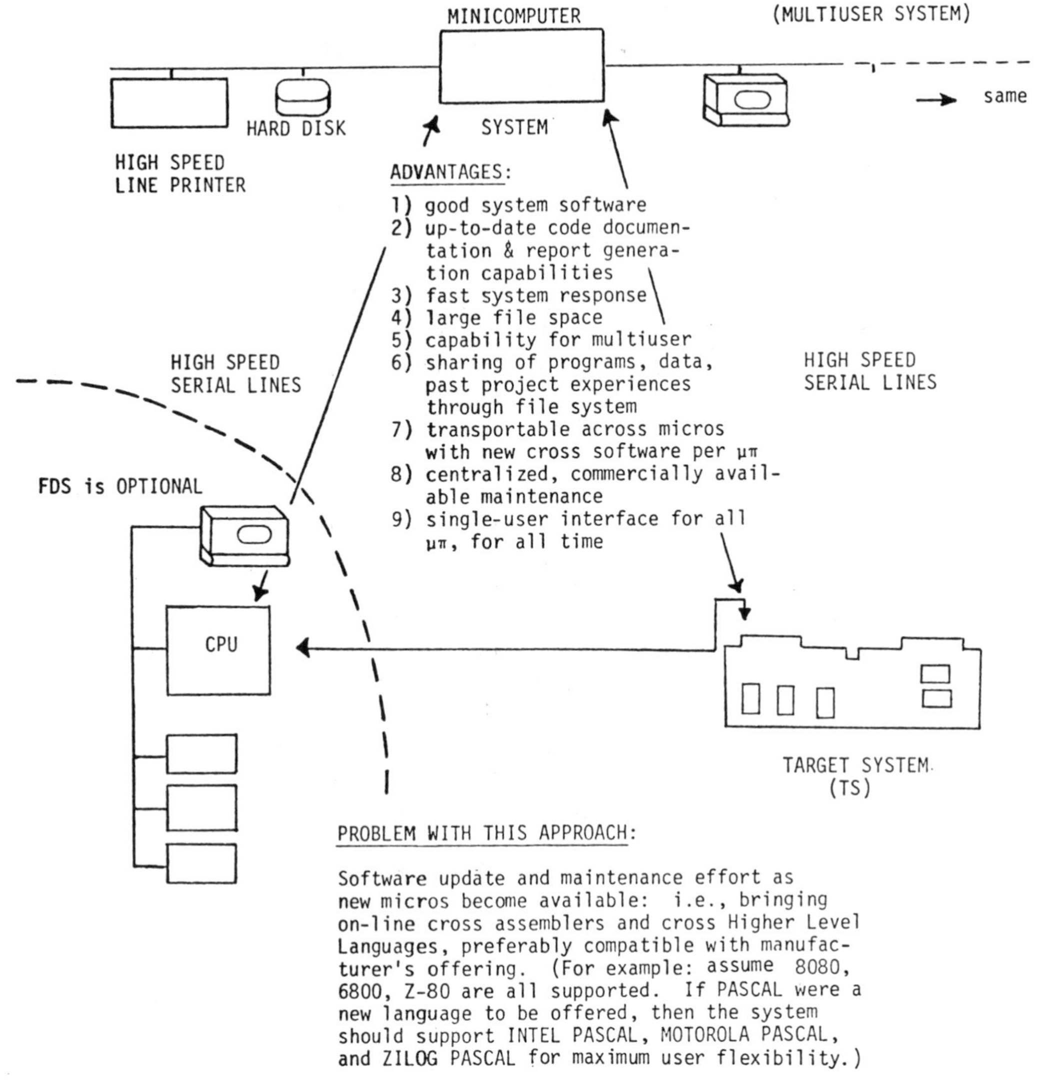

MINICOMPUTER
(MULTIUSER SYSTEM)
same
HARD DISK
SYSTEM
HIGH SPEED
LINE PRINTER
ADVANTAGES:
1) good system software
2) up-to-date code documentation & report generation capabilities
3) fast system response
4) large file space
5) capability for multiuser
6) sharing of programs, data, past project experiences through file system
7) transportable across micros with new cross software per μπ
8) centralized, commercially available maintenance
9) single-user interface for all μπ, for all time
HIGH SPEED
SERIAL LINES
HIGH SPEED
SERIAL LINES
FDS is OPTIONAL
CPU
TARGET SYSTEM
(TS)
PROBLEM WITH THIS APPROACH:
Software update and maintenance effort as new micros become available: i.e., bringing on-line cross assemblers and cross Higher Level Languages, preferably compatible with manufacturer's offering. (For example: assume 8080, 6800, Z-80 are all supported. If PASCAL were a new language to be offered, then the system should support INTEL PASCAL, MOTOROLA PASCAL, and ZILOG PASCAL for maximum user flexibility.)

Figure 3-10. The Triad concept.

can be split among several different parts of an organization. An example of such use of a Triad system for a coordinated microprocessor development within an organization is shown in Figure 3-11. In particular, manufacturing could have its own set of terminal and target systems to exercise products coming down the assembly line. At the same time, engineering could be developing new software or diagnostics for these products with its own set of terminals and Triads. Because both organizations share the same data base, the transferral of updated diagnostics to manufacturing could be immediate. Other parts of the organization, such as field service and marketing, could also share this data base. Naturally, such a time-shared system would implement protection and access mechanisms to allow selective sharing among users as defined by the project management.

Generation 3

Generation 3 development tools will include all the desirable features found in the host multi-user downloading approach of Generation 2 and, in addition, it will elim-

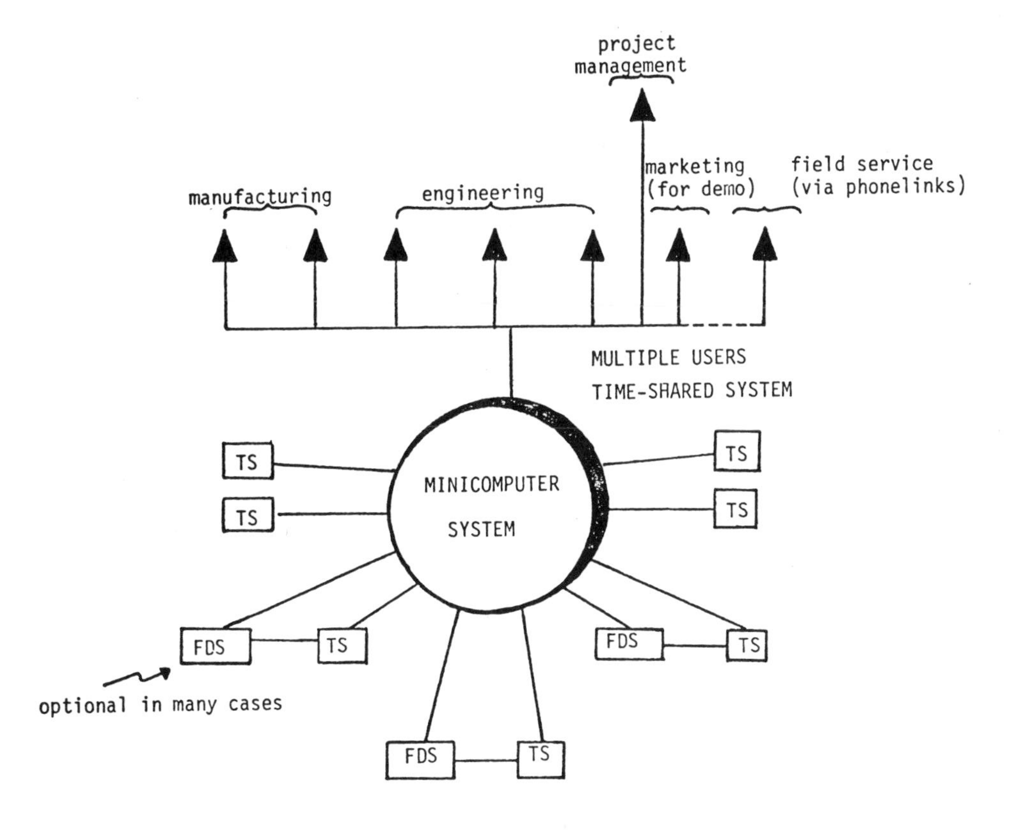

Elaborations of "Triad" system may be warranted in large organizations where a number of microprocessor applications are under development. The system is designed to give one programmer access to several different Triads, each dedicated to a different microprocessor. In the more powerful time-sharing system engineers and programmers can work simultaneously on development efforts involving different microprocessors.

Figure 3-11. Widespread use of a Triad system for coordinated microprocessor development within an organization.

inate the major drawback in terms of software updating and software support of that system. As the Universal Development System, it will give full capability for process support at all levels of languages that are offered by each individual manufacturer. This extensive software support will be capable of being automatically updated to include new offerings of software and hardware support as each individual manufacturer evolves its product line. This capability is crucial, since the growth patterns of the microprocessor market have not been able to standardize on any one processor or on any one language. We expect this phenomena to continue for the foreseeable future. Finally, the third generation development tool must also be as cost-effective as current systems while offering increased capabilities and desirable features that are not existent on any generation 1 or 2 system.

APPLICATIONS TO PROCESS CONTROL

A wide range of microprocessor units are currently employed for local and distributed control tasks. Examples of such systems are the Honeywell TDS-2000 system, the EPTAK controller by Eagle systems and the READAC system from Westinghouse. Current systems employ microprocessors at all levels, ranging from simple operator display functions to actual local control situations involving PID, DDC or CM. Instead of discussing any one manufacturer's system in detail, this section will outline some of the issues in distributed monitor/control systems by using a simplified model. Typically, we find a cost-effective organization for distributed intelligence in such a control configuration to be a hierarchical architecture, as shown in Figure 3-12.

The central issue in the design of such a system is the exact distribution of intelligence within the hierarchy for the particular application. For example, in some cases, it is best to centralize the decision-making and control functions within the SCU. This scheme would place the BCUs and LCUs under complete control of the SCU, with the attendant increase in the communications and computation demands at that level. On the other hand, it is quite conceivable that a complete decentralization of control and monitoring functions would make sense for a particular application. In this case, BCUs could operate as totally independent agents performing local control and only sending updates or alarms to LCUs. Individual LCUs, with their multi-drop line of BCUs could also perform a sub-system control task independent of SCU direction. The trade-off between centralization and decentralization of intelligence in such a system is the key issue that affects both hardware and software systems design. This trade-off is best ascertained following the identification of user

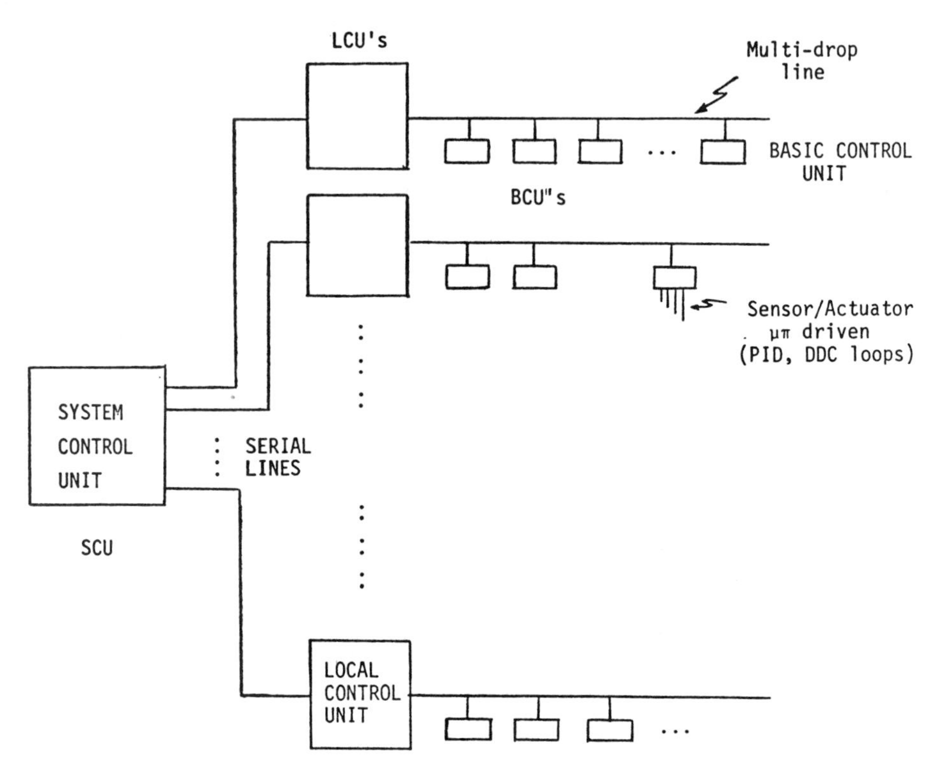

Figure 3-12. Hierarchical architecture for distributed monitor/control systems.

needs and the accompanying system requirements specification. Designing a hierarchical system without such knowledge reflects a bottom-up synthesis approach as discussed earlier in this chapter in the section on microprocessor project management, and can only lead to constantly shifting designs at both the hardware and software levels. The goals of the system must meet the perceived user needs and must be specified before any centralization or decentralization of microprocessors can be made in such a hierarchichal architecture.

Once the system design has been specified, the central issues relating to the operational specifications of the hierarchy can be considered. These issues break down into three major categories: communications, the user interface, and applications specific software. "Communications" covers a wide range of functions, some of which can be identified as follows.

- Physical communication by direct wire, telephone link, microwave and fiber optics.
- Communication protocols for reliability, error checking and security.
- Communication rates compatible with system components such as printers, tape units and CRTs.
- Communications to accommodate system functions that have high and low data rate needs.

Similarly, the user interface must provide a wide range of capabilities, including those listed below.

- Activity reports, logs, alarms and inter-operator message communications.
- Maintenance of a user data base for data analysis, historical archives and general record-keeping.
- Human interfaces such as CRTs, alarms, displays and annunciators.

Finally, the applications software must be designed to execute the specific control and monitoring tasks.

- Traditional control by event and control by time tasks.
- Process control algorithms for PID, DDC, CM and other mechanisms.
- Any other user applications, software specific to the current system tasks.

Given such a hierarchical system architecture, the responsibility for integration of these three functions lies with the overall software control system, the operating system. The form and complexity of this software operating system is also of major importance in the design process. In fact, the operating system itself must undergo the same centralization and decentralization design arguments as outlined above. Thus, two levels of complex design must be accomplished: the operating system design, and then the applications specific software functions. The complexity involved in the execution of such a project indicates why correct project management techniques are so vital to the development of a cost-effective distributed control system.

THE FUTURE

The potential applications of microprocessor technology are so numerous it is hard to visualize any aspect of contemporary life that will escape its impact. Devising a successful microprocessor application, however, is only the first step toward achieving its adoption and acceptance. The advent of the microprocessor presages more than just a technical revolution. It will probably touch more aspects of daily life than have been affected by all previous computer technology. In many instances, society is neither aware of nor prepared for the microprocessor's non-technical impact. For example, the introduction of microprocessors in automobiles to improve fuel economy and reduce exhaust emissions will have a profound effect on tens of thousands of small automobile-repair shops and hundreds of thousands of gasoline service stations. Millions of maintenance workers, who at this moment may never have heard of microprocessors, much less seen one, must quickly become acquainted with them and, to some degree, become familiar with their testing and replacement. Otherwise, the entire network of automobile service will have to be revised.

In other areas, microprocessor technology is likely to move more slowly than one might expect. For example, it would not be difficult to equip every gasoline service station with a microprocessor terminal that would record the details of every credit-card transaction. At the end of each day, the recorded information could be transmitted rapidly by code over an ordinary telephone circuit to computers in the credit-card company's central office, speeding up by several weeks the billings on millions of dollars' worth of sales. Although such systems are eminently practicable, they have not been adopted. One can infer that the innovation is unwelcome because it threatens the existence of entire divisions of credit-card organizations: the optical-character-recognition (OCR) divisions that now transcribe into computer-usable form the transaction data from millions of separate pieces of paper filled out and mailed in by service-station attendants. Entire divisions of companies do not willingly disappear overnight.

These are just two examples of how thousands of business organizations and millions of individuals may be affected by what appears to be a straightforward engineering decision of whether or not to apply the new microprocessor technology. Within the business organization itself, the microprocessor and microcomputer are making more acute the already difficult question of how to distribute computing and information resources for maximum effectiveness. Although many companies are rapidly consolidating and centralizing all aspects of their computing resources, other companies are decentralizing with equal aggressiveness. The elusive "optimal" strategy involves delicate considerations of management control, strategies of system development and operational procedures. Business managers who had to penetrate the mysteries of the $250,000 room-sized computer barely 15 years ago, and of the $25,000 minicomputer 6 or 7 years ago, must now try to weigh the costs and benefits of the $250 microcomputer and the $25 microprocessor.

REFERENCES

1. Huff, S.L. and Madnick, S.E., "An Approach to Constructing Functional Requirement Statements for System Architectural Design," Technical Report #6, Center for Information Systems Research, Massachusetts Institute of Technology, Cambridge, June 1978.
2. "Special Issue on Microprocessor Technology and Applications," *Proceedings of the IEEE* **64** No. 6, June 1976.
3. "Special Issue on Small Scale Computing," *Computer* **10**, *No. 3,* March 1977.
4. Toong, Hoo-min D., "Microprocessors," *Scientific American* **237**, *No. 3,* 1977.
5. Toong, Hoo-min D., "Support Tools for Microprocessor Applications," Center for Information Systems Research, Massachusetts Institute of Technology, Cambridge, December 1978.
6. Madnick, Stuart E., "Trends in Computers and Computing: The Information Utility," *Science* **195**, *No. 4283:* 1191–1199, March 18, 1977.
7. Whisler, Thomas L., *The Impact of Computers on Organizations,* Praeger, New York, 1970.
8. Woolridge, Susan, *Project Management in Data Processing,* Petrocelli/Charter, New York, 1976.

4.
Fractionating Column Control

Aaron R. Kramer

State University of New York,
Maritime College
Bronx, N. Y.

INTRODUCTION

Historically, there have been countless studies, articles, discussions and analyses by the chemical industry concerning the best way to control a fractionating column. The advent of integrated circuit technology and its subsequent development into computer hardware naturally prompted chemical industry investigation into the possibilities of applying this technology to the control of fractionating columns. There are many sources of information available on column control and a similar storehouse of knowledge in mini- and microcomputer technology. Unfortunately, the chemical industry is experiencing difficulty in absorbing and integrating these technologies. How can the chemical industry best utilize mini- and microcomputer technology for the control of fractionating columns? The purpose of this chapter is to attempt to answer this question by discussing 1) potential uses for mini- and microcomputer control on fractionating columns, and 2) the design of a mini- and microcomputer application to column control via systems engineering methods.

POTENTIAL USES FOR MINI- AND MICROCOMPUTERS IN COLUMN CONTROL

Fractionating columns, as their name implies, are designed to separate light chemical components of a feedstock from heavier components in varying degrees. The prime consideration in column control is to obtain maximum product flow with maximum purity, while minimizing energy input and cost. Mini- and microcomputers can assist, and often perform these functions at various levels dictated by design and control philosophical considerations. These levels may be broadly classified as follows.

- Supervisory set point analog control (SPC).
- Direct digital control (DDC).
- Communication interface with a host minicomputer.
- Column optimization and other advanced control techniques.

Supervisory or Set Point Analog Control (SPC)

Supervisory analog control is the process of applying a microcomputer to conventional analog control instrumentation in an "operator assist" mode. The microcomputer is programmed to aid the operator on a local basis or to provide information as to set point variation requirements. In this mode, the computer provides information on changes in environmental conditions which may necessitate operational changes in the column and accumulate data for further analysis. Primarily, this application would be used on a retrofit basis as a learning tool in an established plant, but it is not intended to provide closed-loop automatic control. A second application in SPC analog control is the closed-loop function which the microcomputer can provide. After a suitable learning period, the programmed functions, such as those outlined above, may provide supervisory functions to the analog set points. These set points may be manipulated by the computer as defined in the design and control philosophy. When sufficient hardware reliability and capability is acquired via this learning experience, the computer may become more than a supervisory function and ultimately may replace the analog controllers with equivalent digital control. Direct digital control is achieved when this point has been reached.

Direct Digital Control (DDC)

Direct digital control of a tower can be accomplished in two basic ways:

1. Replace the analog control loops with digital equivalents, in which case all conventional central algorithms (such as proportional derivative) and intergral action are digitally implemented.
2. Perform DDC with adaptive techniques or multivariate control systems of a more sophisticated nature. A DDC replacement of analog controllers does not offer any distinct advantages. The inherent problem of analog controller interval interaction would still exist.

For example:

The proportional, integral and derivative connrol algorithm in analog is

$$\text{output} = K_pE + \tau_I \int Edt + K_0 \frac{dE}{dt}$$

where E = error.

K_p, τ_I and K_0 are proportional, integral and derivative resettable constants. In most cases, the internal mechanics of an analog controller causes interaction between the various modes of control. Changing proportional band may affect the integral and derivative time constants in the controller output. Internal interaction of control elements is very common in most analog controllers. This can be more clearly seen by examining the control algorithm of an analog controller in transfer function mathematical format.[1]

$$\frac{\text{output(s)}}{\text{input(s)}} = \frac{(\tau_r S + 1)(\tau_d S + 1)}{\left(\frac{B}{k_c}\tau_r S + 1\right)(\alpha\tau_d S + 1)}$$

$$\text{where } \tau_r = K_p\tau_I,\ \tau_d = \frac{K_d}{K_p}.$$

Note that any change in K_p changes τ_r and τ_d, indicating interaction between controller modes.

The ideal controller in digital form,[2] using difference techniques, can be written as

$$\text{output} = K_pE_n + \frac{K_I}{\tau}\sum_{i=0}^{n} E_i + \frac{K_D}{T}(E_n - E_{n-1}) + \bar{O}.$$

There are many forms of digital algorithms representing analog controllers in the literature. The principle advantage of digital controllers implementation in microcomputer hardware is the total elimination of internal interaction between control elements of the controller. The implementation of the digital expressions for the various elements are simply the sum of independent quantities of each control mode. Changing K_p, for example, has no effect on K_I or K_d. This simple expression provides the opportunity for better tuning capability of the controller and the process to which it is applied.

Proportional, integral and derivative controllers are not the only modes of control that can be implemented digitally. Quite the contrary, digital implementation of controllers is limited only by the imagination of the control engineer and his knowledge of process control theory. In fact, systems which are digitally represented are often characterized by a series of equations reduced to matrix form, indicating states of the process and its control. When the matrix is solved, the solution brings the existing state of the process to a desired state specified by some constraint.

A brief example of such a set of equations is:

$$\bar{X}(n + 1) = \bar{A}X(n) + \bar{B}u(n).$$

$$\bar{X}(n + 1) = \begin{bmatrix} \dot{X}_1 \\ \dot{X}_2 \\ \dot{X}_3 \end{bmatrix}$$

$$\bar{X} = \begin{bmatrix} X_1 \\ X_2 \\ X_3 \end{bmatrix}$$

$$\bar{A} = \begin{bmatrix} a_{11} a_{12} a_{13} \\ a_{21} a_{22} a_{23} \\ a_{31} a_{32} a_{nn} \end{bmatrix}$$

$$\bar{B} = \begin{bmatrix} b_{11} b_{12} b_{13} \\ b_{21} b_{22} b_{23} \\ b_{31} b_{32} b_{nn} \end{bmatrix}$$

$$u = \begin{bmatrix} u_1 \\ u_2 \\ u_3 \end{bmatrix}$$

$\dot{X}_1, \dot{X}_2, \dot{X}_n$ . . . represent the time rate of change of each variable of the process.

a_{11}, a_{21}, a_{nn} . . . represent a transition matrix (transfer function) of the process.

X_1, X_2, X_n . . . represent the state of the variable at the present time.

b_{11}, b_{12}, b_{nn} . . . represent a control transfer function matrix.

u_1, u_2, u_3 . . . represent the control inputs necessary to drive the process to desired levels.

The solution of the state equations may take several different aspects, depending on the imagination of the control engineer, and limited only by process constraints. These solutions may thus be in the form of optimization of the process through the use of adaptive control, minimizing the main square error, etc. In all cases, it is obvious that implementation of this type of process description and the solution of the control problems can only be implemented by a digital computer.

Communication with Host Computer

The microcomputer can be utilized as an effective off-line operating and management reporting instrument. Among the obvious functions that the micro- and minicomputer configuration can perform are accounting of heat, material and composition balances, alarm conditions, point data trending, maintenance anaylsis, shift data and (currently) optimization of local unit operations. The reporting of various groups of data (status reports) can often lead to better yield prediction and control, minimization of reflux, assistance in maintenance detection, reduction of energy consumption and maintenance of product purity. These status reports are grouped as follows.

1. Shift date (used to assist operators).
2. Management data (used to assist in decision-making).
3. Maintenance data (used to assist maintenance personnel).

Shift Data

Shift data report only that information necessary to operate the tower in accordance with management instructions. Decisions made are limited only to maintain said operation. Figure 4-1 shows a possible format for shift data reporting.

Management Data

Management data report shift data in addition to several measured and calculated parameters necessary to assist management in the determination of optimum yields for given feedstock slates. (See Figure 4-2.)

DATE	TIME	VALUE	ERROR	DURATION OF ERROR
Hydrocarbon feed rate				
Feed temperature				
Composition of feed				
Heavy key				
Light key				
Critical tray temperature				
Overhead production flow				
Bottom production flow				
Reflux flow				
Reflux temperature				
Product composition				
Other selected key variables				

Figure 4-1a. Shift data status report (printed every hour or on demand—typical).

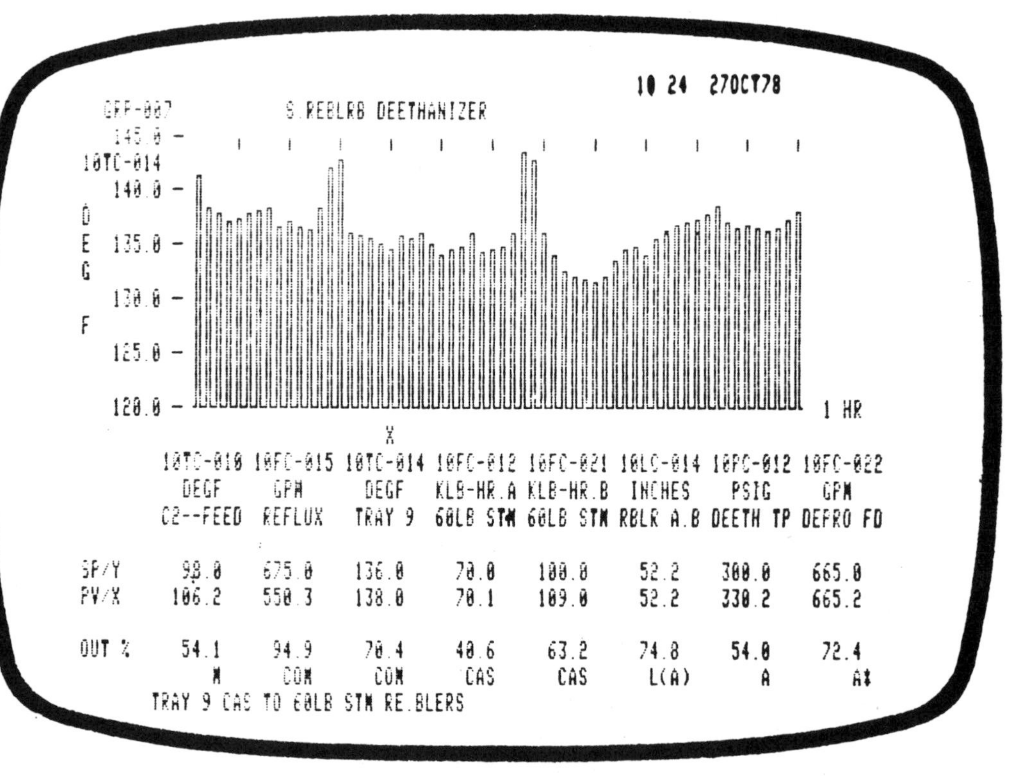

Figure 4-1b. Trend data display (one loop).

DATE	DEMAND TIME	VALUE	VALUE 24 HRS. PREV.	DEVIATION
AVERAGE H/C FEED RATE				
AVERAGE H/C TEMPERATURE				
AVERAGE H/C COMPOSITION				
AVERAGE HEAVY KEY				
AVERAGE LIGHT KEY				
AVERAGE REBOILER LOAD				
AVERAGE PRODUCT FLOW				
AVERAGE BOTTOM FLOW				
AVERAGE TOWER PRESSURE				
AVERAGE ENERGY CONSUMPTION				
AVERAGE PEAK ENERGY CONSUMPTION				
AVERAGE TOWER REFLUX RATIO				
AVERAGE TOWER EFFICIENCY				
" "				
" "				
" "				
OTHER KEY VARIABLES				

Figure 4-2. Management status report (printed every day or on demand).

Maintenance Data

Maintenance data report information primarily as trends and statistics concerning the operation of components of the tower which may affect efficient operation of not only the tower but of other unit operations downstream. To date, however, the use of the mini- and microcomputers in the area of maintenance analysis has been minimal. The ever-increasing cost of repair and replacement of major machinery will ultimately force greater utilization of computer capability in the area of maintenance scheduling, failure prediction and general maintenance analysis (see Figures 4-3 and 4-4). Typically, the host computer interacts with many microcomputers on several unit operations. This chapter deals only with a fractionating column. A typical communication network arrangement between a host and a satellite microcomputer is illustrated in Figure 4-5.

	30 DAY AVERAGE	LAST DAY	TODAY
TOWER			
Pressure drop			
Temperature rise			
Critical tray pressure			
Critical tray temperature			
Other key variables			
REBOILER			
U (heat transfer coefficient)			
Heat supply valve position			
ΔT_m			
Pressure drop			
CONDENSER			
U (heat transfer coefficient)			
Pressure (vacuum)			
ANALYZERS			

Figure 4-3. Maintenance analysis status report (printed every day or on alarm condition—typical).

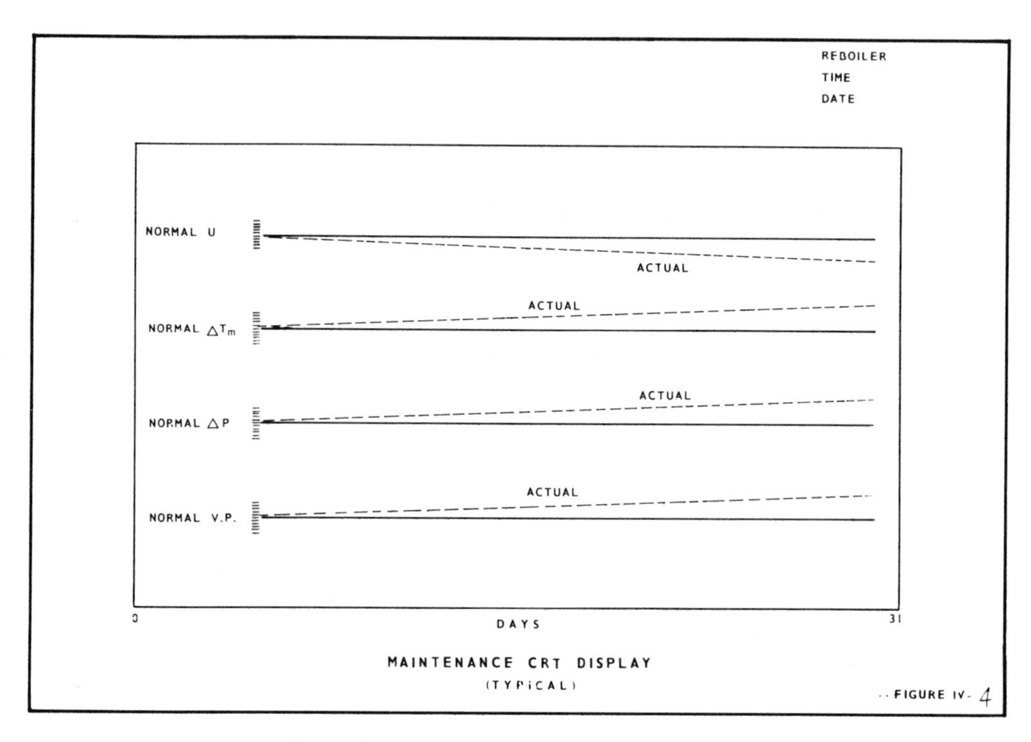

Figure 4-4. Maintenance CRT display (typical).

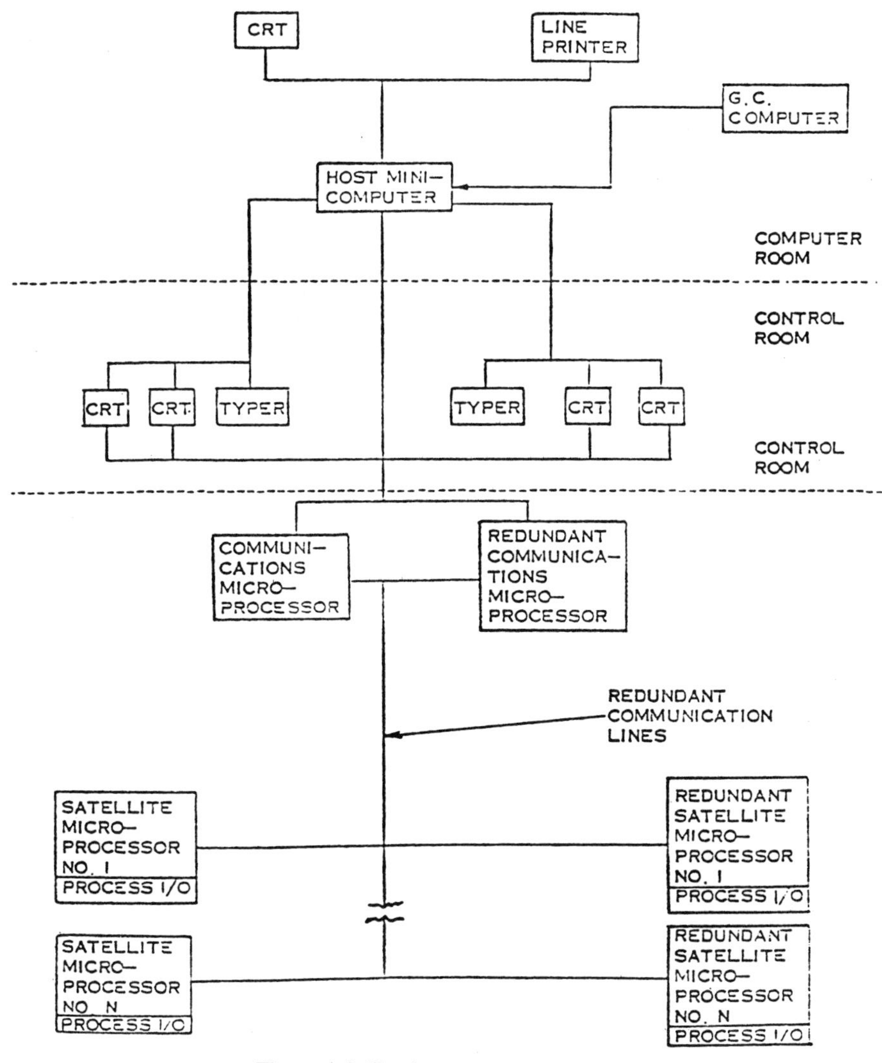

Figure 4-5. Hardware configuration.

Column Optimization and Advanced Control Techniques

Application of mini- and microcomputer technology to on-line process control provides the tools for implementing previously developed, sophisticated mathematical techniques in areas of optimization, multivariable and de-coupling control. Techniques such as the calculus of variations, dynamic programming and lagrangian dynamics are difficult, if not impossible, to manipulate by hand, but can easily be handled by computer.

tion implies that certain variables will be controlled to a "theoretical best" value with respect to other independent variables

Fractionating column control optimiza-
in the process limited by constraints. An optimization application requires, as a minimum, four fundamental steps prior to implementation:

1. A complete mathematical description of the process.
2. A mathematical description of the process constraints.
3. A representation of the variables to

be optimized (the so-called objective function).
4. Criteria for evolution of the optimum point.

Mathematically, the four basic steps may be stated as follows.

1. A process whose dynamics may be given by $x = f(x, u, t)$.
2. Subject to constraints on inputs u and x.
3. A desired reference indicating some output requirements.
4. Criteria or performance function which is to be minimized.

There are several different optimization techniques, a few of which are listed below.

- Unconstrained and constrained minimization.
  1. Gradient methods.
  2. Conjugate gradients.
  3. Minimizing steps using slope information.
  4. Powells method: conjugate directions.
- Direct methods
  1. Lagrange multipliers.
  2. Feasible directions.
  3. Gradient projection.
  4. Linear programming.
  5. Dynamic programming.

Each of the above methods has its own particular advantages and disadvantages too numerous to mention in this text. The subject headings and subheadings listed are standard nomenclature and are found in more detail in references on the subject.[3]

A typical column optimizing scheme, as shown in Figure 4–6, is applied to an ethylene tower. The optimizer is a microcomputer programmed to do many functions, including optimal control. A typical configuration of this optimizer is shown in Figure 4-7a and b.

Multi-variable and de-coupling control are discussed in detail later in this chapter. However, a brief definition of these concepts is offered here for the sake of continuity. A multi-variable control system is a device or set of devices which, by virtue of its intelligence, has the ability to scan key variables of a process and decide on a course of action to achieve the best possible control strategy.

De-coupling control is a device or set of devices which de-couples or isolates the effects of one variable on another. In the case of a tower, it serves to isolate the effects of interaction of several variables on one another.[4]

The mini- and microcomputer, advanced mathematical techniques, optimization, multi-variable control, de-coupling control, feedforward and feedback techniques have reached the level of sophistication where the application of control philosophy must be carried beyond "empirical" engineering. Increasing process complexities, changing world situations in feedstock quantity and quality and changing energy situations require a more engineered and scientific approach. This leads to the subject of engineering a micro- and mini-application of a fractionating column.

THE SYSTEMS ENGINEERING APPROACH

Classical column design is based on four key parameters: feedstock components, purity of product, tray efficiencies and tower capacity. The fractionating column is basically designed to meet these primary requirements. Control philosophy and objectives are largely determined by the control engineer, with input from the process engineers. This as an interface does not work as it might because, in many instances, the process engineer does not consult with the control engineer during the early design phase. The resulting design may provide a column which is difficult to control within required product specification objectives. For example, some columns never seem to reach a steady state

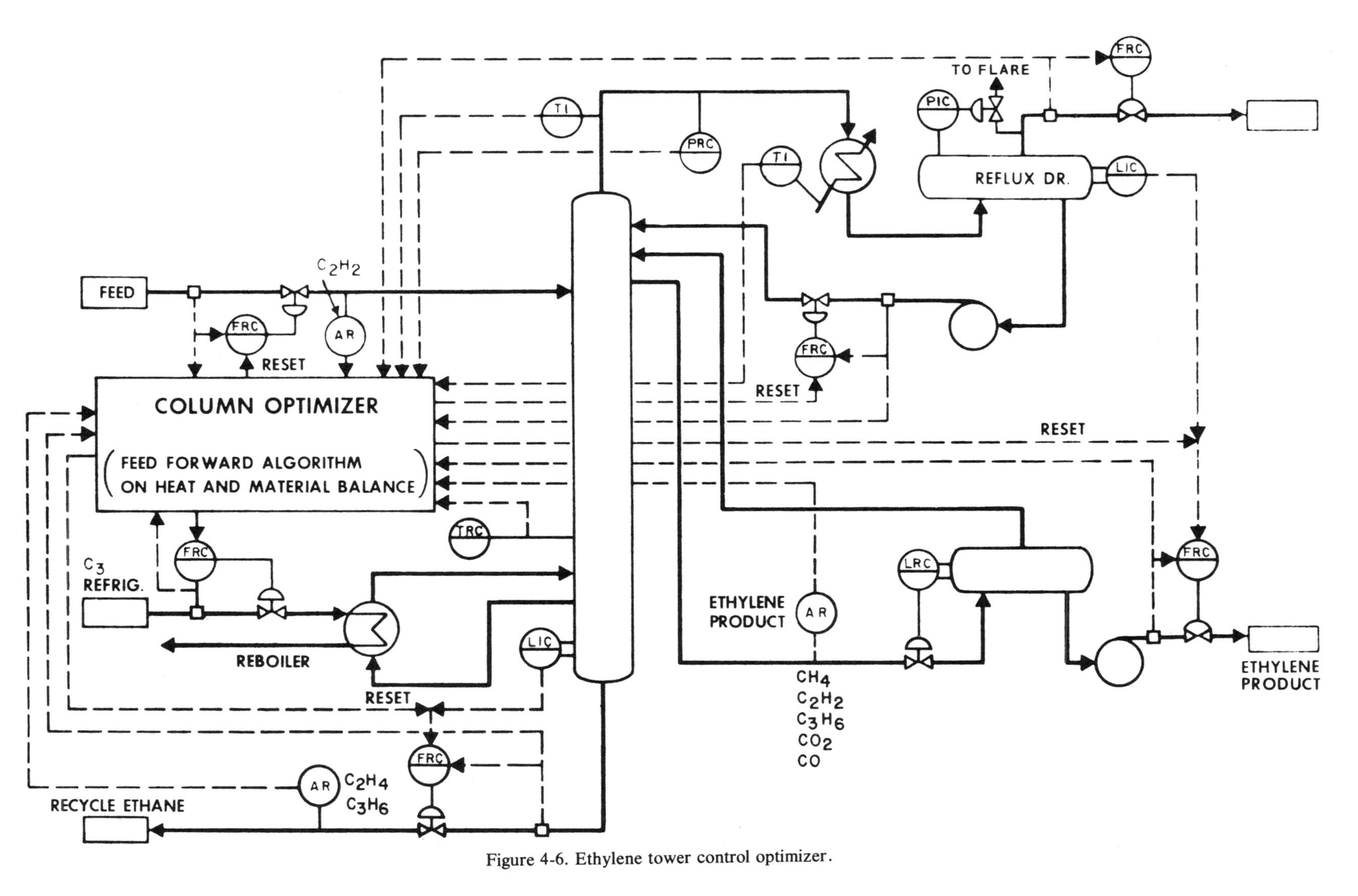

Figure 4-6. Ethylene tower control optimizer.

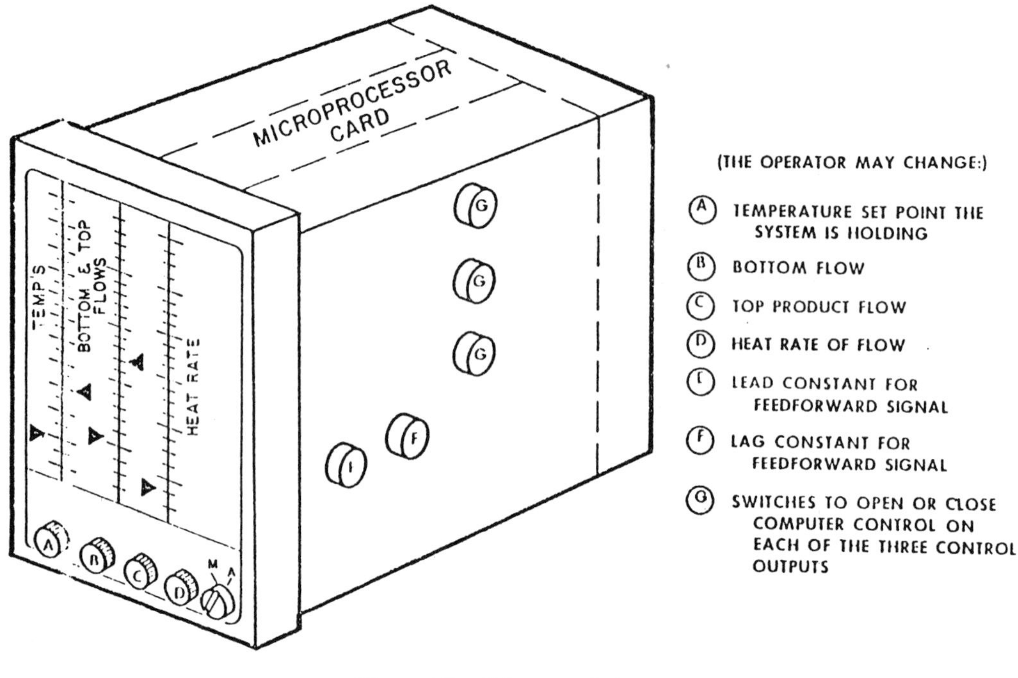

Figure 4-7a. Column optimizer.

INPUTS:

1. FEED RATE (FLOW)
2. LEVEL BOTTOM
3. PRESSURE TOP
4. REFLUX FLOW
5. OVERHEAD PRODUCT FLOW
6. TEMPERATURE TOP
7. TEMPERATURE BOTTOM

OUTPUTS: (CONTROLLED)

A. BOTTOM FLOW
B. TOP PRODUCT FLOW
C. HEAT RATE STEAM FLOW

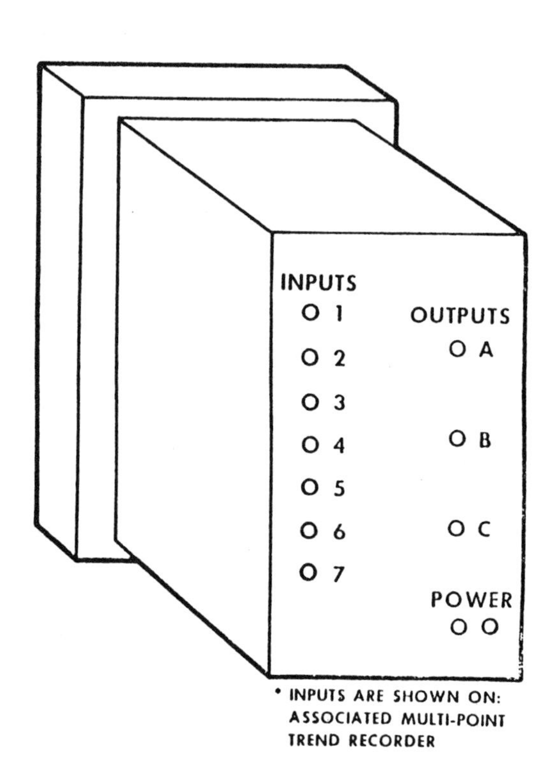

Figure 4-7b. Column optimizer, rear view.

condition even though external disturbances are virtually nonexistent. The term "steady state," in this sense, means that some primary, controlled variable oscillates between some limits with such a large time period that the tower seems to be in a steady state condition but in reality is not. Several questions might be raised about this particular communication problem and its ramifications. Given the problem of designing a column, how best can this column be controlled, how can micro- and/or minicomputer hardware assist in the control scheme and how is the design of the column and the control philosophy verified and evaluated?

The column designer bases his design on established parameters with suitable safety margins and in a steady state time invarient mode. In fact, the parameters selected for the design of the column are assumed to be constant. It is really the only way an industrial column can be designed. The control engineer is then charged with designing a control system to maintain the desired control parameters which are not constant. The dynamic nature of the column (non-steady state) is the result of interaction between many variables, all or some of which may vary in time and which were not compensated for in the original design phases. These time-varying parameters are not predictable from design unless dynamic equations which describe the time-varying action of the column and its components are written, solved and analyzed. There may be a large number of time-dependent differential equations, both linear and non-linear, which must be written in order to describe the dynamic character of a fractionating column adequately. If the questions previously raised are to be answered, one logical method might be the use of a total systems analysis using the scientific method.

The scientific method is used in systems engineering to arrive at solutions to industrial problems. The systems engineering technique consists of four distinct phases:

1. A conceptual phase.
2. A problem definition phase.
3. A possible solutions phase.
4. An evaluation and selection phase.

The systems analysis approach to the control fractionating columns by mini- and microcomputers must follow these phases if a sucessful application of computer technology to fractionating column control is to be realized. To further clarify these phases, a brief explanation is given below.

Conceptual Phase

All too often, the application and selection of a computer takes place before a thorough analysis of the application is carried out. This analysis of the application requires, as a minimum, a definition of the application problem, a list of possible solutions to that problem, a method of evaluating the solutions and a selection of one of the solutions.

Problem Definition Phase

An adequate definition of the problem requires that information from all sources be accumulated to determine the functions for which the computer will be utilized. For example, is the microcomputer to be used for local control, communication to the host computer, optimization or any other applications? Each application requires a unique engineering basis and is subject to process constraints. A microcomputer has a different configuration for optimization applications than it might have for local analog or digital control. Core size, speed and input/output requirements all must be specifically engineered for the process.

Possible Solutions Phase

When all the necessary information has been accumulated, and the facts separated from opinions and hunches, several alternatives for solutions to the problem should be outlined.

Evaluation and Selection Phase

Once the list of possible solutions has been completed, every effort must be made to develop a means of evaluating the various possible solutions.

After evaluation, the selection process ultimately must take place. This process is simply the quantitative or qualitative results of the evaluation.

Analytical Tools

In general, the conceptual, problem definition, possible solutions and evaluation and selection phases are accomplished via various technical and psychological methods governed by external influences and constraints (economics, personnel, etc.). The *actual* solutions to the physical problems require techniques involving analytical tools available to the engineer. One of these analytical tools is the technique of simulation via digital computer.

Simulation as a Method of Solution of Dynamic Control Problems

Simulation may be defined as the imitation of, or a semblance of, that which is to be simulated (copied). In the chemical industry, it may be defined as a system of physical, mathematical relationships resembling or imitating a process or action of a unit operation or unit process.

The Model

Modeling is defined as a preliminary representation of a system serving as the plan from which the final object is to be constructed. A model can be mathematical and may lead to a solution or to a set of equations in order to produce an accepted representation of the physics of the process. A model can also be the construction of a small version of the process (pilot plant) for the purpose of investigating areas of concern prior to large scale construction.

The concepts of simulation and modeling are sometimes used simultaneously, in that a mode of a process must be defined in terms of physical principles (i.e., heat for material balances subsequently simulated or tray-by-tray analysis).

In order to model a process or operation for use by a computer, the process must be rigidly mathematically defined, programmed on a computer and simulated. Finally, the results are analyzed using the computer.

Modeling and Simulation Methods

Modeling and simulation methods may be broadly classified into two categories:

- A steady state (or static) model.
- A dynamic model.

The steady state model and its subsequent simulation can be described simply as a solution to a mathematical expression, the variables of which are the parameters or states of the process condition. If one were to change the independent variable and solve the equation, one could say that a model was derived and various conditions of the process were simulated to determine the result of these changes on the dependent variable. If many equations are involved, some other methods may be required for solution. The use of a digital computer may become necessary in the latter case.

The advent of more sophisticated process and control led to requirements for dynamic studies of these processes. These dynamic simulations and model requirements led to the advent of analog computers and, presently, to digital equivalents of analog computers. The digital equivalents of analog computers manifest themselves in the development of digital simulation and user-oriented modeling languages such that a model may easily be implemented on a digital computer; the solutions yield essentially analog results. Each simulation language has its own character, limitation and advantages, but all require a method to solve time-dependent differential equations (or those equations equivalent to time-dependent differential equations).

It is not our intent here to outline advantages or disadvantages for each simulation language, since one user may prefer one language over the other. Our intention is to list some available dynamic simulation languages and to indicate the sources for these languages as well as to point out those which are indigenous to chemical industry.

Continuous System Simulation Programs

CSMP. Continuous System Modeling Program, users manual, program #360-SC-16X, IBM, 1967.

Mimic. A digital simulator program, H. G. Peterson and F. J. Sansom, SESCA internal memo.

Leans. Lehigh analog simulator, S. M. Morris, Lehigh University, Bethlehem, Pennsylvania.

Gemscope. D. G. Fisher, H. R. Ranstead, Department of Chemical Engineering, University of Alberta, Alberta, Canada.

SL-Mini Point. Interactive simulation language minipoint, R. D. Benham, Interactive Mini Systems, Inc., 5312 West Tucannon, Kennewich, Washington.

Steady State Computer Simulations

System simulates complex towers that have multiple feeds and sidestreams (with or without strippers).

Toray Distillation. Program simulates multicomponent distillation.

Multicomponent Distillation. This program performs tray-by-tray (material balance) calculations using the Thiele-Geddes method for a distillation column.

SP-95 Multi-unit Distillation and Process Simulation. Rigorously simulates separation processes involving vapor-liquid equilibria.

Fractionator Reflux vs Number of Trays, Program 9432. Phillips Petroleum Company, Program Package 1.

Winn Shortcut Distillation Program. This program calculates minimum reflux, minimum stages and product distribution, using the Winn method with Chao-Seader K values.

Underwood-Fenske-Gilliland Shortcut Distillation. Combines the Underwood calculation for minimum reflux, the Fenske minimum tray computation and the Gilliland correlation to give a complete shortcut method of distillation.

SP-07 Distillation Design. Used to simulate the operation of an existing distillation column or to design a new column where the number of trays, feed trays and reflux are unknown.

DR-01 Stream Assay Analysis. Used for the analysis of crude oil and/or other process stream assays. Transforms TBP or ASTM distillation data into pseudocomponents with specified TBP cutpoints.

Fractionator Design, Program 1200, Lewis Matheson. Phillips Petroleum Company, Program Package 1. Program provides rigorous equilibrium stage solution to fractionator problems by Lewis-Matheson method for calculating equilibrium and heat and material balances on each tray.

Batch Distillation. Two computer programs rate batch distillation cycles.

Distillation Curve Comparions. Program performs the following for petroleum distillations.

Shortcut Distillation Column Design Optimizer. This program determines the minimum number of theoretical plates.

General Distillation. Performs tray-to-tray heat, material and equilibrium balances, and calcultes the component split.

SP-10 Distillation Simulation. Calculates performance of a complex distillation column during start-up or during transition from one operating state to another.

Distill. Program calculates distillation separations with simplified English language data input.

Heat and Material Balance for Complex Towers, Tray Sizing and Rating. A combination heat and material balance program and valve tray program.

Complex Column, Program 6029, Theta Convergence. Phillips Petroleum Company, Program Package 1.*

The CSMP Modeling Program

The description of all of the programs listed above would be too cumbersome to be included in this text. However, a brief description of the CSMP (IBM) program is included as an illustration of one general modeling program. This program will be applied later to a particular dynamic study.

The Continuous System Modeling Programm III (CSMP III) allows the digital simulation of continuous processes on large scale digital machines. The program provides an application-oriented language that allows models to be prepared directly and simply from either a block diagram representation or a set of ordinary differential equations. It includes a basic set of functional blocks which can represent the components of a continuous system and accepts application-oriented statements defining the connections between these functional blocks. CSMP III accepts FORTRAN statements, allowing the user to readily handle non-linear and time-variant problems of considerable complexity.

Input and output are simplified by means of user-oriented control statements. Format options are provided for print-plotting in graphic form; scaled plotting in contoured and shaded form; and printing in tabular form. These features permit the user to concentrate upon the phenomenon being simulated, and not the mechanism for implementing the simulation.

The program, when installed in an appropriate computer environment, is a complete system for model development and application, providing means to create "simulation vehicles" capable of performing numerical simulations on a digital computer when supplied with simulation run specifications (for example, model parameters, run parameters and documentation requests).

The process system description may be prepared from a functional representation developed in the tradition of analog computing, as in a block diagram, or from a mathematical representation formulated, perhaps, in ordinary differential equations. In either case, the basic unit of representation is a function.

The program provides both a basic set of functional blocks (also called functions) and the means for defining functions especially suited to the user's particular simulation requirements. Included in the basic set are such conventional analog computer components as integrators and relays, plus many special purpose functions such as delay time, zero-order hold, dead space, limiter functions, variable-flow transport delay, generalized Laplace transform and the arbitrary function of two variables. This complement is augmented by the FORTRAN library, which includes subprograms for logarithmic, exponential, trigonometric and other mathematical functions (see Tables 4-1 and 4-2).

Special functions can be defined either through FORTRAN programming or, more simply, through a macro capability that permits individual existing functions to be combined into a larger functional block. The user is thereby given a high degree of flexibility for different application areas.

Figure 4-8 shows in a block diagram, how mathematical functions are represented by CSMP III language statements. One statement is written for each block in the diagram or for each functional relationship required to define the model. The name of a CSMP III functional block corresponding in behavior to the block or function being represented is placed to the right of the equal sign. The symbolic names of the functional block output are listed to the left of the equal sign in a definite order, each output being uniquely defined according to its position. Once defined, an output need

*The lists of simulation programs were compiled from *Chemical Engineering,* August–September 1973.

Table 4-1. Library of CSMP III function blocks.

CSMP III Statement	Equivalent Mathematical Expression
INTEGRATOR Y = INTGRL (IC, X) where: IC = $y\|_{t=t_s}$ Alternative 'Specification' Form: Y = INTGRL (IC, X, N) where: Y = output array IC = initial condition array X = integrand array N = number of elements in the integrator array (N must be coded as a literal integer constant)	$y(t) = \int_{t_s}^{t} x\,dt + y(t_s)$ where: t_s = start time t = time $\vec{y} = \int_{t_s}^{t} \vec{x}\,dt + \vec{y}(t_s)$ Equivalent Laplace Transfer Function: $\frac{Y(s)}{X(s)} = \frac{1}{s}$
DERIVATIVE Y = DERIV (IC, X) where: IC = $\left.\frac{dx}{dt}\right\|_{t=t_s}$	$y = \frac{dx}{dt}$ Equivalent Laplace Transfer Function: $\frac{Y(s)}{X(s)} = s$
* **1ST ORDER LAG (REAL POLE)** Y = REALPL (IC, P, X) where: IC = $y\|_{t=t_s}$	$p\frac{dy}{dt} + y = x$ Equivalent Laplace Transfer Function: $\frac{Y(s)}{X(s)} = \frac{1}{ps+1}$
* **LEAD-LAG** Y = LEDLAG (P1, P2, X)	$P_2\frac{dy}{dt} + y = P_1\frac{dx}{dt} + x$ Equivalent Laplace Transfer Function: $\frac{Y(s)}{X(s)} = \frac{P_1 s+1}{P_2 s+1}$
* **2ND ORDER LAG (COMPLEX POLE)** Y = CMPXPL (IC1, IC2, P1, P2, X) where: IC1 = $y\|_{t=t_s}$ IC2 = $\left.\frac{dy}{dt}\right\|_{t=t_s}$	$\frac{d^2y}{dt^2} + 2P_1P_2\frac{dy}{dt} + P_2^2 y = x$ Equivalent Laplace Transfer Function: $\frac{Y(s)}{X(s)} = \frac{1}{s^2 + 2P_1P_2 s + P_2^2}$
* **GENERAL LAPLACE TRANSFORM** Y = TRANSF (N, B, M, A, X) where: N and B are, respectively, the degree and coefficient array in the numerator polynomial in S M and A are, respectively, the degree and coefficient array in the denominator polynomial in S Arrays A and B may be set up by statements: STORAGE A(m + 1), B(n + 1) TABLE A(1 - M) = $a_1, a_2, \ldots, a_M$, A(m + 1) = a_0 TABLE B(1 - N) = $b_1, b_2, \ldots, b_N$, B(n + 1) = b_0 where: 1 - M and 1 - N represent subscript ranges and the 0th coefficients follow the highest numbered	$\frac{Y(s)}{X(s)} = \frac{\sum_{i=1}^{m} a_i s^i + a_0}{\sum_{j=1}^{n} b_j s^j + b_0}$ where: $m \leq n$

* The functional blocks, REALPL, CMPXPL, MODINT, PIPE, TRANSF, and LEDLAG, are system Macros and cannot be used in user-defined subprograms.

(Reprinted by permission from IBM Manual: Continuous System Modelling Program III (COSMP III) (Form # 19–7001–3), by International Business Machines Corporation.)

Table 4-1. (*continued*)

CSMP III Statement	Equivalent Mathematical Expression
DEAD TIME (DELAY) Y = DELAY (N, P, X) where: P = delay time N = number of points sampled in interval p (integer constant) and must be ≥ 3, and ≤ 16,378	$y = x(t-p)\ ;\ t \geq p$ $y = 0\ ;\ t < p$ Equivalent Laplace Transfer Function: $\frac{Y(s)}{X(s)} = e^{-Ps}$
ZERO-ORDER HOLD Y = ZHOLD (X1, X2)	$y = x_2\ ;\quad x_1 > 0$ $y =$ last output $;\quad x_1 \leq 0$ $y\|_{t=t_s} = 0$ Equivalent Laplace Transfer Function $\frac{Y(s)}{X(s)} = \frac{1}{s}(1 - e^{st})$
* **MODE-CONTROLLED INTEGRATOR** Y = MODINT (IC, X1, X2, X3)	$y = \int_{t_s}^{t_f} x_3 dt + ic\ ;\ x_1 > 0$, any x_2 $y = ic\ ;\quad x_1 \leq 0,\ x_2 > 0$ $y =$ last output $;\quad x_1 \leq 0,\ x_2 \leq 0$
* **VARIABLE FLOW TRANSPORT DELAY** Y = PIPE (N, IC, P, X1, X2, ND) where: N = number of intervals necessary to define holdup quantity (integer constant) IC = initial condition of entire pipeline P = holdup quantity X1 = flow rate X2 = delayed characteristic ND = degree of interpolation for retrieving delayed characteristic. May be 1 or 2 (integer variable or constant).	$fv = \int_{t_s}^{t} x_1\ dt$ $q = \int_{t_s}^{t} x_2 x_1\ dt$ $y = ic\ ;\ fv < p$ $y = \frac{q(fv-p) - q(fv(t-\Delta t)-p)}{fv - fv(t-\Delta t)}\ ;\ fv \geq p$ where: fv = flow volume q = weighted volume
IMPLICIT FUNCTION Y = IMPL (IC, P, FOFY) where: IC = first guess P = error bound FOFY = output name from final statement in algebraic loop definition	$y = f(y)$ $\|y - f(y)\| \leq p\|y\|$

*The functional blocks, REALPL, CMPXPL, MODINT, PIPE, TRANSF, and LEDLAG, are system Macros and cannot be used in user-defined subprograms.

Table 4-1. (*continued*)

CSMP III Statement	Equivalent Mathematical Expression
ARBITRARY FUNCTION GENERATOR (LINEAR INTERPOLATION) Y = AFGEN (FUNCT, X)	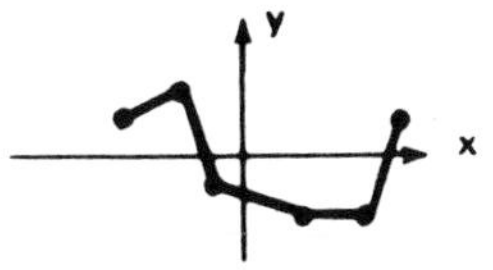 y = f(x)
ARBITRARY FUNCTION GENERATOR (QUADRATIC INTERPOLATION) Y = NLFGEN (FUNCT, X)	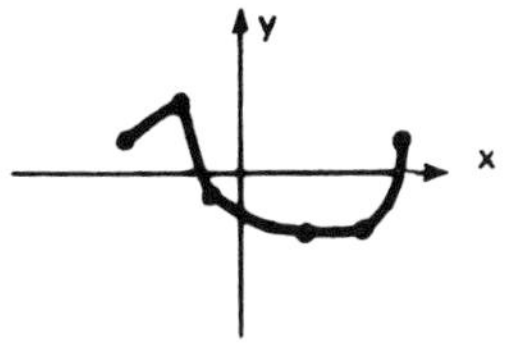 y = f(x)
FUNCTION GENERATOR WITH DEGREE OF INTERPOLATION CHOSEN BY USER Y = FUNGEN (FUNCT, N, X) where: FUNCT = function name N = degree of interpolation to be used. May be 1, 2, 3, 4, or 5 X = value of abscissa	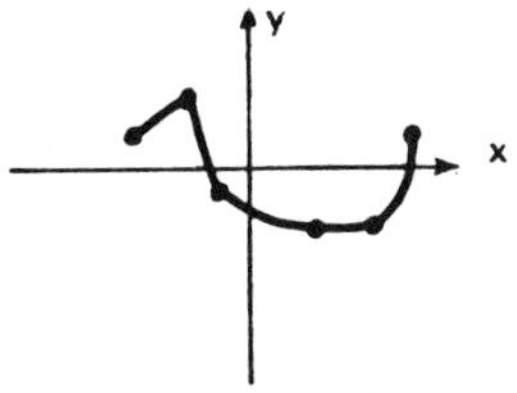 y = f(x)
ARBITRARY FUNCTION OF 2 VARIABLES Y = TWOVAR (FUNCT, X, Z)	y = f(x, z)
SLOPE OF A CURVE Y = SLOPE (FUNCT, N, X) where: FUNCT = name of curve N = degree of interpolation to be used X = value of abscissa	$y = \frac{df}{dx}$ at x

Table 4-1. (*continued*)

CSMP III Statement	Equivalent Mathematical Expression
LIMITER Y = LIMIT (P1, P2, X)	$y = p_1 \; ; x < p_1$ $y = p_2 \; ; x > p_2$ $y = x \; ; p_1 \leq x \leq p_2$ 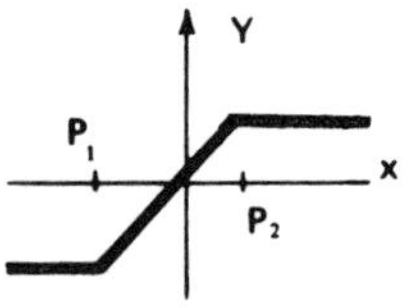 Y, X, P_1, P_2
DEAD SPACE Y = DEADSP (P1, P2, X)	$y = 0 \; ; p_1 \leq x \leq p_2$ $y = x - p_2 \; ; x > p_2$ $y = x - p_1 \; ; x < p_1$ Y, X, P_1, P_2, 45°, 45° 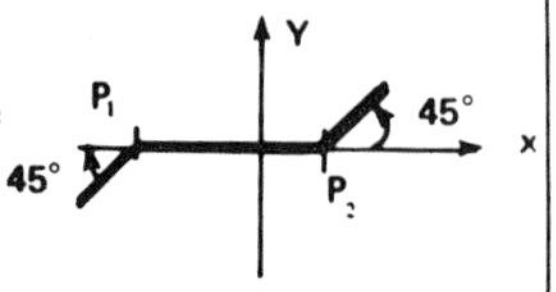
QUANTIZER Y = QNTZR (P, X)	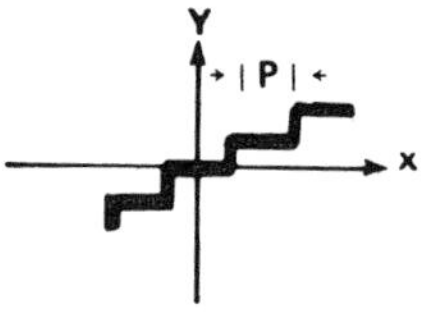 Y, X, | P | $y = kp \; ; (k - ½)\, p < x \leq (k + ½)\, p$ $k = 0, \pm 1, \pm 2, \pm 3, \ldots$
HYSTERESIS LOOP Y = HSTRSS (IC, P1, P2, X) where: $IC = y\|_{t_s}$	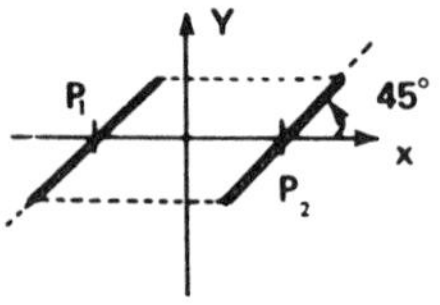 Y, X, P_1, P_2, 45° $y = x - p_2 \; ; (x - x(t - \Delta t)) > 0$ and $y(t - \Delta t) \leq (x - p_2)$ $y = x - p_1 \; ; (x - x(t - \Delta t)) < 0$ and $y(t - \Delta t) \geq (x - p_1)$ otherwise: $y = y(t - \Delta t)$

Table 4-1. *(continued)*

CSMP III Statement	Equivalent Mathematical Expression
STEP FUNCTION Y = STEP (P)	$y = 0 \; ; t < p$ $y = 1 \; ; t > p$ 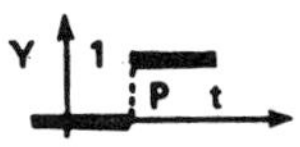
RAMP FUNCTION Y = RAMP (P)	$y = 0 \; ; t < p$ $y = t - p \; ; t > p$ 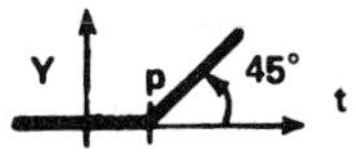
IMPULSE GENERATOR Y = IMPULS (P1, P2) where: P1 = time of first pulse P2 = interval between pulses	$y = 0 \; ; t < p_1$ $y = 1 \; ; (t - p_1) = kp_2$ $y = 0 \; ; (t - p_1) \neq kp_2$ $k = 0, 1, 2, 3, \ldots$ 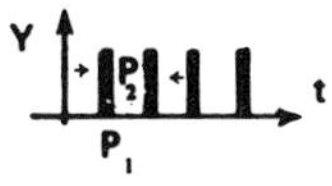
PULSE GENERATOR (WITH X > 0 AS TRIGGER) Y = PULSE (P, X) where: P = minimum pulse width	$y = 1 \; ; t_t \leq t < (t_t + p)$ or $x > 0$ $y = 0$; otherwise (t_t = time of trigger) 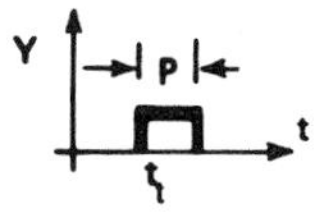
TRIGONOMETRIC SINE WAVE WITH DELAY, FREQUENCY AND PHASE PARAMETERS Y = SINE (P1, P2, P3) where: P1 = delay P2 = frequency (in radians per unit time) P3 = phase shift in radians	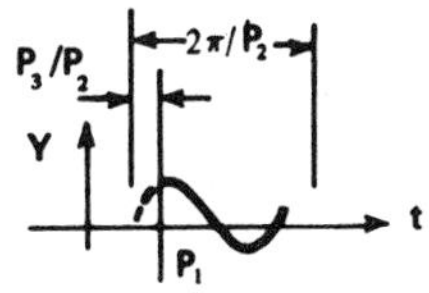 $y = 0 \; ; \; t < p_1$ $y = \sin(p_2 (t - p_1) + p_3) \; ; \; t > p_1$
NOISE (RANDOM NUMBER) GENERATOR WITH NORMAL DISTRIBUTION Y = GAUSS (S, P1, P2) where: S = seed (S is first truncated to an integer if real valued, then its absolute value is used once to initialize the generator) P1 = desired mean of values of Y P2 = desired standard deviation of values of Y	Normal Distribution of Variable y p(y) = probability density function 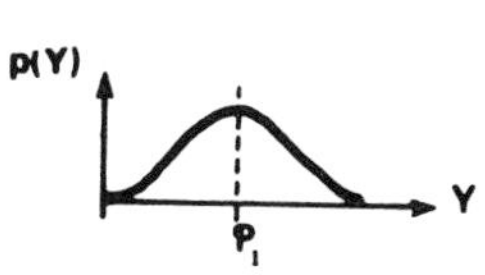
NOISE (RANDOM NUMBER) GENERATOR WITH UNIFORM DISTRIBUTION Y = RNDGEN (S) where: S = seed (S is first truncated to an integer if real valued, then its absolute value is used once to initialize the generator)	Uniform Distribution of Variable y p(y) = probability density function 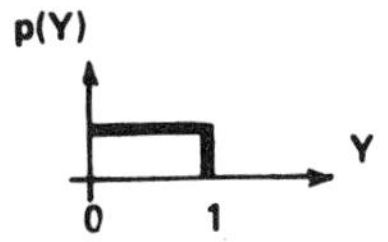

Table 4-1. (*continued*)

CSMP III Statement	Equivalent Mathematical Expression
SAMPLING INTERVAL SWITCH Y = SAMPLE (P1, P2, {P3 / N}) where: P_1 = start time for sampling to occur P_2 = last time for sampling to occur P_3 = time interval between samples if entered as a floating-point number N = number of sampling intervals if entered as a fixed-point number	$y = 1.0$; $TIME = p_1 + k\, p_3 < p_2$ $k = 0, 1, 2, \ldots$ or $y = 1.0$; $TIME = p_1 + k(p_2 - p_1)/n \le p_2$ $k = 0, 1, 2, \ldots$ $y = 0.0$ otherwise

CSMP III Statement	Equivalent Mathematical Expression
SCALAR-TO-ARRAY CONVERTOR CALL ARRAY (V1, V2, . . . , VN, X)	x(1) = v1 x(2) = v2 x(3) = v3 . . . x(n) = vn
ARRAY-TO-SCALAR CONVERTOR Y1, Y2, . . . , YM = SCALAR (X(2))	y1 = x(2) y2 = x(3) . . . ym = x(m + 1)
DOUBLE PRECISION FLOATING-POINT TO SINGLE PRECISION Y = ZZRND (DNUMBR) where: DNUMBR is a double precision floating-point number. Y = rounded single precision value of DNUMBR	To convert the double precision floating-point number, DNUMBR, to single precision, rounding to hexadecimal digit.

Table 4-1. *(continued)*

CSMP III Statement	Equivalent Mathematical Expression	
COMPARATOR		
Y = COMPAR (X1, X2)	$y = 0;$	$x_1 < x_2$
	$y = 1;$	$x_1 \geq x_2$
OUTPUT SWITCH		
Y1, Y2 = OUTSW (X1, X2)	$y_1 = x_2, y_2 = 0;$	$x_1 < 0$
	$y_1 = 0, y_2 = x_2;$	$x_1 \geq 0$
INPUT SWITCH RELAY		
Y = INSW (X1, X2, X3)	$y = x_2;$	$x_1 < 0$
	$y = x_3;$	$x_1 \geq 0$
RESETTABLE FLIP-FLOP		
Y = RST (X1, X2, X3)	$y = 0;$	$x_1 > 0$
	$y = 1;$	$x_2 > 0, x_1 \leq 0$
	$y = 0;$	$x_1 \leq 0, x_2 \leq 0$, $x_3 > 0, y(t - \Delta t) = 1$
	$y = 1;$	$x_1 \leq 0, x_2 \leq 0$, $x_3 > 0, y(t - \Delta t) = 0$
	$y = 0;$	$x_1 \leq 0, x_2 \leq 0$, $x_3 \leq 0, y(t - \Delta t) = 0$
	$y = 1;$	$x_1 \leq 0, x_2 \leq 0$, $x_3 \leq 0, y(t - \Delta t) = 1$
FUNCTION SWITCH		
Y = FCNSW (X1, X2, X3, X4)	$y = x_2;$	$x_1 < 0$
	$y = x_3;$	$x_1 = 0$
	$y = x_4;$	$x_1 > 0$

Table 4-1. (*continued*)

CSMP III Statement	Equivalent Mathematical Expression	
NOT Y = NOT (X)	$y = 1;$ $y = 0;$	$x \leq 0$ $x > 0$
AND Y = AND (X1, X2)	$y = 1;$ $y = 0;$	$x_1 > 0, x_2 > 0$ otherwise
NOT AND Y = NAND (X1, X2)	$y = 0;$ $y = 1;$	$x_1 > 0, x_2 > 0$ otherwise
INCLUSIVE OR Y = IOR (X1, X2)	$y = 0;$ $y = 1;$	$x_1 \leq 0, x_2 \leq 0$ otherwise
EXCLUSIVE OR Y = EOR (X1, X2)	$y = 1;$ $y = 1;$ $y = 0;$	$x_1 \leq 0, x_2 > 0$ $x_1 > 0, x_2 \leq 0$ otherwise
NOT OR Y = NOR (X1, X2)	$y = 1;$ $y = 0;$	$x_1 \leq 0, x_2 \leq 0$ otherwise
EQUIVALENT Y = EQUIV (X1, X2)	$y = 1;$ $y = 1;$ $y = 0;$	$x_1 \leq 0, x_2 \leq 0$ $x_1 > 0, x_2 > 0$ otherwise

Table 4-2. FORTRAN functions.

CSMP III Statement	Equivalent Mathematical Expression
EXPONENTIAL Y = EXP(X)	$y = e^x$
NATURAL LOGARITHM Y = ALOG(X)	$y = \ln(x)$
COMMON LOGARITHM Y = ALOG10(X)	$y = \log_{10}(x)$
ARCTANGENT Y = ATAN (X)	$y = \tan^{-1}(x)$
TRIGONOMETRIC SINE Y = SIN (X)	$y = \sin(x)$
TRIGONOMETRIC COSINE Y = COS (X)	$y = \cos(x)$
SQUARE ROOT Y = SQRT (X)	$y = \sqrt{x}$
HYPERBOLIC TANGENT Y = TANH (X)	$y = \tanh(x)$
ABSOLUTE VALUE (REAL ARGUMENT AND OUTPUT) Y = ABS (X)	$y = \|x\|$
ABSOLUTE VALUE (INTEGER ARGUMENT AND OUTPUT) Y = IABS (X) also specify FIXED IABS	$y = \|x\|$
TRANSFER OF SIGN (REAL ARGUMENT AND OUTPUT) Y = SIGN (X1, X2)	$y = \|x_1\|$; $x_2 \geq 0$ and $x_1 \neq 0$ $y = -\|x_1\|$; $x_2 < 0$ and $x_1 \neq 0$ $y = 0$; $x_1 = 0$
TRANSFER OF SIGN (INTEGER ARGUMENT AND OUTPUT) Y = ISIGN (X1, X2) also specify FIXED ISIGN	$y = \|x_1\|$; $x_2 \geq 0$ and $x_1 \neq 0$ $y = -\|x_1\|$; $x_2 < 0$ and $x_1 \neq 0$ $y = 0$; $x_1 = 0$

(Reprinted by permission from IBM Manual "Continuous System Modelling Program III (COSMP III) (Form # 19–7001–3), by International Business Machines Corporation.)

Table 4-2. (*continued*)

CSMP III Statement	Equivalent Mathematical Expression
LARGEST VALUE (INTEGER ARGUMENTS AND REAL OUTPUT) Y = AMAX0 (X1, X2, . . . , XN)	$y = \max(x_1, x_2, \ldots, x_n)$
LARGEST VALUE (REAL ARGUMENTS AND OUTPUT) Y = AMAX1 (X1, X2, . . . , XN)	$y = \max(x_1, x_2, \ldots, x_n)$
LARGEST VALUE (INTEGER ARGUMENTS AND OUTPUT) Y = MAX0 (X1, X2, . . . , XN) also specify FIXED MAX0	$y = \max(x_1, x_2, \ldots, x_n)$
LARGEST VALUE (REAL ARGUMENTS AND INTEGER OUTPUT) Y = MAX1 (X1, X2, . . . , XN) also specify FIXED MAX1	$y = \max(x_1, x_2, \ldots, x_n)$
SMALLEST VALUE (INTEGER ARGUMENTS AND REAL OUTPUT) Y = AMIN0 (X1, X2, . . . , XN)	$y = \min(x_1, x_2, \ldots, x_n)$
SMALLEST VALUE (REAL ARGUMENTS AND OUTPUT) Y = AMIN1 (X1, X2, . . . , XN)	$y = \min(x_1, x_2, \ldots, x_n)$
SMALLEST VALUE (INTEGER ARGUMENTS AND OUTPUT) Y = MIN0 (X1, X2, . . . , XN) also specify FIXED MIN0	$y = \min(x_1, x_2, \ldots, x_n)$
SMALLEST VALUE (REAL ARGUMENTS AND INTEGER OUTPUT) Y = MIN1 (X1, X2, . . . , XN) also specify FIXED MIN1	$y = \min(x_1, x_2, \ldots, x_n)$

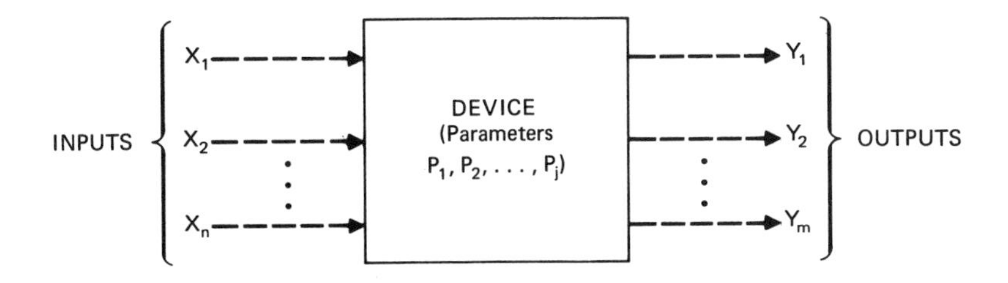

MATHEMATICAL EXPRESSION

$$Y_1, Y_2, \ldots, Y_m = f(P_1, P_2, \ldots, P_j, X_1, X_2, \ldots, X_n)$$

EQUIVALENT CSMP III STATEMENT

Y1, Y2, . . . , YM = DEVICE (P1, P2, . . . , PJ, X1, X2, . . . , XN)

Figure 4-8. CSMP functional blocks.

not be defined again; that is, its symbolic name does not ordinarily appear to the left of the equal sign in any other statement.

To complete the statement, block parameters and inputs to the block are listed in a definite order and enclosed in parentheses to the right of the functional block name. A constant input may be given there as either a numerical constant or the symbolic name of a constant. Other inputs given will be outputs of other (or sometimes the same) structure statements, or expressions composed of outputs and constants.

While the use of an output of one structure statement as an input to another expresses a block-to-block connection (or a link between mathematical formulae), parameters internally modify the functional block. Some functional blocks require initial conditions of their outputs to be supplied. Like constants, parameters and initial conditions may be either explicit numerical constants or symbolic variables. Symbolically named constants, parameters and initial conditions are variable in the sense that their values may be set each time the model is used, without change to the model.

Structure of the Model

The nucleus of a continuous system simulation language (CSSL) is a mechanism for solving the differential equations that represent the dynamics of the model. If the simulation requires no auxiliary computation, the simple implicit structure of CSMP III is sufficient. Usually, there are computations to be performed at the beginning and end of each run. The explicit formulation of model structure is designed for these more complex requirements. For example, certain parameters of a model might be considered basic; secondary parameters and initial conditions are often expressed as functions of these basic parameters. Evaluation of these functions is desired just once per run. Frequently, one needs to perform some terminal evaluation of a solution. In a design study, for instance one might compute some "figure of merit" for each run of a parameter search.

To satisfy these requirements, the CSMP III formulation of a model is divided into three segments—initial, dynamic, and terminal—that describe, respectively, the computations to be performed before, during and after each solution. These segments represent the highest level of the structural hierarchy. Each of the segments may comprise one or more sections. These, in turn, contain the structure statements that specify model dynamics and associated computations.The sections represent rational groupings of structure statements and may be processed as either parallel or procedural entities. The overall structural hierarchy is illustrated in Figure 4-9

Initial, dynamic and terminal segments. The initial segment is intended exclusively for the computation of initial condition values and those parameters that the user prefers to express in terms of more basic parameters. Thus, if a model repeatedly makes use of the cross-sectional area of a cylindrical member, the initial segment might contain the statement

AREA = 3.1416 * (R **2),

with radius R, which might be considered the more basic parameter, specified on a data card as follows.

PARAMETER R = 7.5

If initial conditions or parameters are to be computed, rather than specified as constant values, then they must be computed in the initial segment.

The initial segment is optional.

The dynamic segment is normally the most extensive in the model. It contains the complete description of the systems dynamics, together with any other computations required during the solution of the system. The dynamic segment is analogous to the block diagram representation or to the ordinary differential equation representation of the system dynamics. The structure statements within this segment are generally a mixture of CSMP III and FORTRAN statements. The following might be considered representative.

DRAG = 0.5*RHO*S*CD*(V**2)

VX=INTGRL [VZERO, (THRUST-DRAG)/MASS]

For most models, the dynamic segment consists of a single section. For complicated systems, it is often desirable, and sometimes required, that it be divided into several sections. In modeling an industrial process, for example, it might be desirable to separate the various unit process models into separate sections, simply because the physical units are distinct.

The dynamic segment is required. This segment may be declared explicitly by a DYNAMIC statement, or implicitly by the

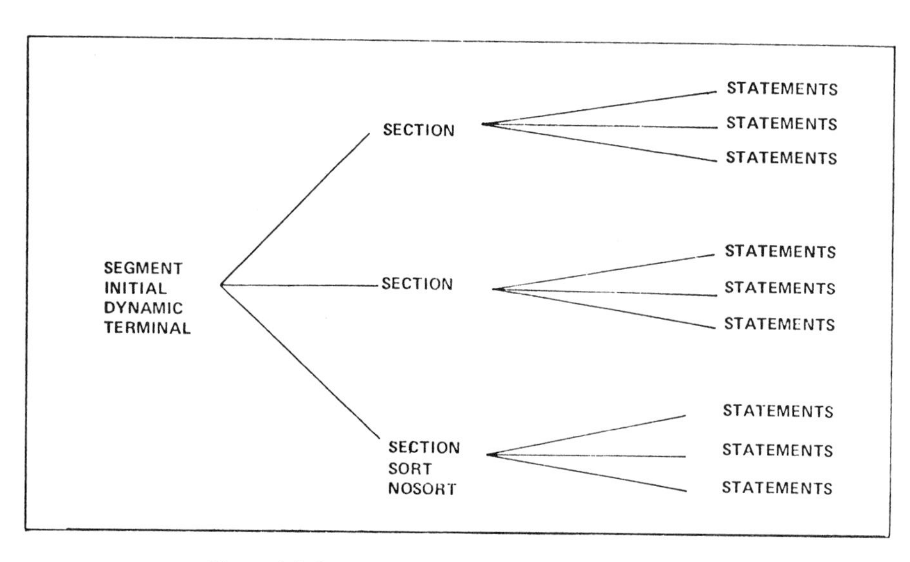

Figure 4-9. Structural hierarchy of CSMP III language.

absence of INITIAL, DYNAMIC or TERMINAL statements.

The terminal segment is used for those computations required at the end of the run, after completion of the solution. This can be calculated based on the final value of one or more model variables, or one might incorporate an optimization algorithm that will modify the values of critical system parameters.

The terminal segment is optional.

Segmentation, the explicit division of the model into computations to be performed before, during and after each solution, is provided by the control statements: INITIAL, DYNAMIC, TERMINAL and END. In the general case, as illustrated in Figure 4-10, INITIAL is the first statement of the specification of model structure (it is not normally the first statement of the composite CSMP III data deck, since certain other translation control statements and all MACRO definitions must be placed before any structure statements). The DYNAMIC statement separates the initial segment from the dynamic segment. Similarly, the TERMINAL statement separates the dynamic segment from the first statement of the terminal segment. The first use of the END control statement is mandatory to complete both the model specifications and

```
          -----------
          -----------
INITIAL
          -----------
          -----------
DYNAMIC
          -----------
          -----------
          -----------
TERMINAL
          -----------
          -----------
END
```

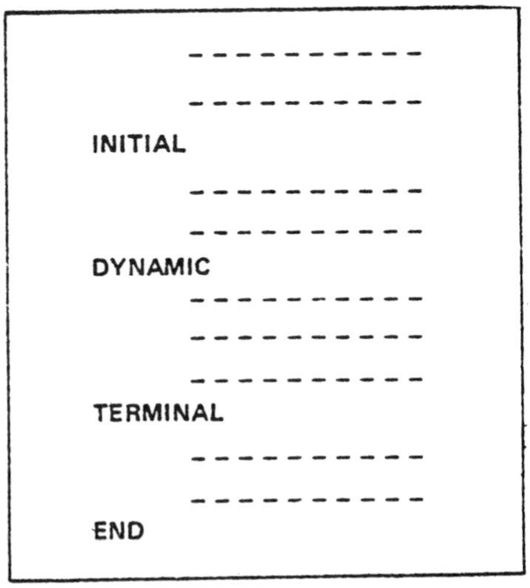

Figure 4-10. Segmentation of model.

the data and execution control for the first case. Subsequent END statements must be used to de-limit data and execution control input for all additional cases, if any. A detailed example of the use of the CSMP modeling program is shown later in this section.

How Simulation and Modeling fit into the Chemical Industry

Increasing process complexities, the requirements for better energy management and improved product quality are forcing implementation of more sophisticated control strategies and a re-orientation of their methods of analysis. Prior to the current widespread acceptance of computers in the chemical and petroleum industry, complete understanding of a process both from a design and a dynamic operation standpoint was only based on experience and judgment. The current trend toward larger plants and the variable throughput operation of existing plants require engineers to investigate, more analytically, various methods of efficient and safer operation. Once these methods are formulated, they must be verified, implemented and controlled. The verification is best accomplished by the solution of a computer model. The results then can be analyzed and corrections made prior to a construction phase (in the case of new plant). In the case of existing plant, different operating parameters can be instituted.

Modeling and simulation of a unit process and, eventually, the entire train requires, as a minimum:

- A thorough understanding of the physical and chemical principles of the process.
- A reasonable ability to translate the process into mathematics.
- An ability to forecast results without actual implementation, so that the model resembles the process.
- An ability to transform the mathematics into computer language.

- An ability to translate computer outputs or results into usable information.

Very often, the mathematical exercise of formulating a model, along with the application of the items listed above, result in a better understanding of the process. Model formulation may sometimes negate the requirements for further simulation, in those cases when a solution has been found as a result of this mathematical exercise.

There are some modeling and simulation applications which should be carried through to a solution on the computer in order to completely understand the program. They may be broadly classified as follows.

- New process dynamics.
- Control system design.
- Optimization.
- Logic system check-out.
- Training.
- Energy management.
- Miscellaneous.

Modeling and simulation as applied to these items offer alternate solutions to situations without the usual trial and error often required during construction and start-up of the actual plant. That is not to say that some experimentation after construction may not still be required, but modeling and simulation at least help to define the areas of concern.

The Interaction of a Microcomputer and the Simulation Process

The development of microcomputer hardware has enabled the implementation of advanced control technology at key unit operations located throughout a plant. The simulation and modeling of a chemical column when completed on an off-line computer provides solutions to problems previously defined. The implementation of the solutions may in some cases require design changes to the column and/or its associated equipment, but in all cases, some modification in the control systems components is required. The advanced control techniques necessary to meet the control objectives (optimization, energy management, tray analysis, composition analysis, etc.) require a digital computer. A microcomputer, when communicating with a host minicomputer, can be loaded directly from the host previously supplied with the simulation results. A typical development system showing a hardware configuration of a host minicomputer and a dedicated microcomputer is shown in Figure 4-11. The process input/output interface is used in this case to load the advanced control and algorithms into the system controller. Verification of the operation can be accomplished with a continuous updating of the model and subsequent updating of the control algorithms.

Computer simulation of process control systems and the integration of microcomputers in the chemical industry provides the tools that enable solutions to the ever-increasing problems of the "real world."

Computer Model Development

This section outlines, in some detail, the development, analysis, solution and conclusion of a control philosophy application on a typical depropanizer column in an ethylene plant.

Mathematical Development

The mathematical development is limited to first order lumped parameter equations. Where necessary, non-linear functions are either linearly approximated or the actual function is generated.

The mathematical model is developed for the column, condenser, reflux drum and associated controls. Each component was modeled separately and then combined to form the system.

Assumptions

It was assumed that a multicomponent mixture did not affect column dynamics, since interest was focused on column stability and not on composition.

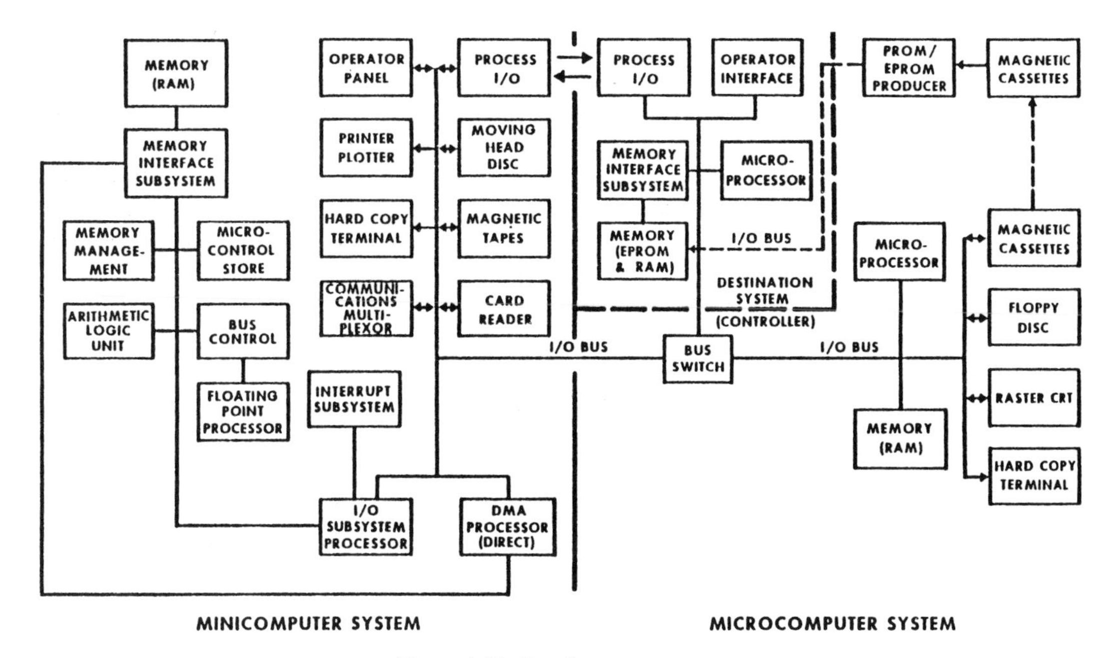

Figure 4-11. Development system.

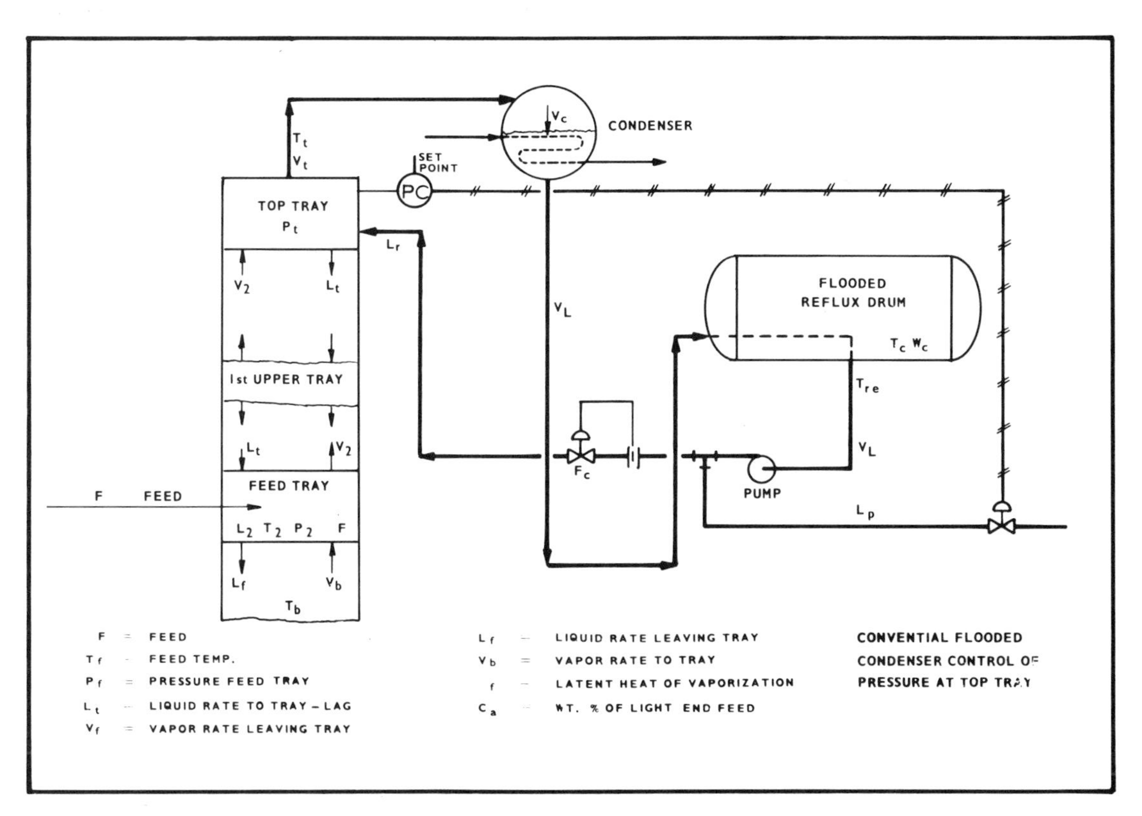

Figure 4-12. Conventional flooded condenser control of pressure at top tray.

It was also assumed (1) that only two trays existed—feed tray and top tray—such that first order lags between the two trays simulated the intervening trays (the first order lag for the vapor boil-up was considered much faster than the first order lag for liquid coming down from the top tray); and (2) that all vapor equations followed the perfect gas law at these pressures and temperatures.

The column simulation began at the feed tray. The re-boiler and bottom section were not considered.

The reflux flow was at a constant ratio to feed flow.

Mathematical derivation

Feed tray. (Figure 4-13.) The feed tray mathematical model consists of an energy balance resulting in calculation of feed tray temperature, material balance and vapor boil-up rate. The vapor boil-up is modeled using the well known "Virial" approximation.

Top tray. (Figure 4-14.) The top tray mathematical model consists of an energy balance, a material balance and equations describing the top tray pressure variations and vapor generation.

The generation of vapor is accomplished by using an expression which is a function of the temperature difference between the top tray gas and liquid temperatures:

$$V_E = \frac{\alpha_E}{P_T}(T_T - T_E)$$

where

V_E = vaporization rate

α_E = constant

P_T = top tray pressure (psia)

T_T = top tray temperature, liquid (°F)

T_E = top tray temperature, gas (°F).

Note that the vaporization rate is zero when $T_T = T_E$ and that vapor is generated when $T_T > T_E$. This concept and the equation are well documentated in the literature.[1] The top tray pressure equation is the dynamic equivalent of the perfect gas law:

ENERGY BALANCE $$\frac{dT_2}{d\theta} = \frac{\lambda FCA}{C_2} + \frac{C_{pf}F[T_f - T_r]}{C_2} + \frac{L_t' C_{pt}[T_t - T_2]}{C_2} + \frac{V_b C_b [T_b - T_2]}{C_2}$$

MATERIAL BALANCE $$L_f = L_t + F - V_2 + V_b$$

VAPOR EQUATION $$V_2 = \frac{\alpha b T_2}{P_2}$$

LAG EQUATION $$\frac{\tau\, dL_t'}{d\theta} + L_t' = L_t$$

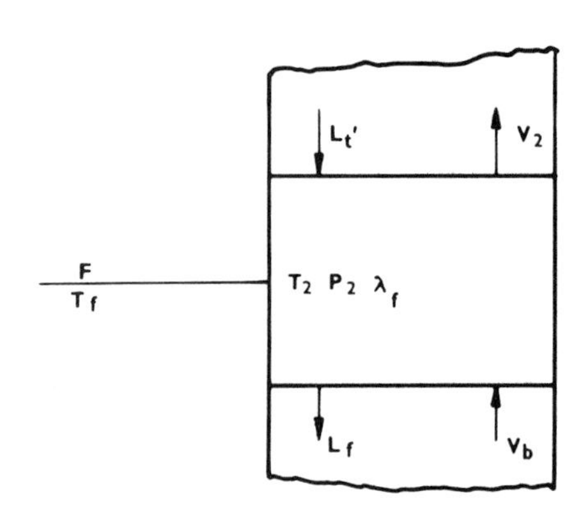

Figure 4-13. Feed tray.

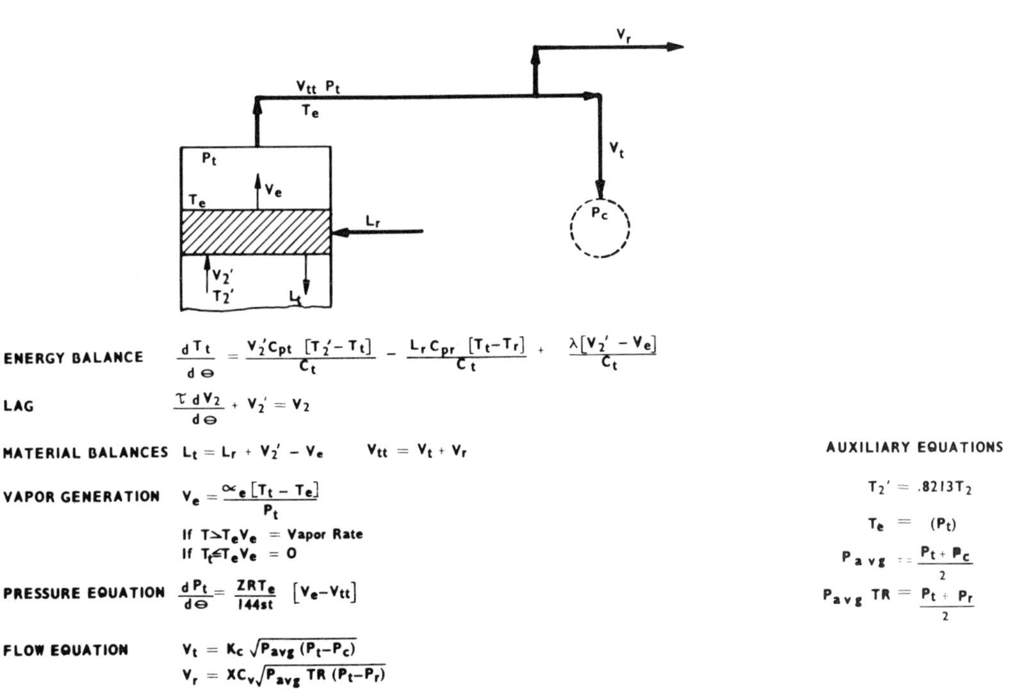

V_r
V_{tt} P_t
T_e
V_t
P_t
V_e
T_e
L_r
P_c
V_2'
T_2'
L_t
ENERGY BALANCE $\frac{dT_t}{d\theta} = \frac{V_2' C_{pt}[T_2' - T_t]}{C_t} - \frac{L_r C_{pr}[T_t - T_r]}{C_t} + \frac{\lambda[V_2' - V_e]}{C_t}$
LAG $\frac{\tau\, dV_2}{d\theta} + V_2' = V_2$
MATERIAL BALANCES $L_t = L_r + V_2' - V_e$ $V_{tt} = V_t + V_r$
VAPOR GENERATION $V_e = \frac{\alpha_e[T_t - T_e]}{P_t}$
If $T > T_e V_e$ = Vapor Rate
If $T_t \le T_e V_e = 0$
PRESSURE EQUATION $\frac{dP_t}{d\theta} = \frac{ZRT_e}{144st}[V_e - V_{tt}]$
FLOW EQUATION $V_t = K_c\sqrt{P_{avg}(P_t - P_c)}$
$V_r = XC_v\sqrt{P_{avg}\,TR\,(P_t - P_r)}$
AUXILIARY EQUATIONS
$T_2' = .8213T_2$
$T_e = (P_t)$
$P_{avg} = \frac{P_t + P_c}{2}$
$P_{avg}\,TR = \frac{P_t + P_r}{2}$

Figure 4-14. Top tray.

$$\frac{dP_T}{dt} = \frac{Z_R T_E}{144 V_{ST}} (V_E - V_{TT})$$

where

P_T = pressure at top tray (psia)

Z_R = constant

T_E = top tray temperature, gas (°F)

V_{ST} = vapor volume in top tray cavity (ft^3)

V_E = vaporization rate

V_{TT} = overhead product flow

t = time (hr).

It can be seen that at steady state condition, $V_E = V_T$, and the pressure remains constant at design pressure. This equation will be used to simulate the top tray pressure.

The remaining equations describe the flow through piping and valves.

Condenser. (Figure 4-15.) The condenser model consists of an energy balance, a material balance in the liquid and vapor phase and several expressions relating to condensing tube surface area, liquid volume, liquid surface area and gas volume, all of which are variables.

The pressure in the condenser is described by the dynamic equivalent of the perfect gas equation, with the condensation rate as a function of the tubes surface area. It should be noted here that condensation rate is not necessarily equal to the condensate flow. This occurs only at the steady state condition when the number of exposed tubes is sufficient to condense the incoming vapor. Should the incoming vapor increase via an increase in feed or boil-up through the tower, the pressure in the condenser will increase until tubes are exposed and an increase in condensation rate occurs.

The pressure equation is

$$\frac{dP_C}{dt} = \frac{Z_R T_E}{144 V_R} (V_T - V_C);$$

the level equation is

$$\frac{dH_C}{dt} = \frac{V_C - V_L}{\rho_L A};$$

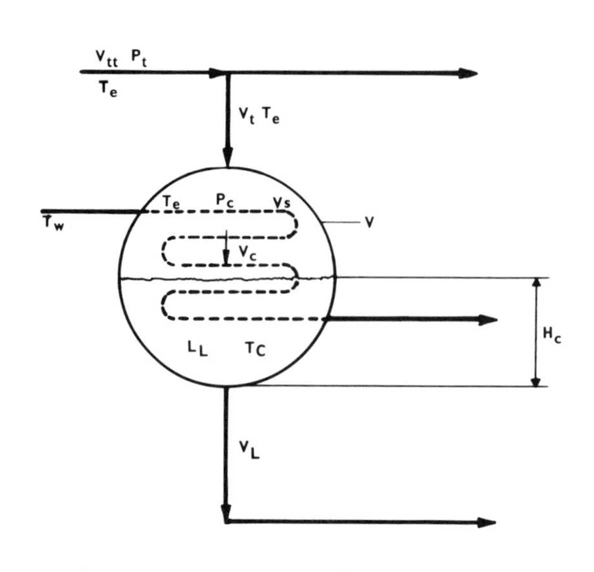

ENERGY BALANCE

$$\frac{dT_c}{d\Theta} = \frac{\lambda V_c}{C_c} + \frac{C_p(V_t)[(T_e - T_c)]}{C_c} - \frac{UA_c[T_c - T_w]}{C_c}$$

PRESSURE EQUATION

$$\frac{dPc}{d\Theta} = \frac{RATE}{144V_s}[(V_t - V_c)]$$

MATERIAL BALANCE

$$\frac{dhc}{d\Theta} = \frac{V_c' - V_L}{\rho_L A}$$

FLOW EQUATION

$$V_L = K_L\sqrt{P_{avg\,cr} \quad (P_c - P_r)}$$

AUXILIARY EQUATION

$$V_c = f(A_c) = 3842.187A_h$$

$$A_h = f(H_c) \quad \text{TABLE I}$$

$$A = f(H_c) = 31\,\frac{[3600 - (60 - H_c)^2]^{1/2} \times 2}{12}$$

$$A_c = 512.292 \times A_h$$

$$V_s - V - L_L = 2435 - 31[(78.549 - A_h)]$$

$$P_{avg\,cr} = \frac{P_c + P_r}{2}$$

Figure 4-15. Condenser.

and the condensation rate equation is

$$V_C = f(A_C) \text{(from Table 4-3)};$$

where

P_C = condenser pressure (psia)

Z_R = constant

T_E = vapor overhead temperature (°F)

t = time

V_T = total overhead product vapor

V_C = condensation rate

A_C = exposed area of condenser tubes (ft^2)

H_C = level in condenser (ft)

ρ_L = liquid density (ft^3)

A = surface area of liquid in condenser (ft^2)

V_L = condensate flow.

Reflux drum. Since the reflux drum is completely flooded it has no dynamic effect upon this simulation.

Control system. (Figure 4-16.) The control system consists of a valve and pressure controller. The valve is modeled such that it is capable of passing 75% of flow at 50% opening. The pressure controller is a three mode type.

Results and Conclusions of the Simulation Exercise

(Figure 4-17 shows the results of the dynamic simulation.)

The computer model development for this column illustrates the application of the systems analysis approach.

APPLICATIONS OF MICROCOMPUTER SYSTEMS TO COLUMN CONTROL

There are many different methods for the control of fractionating columns. The one common denominator contained in all control philosophies is the use of either conventional single loop feedback control, or a feedforward and feedback application. Historically, engineers have designed control systems for columns based on experience and judgment, and some still contend

Table 4-3. Variables.

FEED TRAY

T_2	=	Feed tray temperature
P_2	=	Feed tray pressure
τ_F	=	Latent heat of vaporization
L_F	=	Liquid from feed tray
V_B	=	Vapor from bottoms
F	=	Feed rate
L_T	=	Liquid on feed tray from top tray
V_2	=	Vapor rate from feed tray
C_A	=	Weight % overhead in feed
C_2	=	Thermal capacitance of material on tray
C_{PF}	=	Specific heat at constant pressure
T_F	=	Temperature of feed
C_{PT}	=	Heat capacity of material from top tray
T_T	=	Top tray temperature
T_B	=	Bottom tray temperature into feed tray
C_B	=	Heat capacity of materials from bottom tray
V_2	=	Boil-up rate from feed tray
α_B	=	Vaporization constant
τ_2	=	Time constant

REFLUX DRUM

T_R	=	*Temperature of reflux liquid*
$\mathbf{T}_X$	=	*Temperature of film*
$\mathbf{T}_E$	=	*Temperature of vapor*
$\mathbf{T}_C$	=	*Temperature of condensate*
$\mathbf{V}_L$	=	*Condensate flow from condenser*
$\mathbf{Z}_R$	=	*Compressibility and gas constant*

TOP TRAY

T_T	=	Top tray temperature liquid
T_E	=	Top tray temperature, vapor
α_E	=	Constant
V_E	=	Evaporation rate
P_T	=	Top tray pressure
V_{TT}	=	Overhead product flow
V_T	=	Flow into condenser
V_2'	=	Lag from vapor rate from feed tray
L_T	=	Liguid from top tray down
L_R	=	Reflux
C_T	=	Heat capacity of liquid in tray
C_P	=	Specific heat in tray
C_{PR}	=	Specific heat of reflux
Z_R	=	Compressibility and gas constant
t	=	Time
V_{ST}	=	Vapor space in top tray overhead
P_C	=	Pressure in condenser
λ_T	=	Heat of vaporization
T_2'	=	Fictitious temperature from bottom (= 3213 T_2)
T_1	=	Lag on vapor boil-up
V_R	=	Vapor bypass flow coefficient
K_C	=	Condenser flow coefficient
X_{CV}	=	Valve flow coefficient for flow to reflux drum, vapor
X	=	Position

CONDENSER DATA

T_C	=	Condenser temperature
T_W	=	Water temperature

Table 4-3. (*continued*)

T_E	=	Overhead temperature, top tray
λ_e	=	Heat of condensation
L_L	=	Liquid volume
H_C	=	Height of liquid in condenser
P_C	=	Condenser pressure
C_P	=	Specific heat
Z_R	=	Compressibility and gas constant
ρ_g	=	Density, gas
V_S	=	Volume of vapor space
A_H	=	Cross-sectional area of drum exposed
V_t	=	Overhead vapor rate
V_R	=	Reflux drum hot vapor bypass flow
V_c	=	Condensation rate
A	=	Area of liquid surface
A_C	=	Surface area of tubes exposed
A_H	=	Cross-section at area of drum exposed
C_C	=	Thermal capacitance of liquid in condenser
V_L	=	Condensate flow
P_R	=	Reflux drum pressure
K_L	=	Flow coefficient from condenser to reflux drum

that control of a column is an art. Microcomputer technology, and computer technology in general, offer the opportunity to remove column control from the "art" category and to make use of experience and judgment in addition to science and engineering.

The purpose of this section is to discuss the conventional methods of control and introduce the subject of multi-variable and interactive control. This requires a stringent application of the science and engineering approach.

A typical simple binary column with a steam re-boiler, water-cooled condenser and reflux drum was shown, with a typical

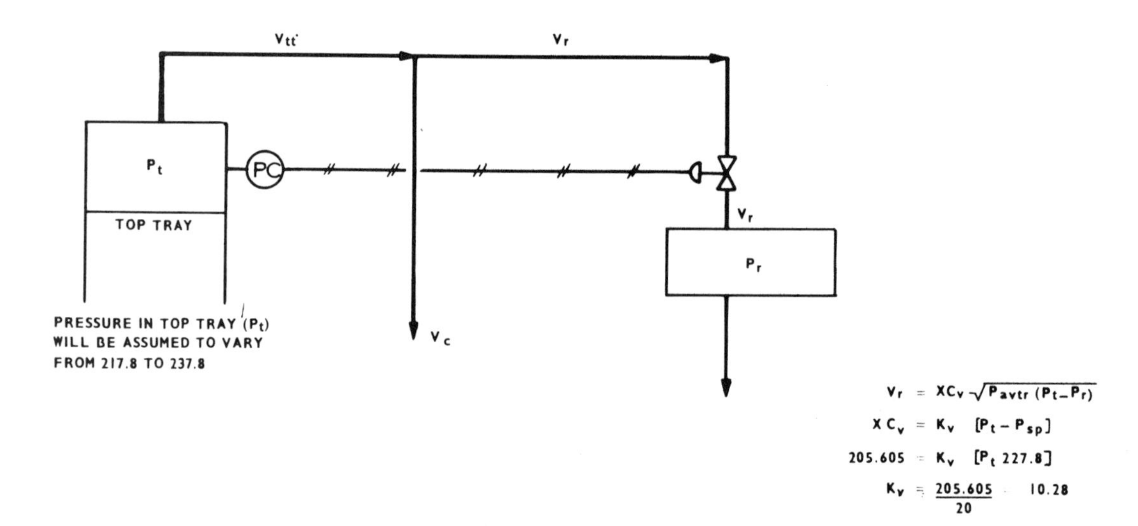

Figure 4-16. Control system.

control scheme, in Figure 4-18. Conventional single loop control is used throughout. In this type of column, there are key variables which greatly affect the operation of the column. Some of these variables can be manipulated; others cannot. Among the manipulated variables are column pressure, feed flow rate, feed temperature, energy input, bottom product flow rate, energy removal and product flow rate. Variables which cannot be manipulated are ambient barometric pressure, ambient temperature, feed composition, cooling water temperature and inert vapor generation.

The discussion that follows is very general and is limited to one column.

Conventional Control

Single Loop Feedback

The single loop feedback control of a binary distillation column is primarily a single variable (single input, single output) philosophy. Consider the typical fractionating column shown in Figure 4-18. This particular column is pressure controlled and compensates for inerts generated through the split range pressure control at the overheads.

Note that each loop is essentially independently operated as a feedback device and that no provision is made to compensate for any interaction among the controlled variables. The standard analog controller, discussed earlier in this chapter, has inherent internal interaction. Should retuning of a controller become necessary, the variable on which this controller is applied may interact with other process variables, which would necessitate retuning of other variables.

The application of a microcomputer to this system, with the conventional control shown, serves no purpose unless the computer acts to change set points, optimizes, de-couples or, more important, totally replaces the analog conventional control with a digital equivalent. Figure 4-19 shows a possible DDC configuration. DDC can be applied with a microcomputer and can perform functions on this column (as was discussed earlier). However, an economic and engineering study should be implemented prior to a decision regarding the inclusion of a microcomputer on a conventional column control system.

Feedforward and Feedback Control

Feedback control may be defined as the process of maintaining a controlled variable through the action of a device which compares the deviation between a set point and measurement of the controlled variable and which tends to drive that deviation to zero. One important disadvantage of feedback control is that the error or deviation must exist before corrective action takes place.

Feedforward control is a process whereby the magnitude of the error is anticipated and corrective action is taken prior to the occurrence of the error.[5]

Feedforward/feedback control is a combination of both. The process of combining the attributes of both concepts results in error anticipation and corrective action followed by readjustment. Feedforward/feedback control can best be illustrated by Figure 4-20.

Figure 4-21 shows the application of this concept to a column. In this case, the feedforward components of control consist of a feed analyzer and feed flow transmitter. These two components analyze and sense the disturbances due to a feed change and composition. If the feed and/or composition changes, the output manipulates the product draw and bottoms drain-off to approximately the correct or desired quantity. The product and bottoms analyzers then correct the quantity of product and bottoms draw to the desired quantity. The lead lag functions on the feed flow and bottoms flow should be carefully set with respect to the dynamics of the column and are not the same for the overhead and bottoms. The reader is referred to the literature for methods of tuning lead lag con-

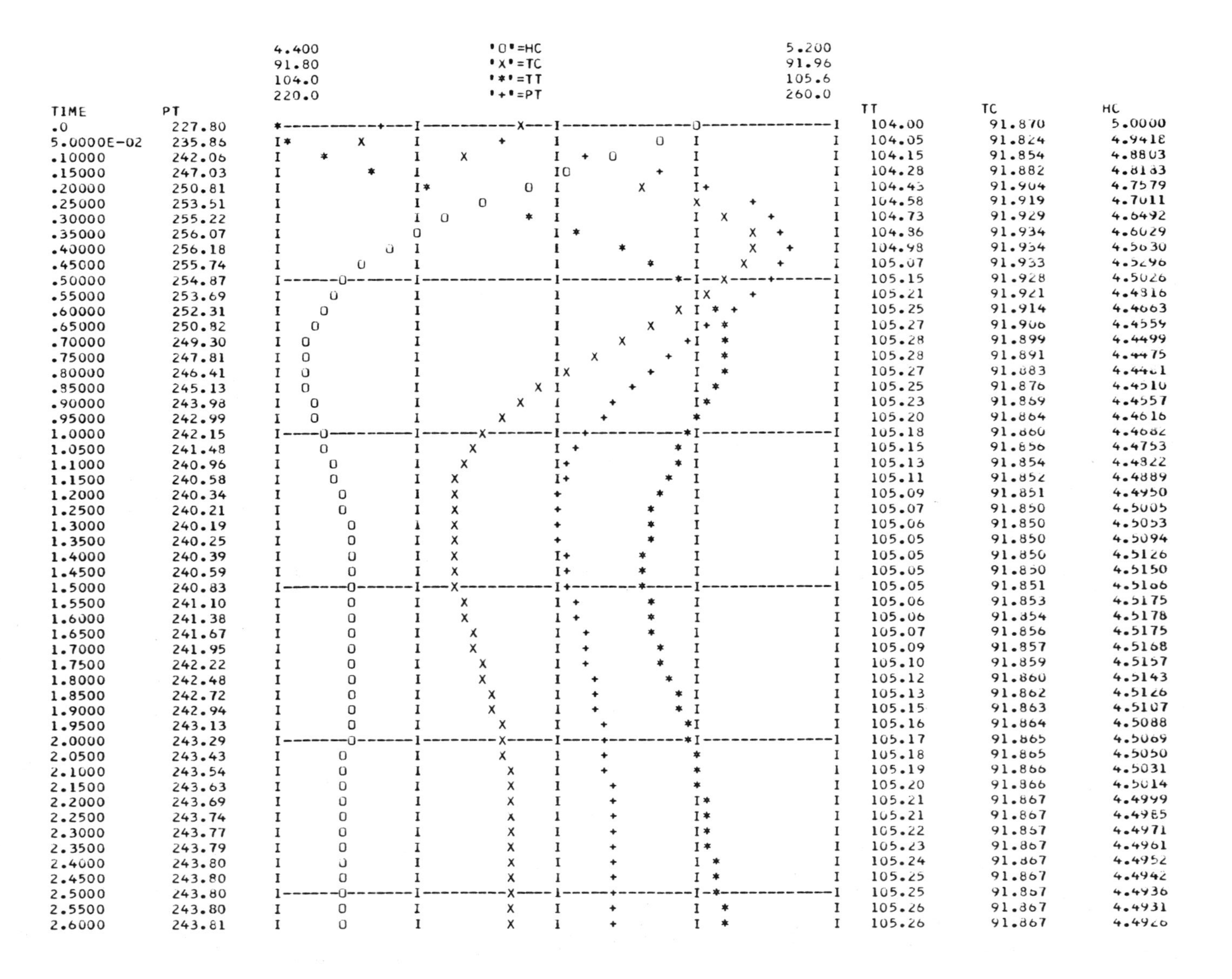

TIME	PT	TT	TC	HC
.0	227.80	104.00	91.870	5.0000
5.0000E-02	235.86	104.05	91.824	4.9418
.10000	242.06	104.15	91.854	4.8803
.15000	247.03	104.28	91.882	4.8183
.20000	250.81	104.43	91.904	4.7579
.25000	253.51	104.58	91.919	4.7011
.30000	255.22	104.73	91.929	4.6492
.35000	256.07	104.86	91.934	4.6029
.40000	256.18	104.98	91.934	4.5630
.45000	255.74	105.07	91.933	4.5296
.50000	254.87	105.15	91.928	4.5026
.55000	253.69	105.21	91.921	4.4816
.60000	252.31	105.25	91.914	4.4663
.65000	250.82	105.27	91.906	4.4559
.70000	249.30	105.28	91.899	4.4499
.75000	247.81	105.28	91.891	4.4475
.80000	246.41	105.27	91.883	4.4461
.85000	245.13	105.25	91.876	4.4510
.90000	243.98	105.23	91.869	4.4557
.95000	242.99	105.20	91.864	4.4616
1.0000	242.15	105.18	91.860	4.4682
1.0500	241.48	105.15	91.856	4.4753
1.1000	240.96	105.13	91.854	4.4822
1.1500	240.58	105.11	91.852	4.4889
1.2000	240.34	105.09	91.851	4.4950
1.2500	240.21	105.07	91.850	4.5005
1.3000	240.19	105.06	91.850	4.5053
1.3500	240.25	105.05	91.850	4.5094
1.4000	240.39	105.05	91.850	4.5126
1.4500	240.59	105.05	91.850	4.5150
1.5000	240.83	105.05	91.851	4.5166
1.5500	241.10	105.06	91.853	4.5175
1.6000	241.38	105.06	91.854	4.5178
1.6500	241.67	105.07	91.856	4.5175
1.7000	241.95	105.09	91.857	4.5168
1.7500	242.22	105.10	91.859	4.5157
1.8000	242.48	105.12	91.860	4.5143
1.8500	242.72	105.13	91.862	4.5126
1.9000	242.94	105.15	91.863	4.5107
1.9500	243.13	105.16	91.864	4.5088
2.0000	243.29	105.17	91.865	4.5069
2.0500	243.43	105.18	91.865	4.5050
2.1000	243.54	105.19	91.866	4.5031
2.1500	243.63	105.20	91.866	4.5014
2.2000	243.69	105.21	91.867	4.4999
2.2500	243.74	105.21	91.867	4.4985
2.3000	243.77	105.22	91.857	4.4971
2.3500	243.79	105.23	91.867	4.4961
2.4000	243.80	105.24	91.867	4.4952
2.4500	243.80	105.25	91.867	4.4942
2.5000	243.80	105.25	91.857	4.4936
2.5500	243.80	105.26	91.867	4.4931
2.6000	243.81	105.26	91.867	4.4926

```
2.7000   243.80   I     0       I         X    I     +       I *          I   105.27   91.867   4.4917
2.7500   243.79   I     0       I         X    I     +       I *          I   105.27   91.867   4.4913
2.8000   243.77   I     0       I         X    I     +       I *          I   105.27   91.867   4.4913
2.8500   243.77   I     0       I         X    I     +       I *          I   105.27   91.867   4.4913
2.9000   243.76   I     0       I         X    I     +       I *          I   105.27   91.867   4.4913
2.9500   243.76   I     0       I         X    I     +       I *          I   105.27   91.867   4.4913
3.0000   243.76   I-----0-------I---------X----I-----+-------I-*----------I   105.27   91.867   4.4913
3.0500   243.77   I     0       I         X    I     +       I *          I   105.27   91.867   4.4913
3.1000   243.77   I     0       I         X    I     +       I *          I   105.27   91.867   4.4913
3.1500   243.78   I     0       I         X    I     +       I *          I   105.28   91.867   4.4913
3.2000   243.80   I     0       I         X    I     +       I *          I   105.28   91.867   4.4912
3.2500   243.81   I     0       I         X    I     +       I *          I   105.28   91.867   4.4910
3.3000   243.83   I     0       I         X    I     +       I *          I   105.28   91.867   4.4908
3.3500   243.84   I     0       I         X    I     +       I *          I   105.28   91.867   4.4906
3.4000   243.86   I     0       I         X    I     +       I *          I   105.29   91.867   4.4905
3.4500   243.87   I     0       I         X    I     +       I  *         I   105.29   91.868   4.4903
3.5000   243.88   I-----0-------I---------X----I-----+-------I--*---------I   105.29   91.868   4.4901
3.5500   243.89   I     0       I         X    I     +       I  *         I   105.29   91.868   4.4901
3.6000   243.90   I     0       I         X    I     +       I  *         I   105.29   91.868   4.4900
3.6500   243.91   I     0       I         X    I     +       I  *         I   105.29   91.868   4.4899
3.7000   243.91   I     0       I         X    I     +       I  *         I   105.29   91.868   4.4898
3.7500   243.92   I     0       I         X    I     +       I  *         I   105.29   91.868   4.4898
3.8000   243.92   I     0       I         X    I     +       I  *         I   105.29   91.868   4.4898
3.8500   243.92   I     0       I         X    I     +       I  *         I   105.29   91.868   4.4898
3.9000   243.92   I     0       I         X    I     +       I  *         I   105.29   91.868   4.4898
3.9500   243.92   I     0       I         X    I     +       I  *         I   105.29   91.868   4.4898
4.0000   243.92   I-----0-------I---------X----I-----+-------I--*---------I   105.29   91.868   4.4898
4.0500   243.92   I     0       I         X    I     +       I  *         I   105.29   91.868   4.4898
4.1000   243.92   I     0       I         X    I     +       I  *         I   105.29   91.868   4.4898
4.1500   243.92   I     0       I         X    I     +       I  *         I   105.29   91.868   4.4898
4.2000   243.92   I     0       I         X    I     +       I  *         I   105.29   91.868   4.4898
4.2500   243.92   I     0       I         X    I     +       I  *         I   105.29   91.868   4.4898
4.3000   243.92   I     0       I         X    I     +       I  *         I   105.29   91.868   4.4898
4.3500   243.92   I     0       I         X    I     +       I  *         I   105.29   91.868   4.4898
4.4000   243.92   I     0       I         X    I     +       I  *         I   105.29   91.868   4.4898
4.4500   243.92   I     0       I         X    I     +       I  *         I   105.29   91.868   4.4898
4.5000   243.92   I-----0-------I---------X----I-----+-------I--*---------I   105.29   91.868   4.4898
4.5500   243.92   I     0       I         X    I     +       I  *         I   105.29   91.868   4.4898
4.6000   243.92   I     0       I         X    I     +       I  *         I   105.29   91.868   4.4898
4.6500   243.92   I     0       I         X    I     +       I  *         I   105.29   91.868   4.4898
4.7000   243.92   I     0       I         X    I     +       I  *         I   105.29   91.868   4.4898
4.7500   243.92   I     0       I         X    I     +       I  *         I   105.29   91.868   4.4898
4.8000   243.92   I     0       I         X    I     +       I  *         I   105.29   91.868   4.4898
4.8500   243.92   I     0       I         X    I     +       I  *         I   105.29   91.868   4.4898
4.9000   243.92   I     0       I         X    I     +       I  *         I   105.29   91.868   4.4898
4.9500   243.92   I     0       I         X    I     +       I  *         I   105.29   91.868   4.4898
5.0000   243.92   I-----0-------I---------X----I-----+-------I--*---------I   105.29   91.868   4.4898
5.0500   243.92   I     0       I         X    I     +       I  *         I   105.29   91.868   4.4898
5.1000   243.92   I     0       I         X    I     +       I  *         I   105.29   91.868   4.4898
5.1500   243.92   I     0       I         X    I     +       I  *         I   105.29   91.868   4.4898
5.2000   243.92   I     0       I         X    I     +       I  *         I   105.29   91.868   4.4899
5.2500   243.92   I     0       I         X    I     +       I  *         I   105.29   91.868   4.4898
5.3000   243.92   I     0       I         X    I     +       I  *         I   105.29   91.868   4.4898
5.3500   243.92   I     0       I         X    I     +       I  *         I   105.29   91.868   4.4898
5.4000   243.92   I     0       I         X    I     +       I  *         I   105.29   91.868   4.4898
5.4500   243.92   I     0       I         X    I     +       I  *         I   105.29   91.868   4.4898
5.5000   243.92   I-----0-------I---------X----I-----+-------I--*---------I   105.29   91.868   4.4898
5.5500   243.92   I     0       I         X    I     +       I  *         I   105.29   91.868   4.4898
5.6000   243.92   I     0       I         X    I     +       I  *         I   105.29   91.868   4.4898
```

Figure 4-17. Flood drum computer output.

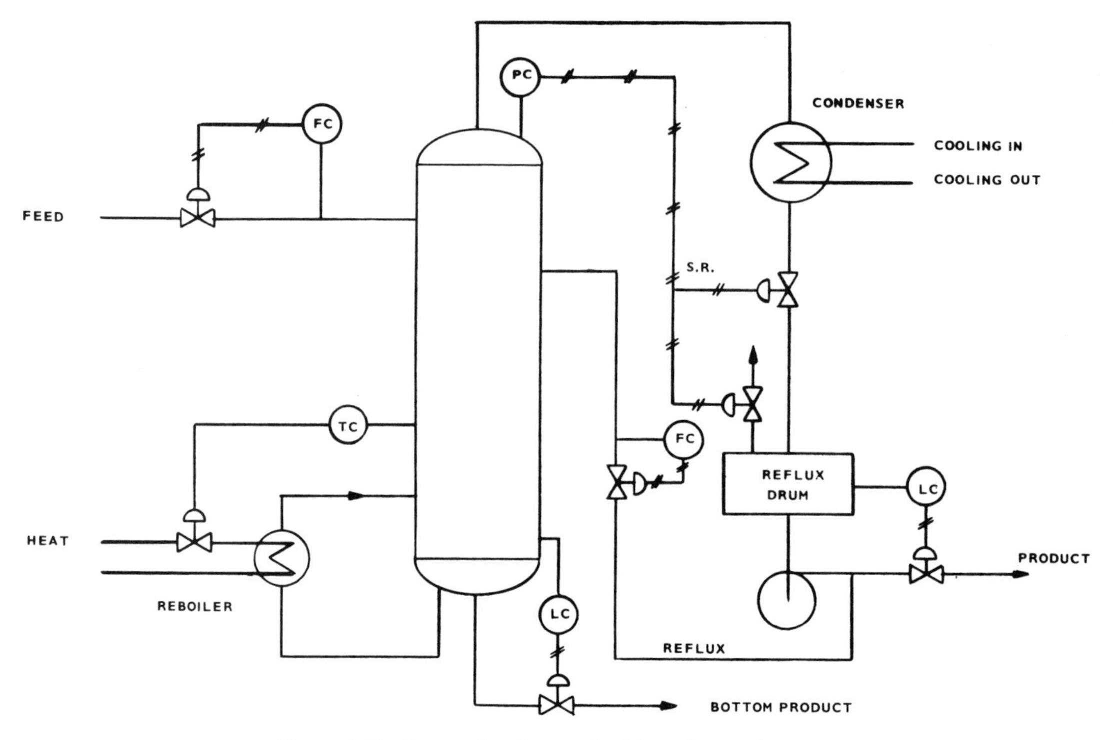

Figure 4-18. Conventional control with analog equipment.

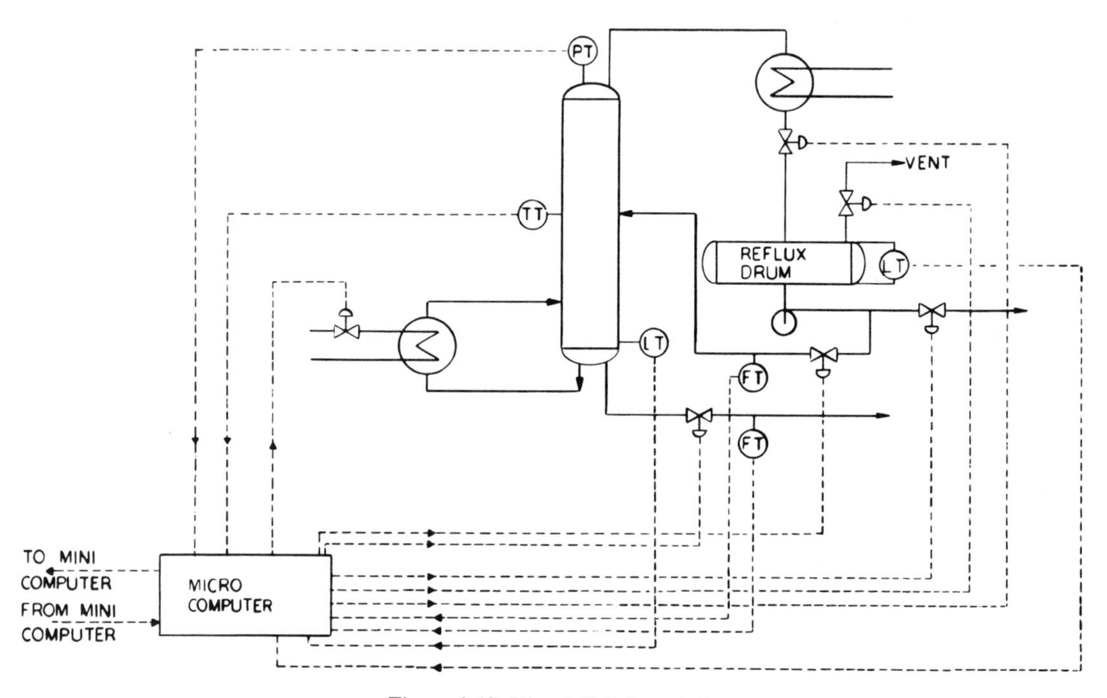

Figure 4-19. Direct digital control.

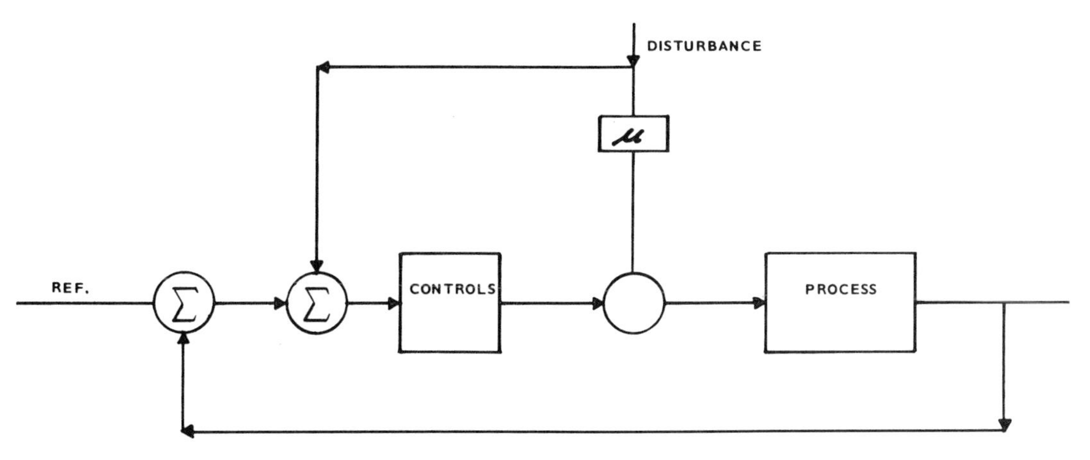

Figure 4-20. Feedforward/feedback control.

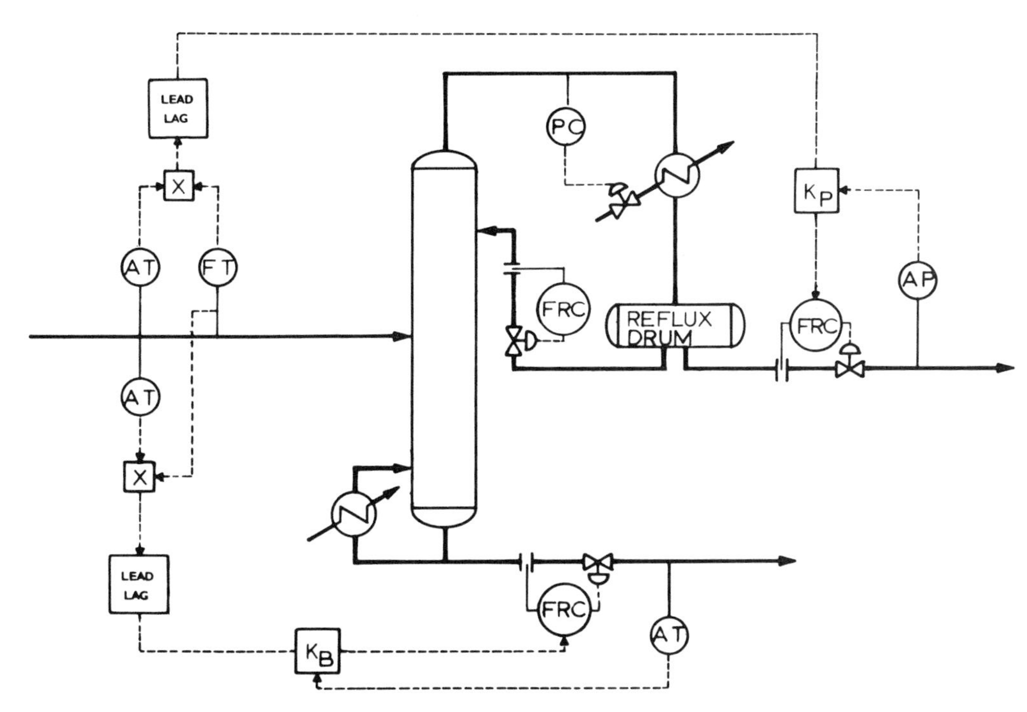

PRODUCT FLOW = (LIGHT KEY COMPUTED) (FEED) (LAG 1) (KP)

K_p = K_1 (PRODUCT ANALYSIS S.P. – PRODUCT ANALYSIS)

BOTTOMS FLOW = (HEAVY KEY COMPUTED) (FEED) (LAG2) (KB)

K_b = K_2 (BOTTOM ANALYSIS S.P. – BOTTOM ANALYSIS)

Figure 4-21. Feedforward column control.

troller and for various feedforward and feedback control applications.[6]

Multi-variable and De-coupling Control

Most industrial process control applications over the past half-century have been based only on simple feedback or feedforward/feedback philosophies. An alternate method of describing this type of control is single variable or single input/single output control, as discussed earlier.

The complexities of modern process technologies, combined with requirements of closer constraints, smaller hold-up times, higher throughput and energy saving systems offer a challenge to the unit operator so complex that he cannot hope to cope with it by using only conventional analog control instrumentation.

All of these factors, plus the realization that manipulation of one variable affects others, quite naturally lead to the study of how these interactions take place and how they can be controlled.

If such a process is multi-variable and interactive, then naturally we must define a multi-variable control system. This is a device or set of devices which has some built-in intelligence to simultaneously look at many variables and to choose, based on a given situation, the best of several programmed control strategies. Consider the effects on the control of a column as a result of varying the steam flow, differential pressure across the column, feed rate, bottom flow rate, reflux rate and other possible variables (product composition, for example). Each of these variables quite commonly affects the others. These effects can be broadly defined as interactive. It can be further stated that conventional control systems do not compensate for the interaction among these variables.

The complex requirements for control and operation, along with the interactive nature of the variables in the column, naturally lead to this question: What can be done to improve what needs to be improved? A method of attack based on these considerations should be as follows.

- Identify the areas to be improved.
- Identify those variables which interact.
- Develop a design philosphy not limited to the single input/single output syndrome.
- Identify the static and dynamic nature of the interaction.

In the case of a binary distillation column, some possible areas of improvement are the following.

- Improvement in product quality.
- Improvement in product flow rate (throughput).
- Minimization of energy input.
- Improvement of dynamic response disturbances.
- Minimization of excessive reflux.

Some variables which interact are given below.

- Steam flow.
- Column differential pressure.
- Column pressure.
- Feed rate.
- Feed composition (one way only).
- Reflux.
- Column temperature.

A design philosophy allowing for interacting variables is, perhaps, the most difficult to institute or engineer because of the inertia of the generally accepted single input/single output syndrome. Proper design should be based on, as a minimum, these factors:

- Heat and material balances.
- Product, flow rate and quality controls.
- Safety controls.
- Feedforward and feedback controls.
- Interaction elimination.
- Interaction compensation.

Identification of the static and dynamic nature of the interaction requires the use of either simulation or field experience on the column in question.

Consider the following generalized example: Figure 4-22 shows multi-variable interaction between manipulated and controlled variables; Figure 4-23 indicates a generalized de-coupler that minimizes the interactive processes and thus helps to improve the operation.

Consider the example by Buckley,[7] reproduced with permission of *Control Engineering*. The example illustrates how a

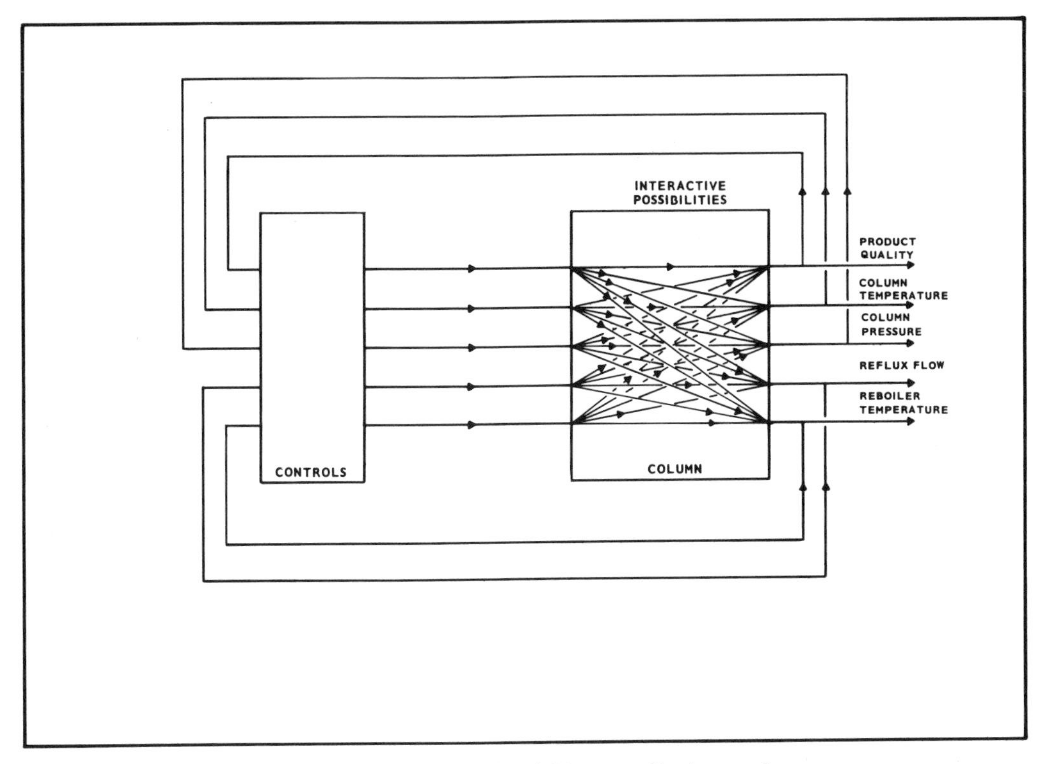

Figure 4-22. Interactive multi-variable generalized control.

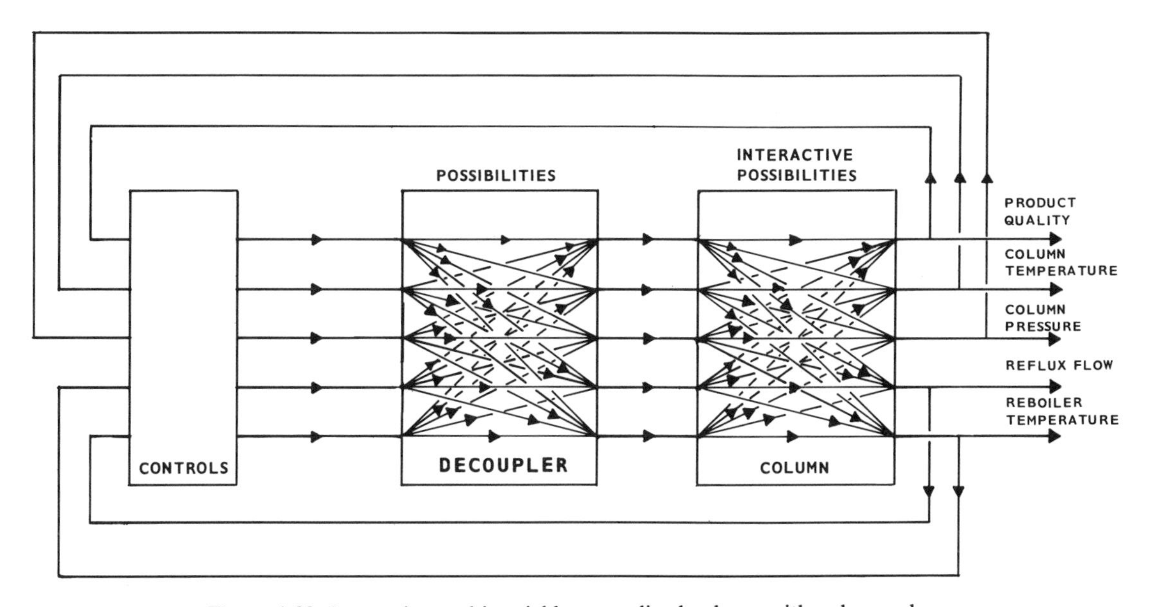

Figure 4-23. Interactive multi-variable generalized column with a de-coupler.

binary distillation column is de-coupled. Governing equations are shown.

$$Y_T(s) = \left[\frac{Y_T(s)}{Z_F(s)}\right] Z_F + \left[\frac{Y_T(s)}{F(s)}\right] F(s) + \left[\frac{Y_T(s)}{L_R(s)}\right] L_R(s) + \left[\frac{Y_T(s)}{V(s)}\right] V(s) + \left[\frac{Y_T(s)}{P(s)}\right] P(s)$$

where

Y_T = mol fraction of more volatile component in vapor leaving top tray

Z_F = mol fraction of volatile component in feed

F = column feed rate

L_R = reflux rate

V = vapor boil-up

P = factor characterizing thermal conduction of feed.

Consider a possible design of a de-coupling control which, for example, reduces the effects of a feed rate change on Y_T. Since $Yy_T(s) = f\,(Z_F, F, L_R, V, P)\,(s)$, it is necessary to negate the effect on Y_T of feed rate change or cause $Y_T(s) = 0$ for all changes in F. In a similar fashion, the bottom fraction of the more volatile key is to be independent of F; or $X_B(s) = 0$.

Given a block diagram of the composition control of a distillation column, Figure 4-24,[7] and noting that a de-coupling device must result in cancellation of the effects of feed, the de-coupling controller may be derived as indicated.

For effects on overhead product and bottom product volatile components, due to change in feed (F), the de-coupling controller is derived directly from the diagram.

For $Y_T(s)$

$$F(s)\left\{\left[\frac{Y_T(s)}{F(s)}\right] + K_{F4}G_{F4}\frac{L_R}{W_R K_{mFR}}\left[\frac{Y_T(s)}{L_R(s)}\right] + \left[\frac{Y_T(s)}{V(s)}\right] K_{F2}G_{F2}\frac{V}{W_S K_{mFS}}\right\} = 0$$

For $X_B(s)$

$$F(s)\left\{\left[\frac{X_B(s)}{F(s)}\right] + K_{F2}G_{F2}\frac{V}{W_s K_{mFS}}\left[\frac{X_B(s)}{V(s)}\right] + \left[\frac{X_B(s)}{L_R(s)}\right] K_{F4}G_{F4}\frac{L_R}{W_R K_{mFR}}\right\} = 0$$

The simultaneous solution of these equations for the controllers $K_{F4}G_{F4}$ and $K_{F2}G_{F2}$ yield the de-coupling control for volatile components in the tops and bottoms as a function of feed. Where $K_{F4}G_{F4}$ is the controller for the top and $K_{F2}G_{F2}$ is the controller for the bottoms:

$$K_{F2}G_{F2}(s) = \frac{\left[\frac{Y_T(s)}{F(s)}\right]\left[\frac{X_B(s)}{L_R(s)}\right] - \left[\frac{(Y_T(s)}{(L_R(s)}\right]\left[\frac{X_B(s)}{F(s)}\right]}{\frac{V}{W_S K_{mFS}}\left\{\left[\frac{Y_T(s)}{L_R(s)}\right]\left[\frac{X_B(s)}{V(s)}\right] - \left[\frac{X_B(s)}{L_R(s)}\right]\left[\frac{Y_T(s)}{V(s)}\right]\right\}}$$

$$K_{F4}G_{F4} = \frac{\left[\frac{X_B(s)}{F(s)}\right]\left[\frac{Y_T(s)}{V(s)}\right] - \left[\frac{Y_T(s)}{F(s)}\right]\left[\frac{X_B(s)}{V(s)}\right]}{\frac{L_R}{W_R K_{mFR}}\left\{\left[\frac{Y_T(s)}{L_R(s)}\right]\left[\frac{X_B(s)}{V(s)}\right] - \left[\frac{X_B(s)}{L_R(s)}\right]\left[\frac{Y_T(s)}{V(s)}\right]\right\}}.$$

The development of an interactive controller, in general, may be thought of as deriving an inverse image of the processes which interact and inserting the controller into the interaction path.

Some of the advantages of the microcomputer for multi-variable and de-coupling control in fractionating column control may be summarized as follows.

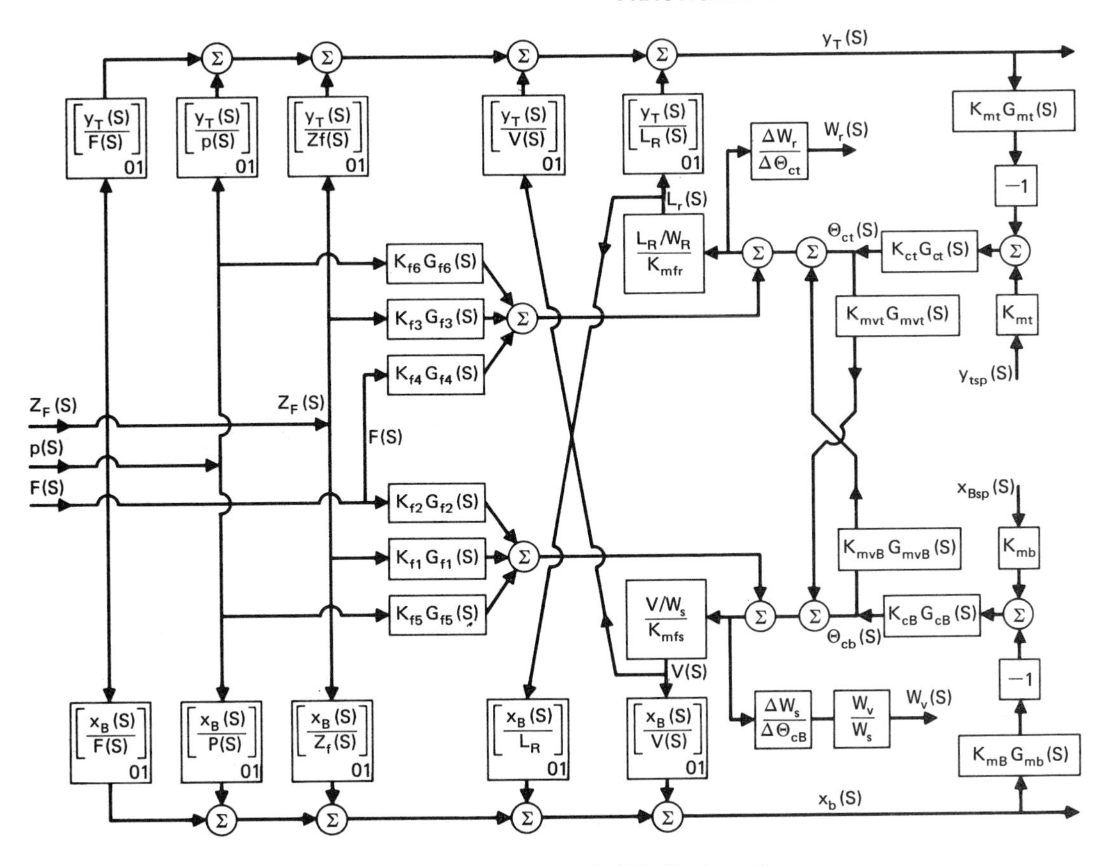

Figure 4-24. Composition control of distillation columns.

1. An on-line device, if designed properly, will improve closed-loop performance.
2. It provides a more simplified and accurate method of controller tuning, particularly for single loop operation where it exists.
3. It tends to minimize the propagation of disturbances from loop to loop.

ENERGY CONSERVATION METHODS IN FRACTIONATING COLUMN CONTROL

Worldwide energy shortages projected over the next 20 years and the subsequent rise in the cost of energy and feedstock have forced oil companies to pay more attention to operating efficiency of unit operations.

Recent studies[8,9] indicate fractionating columns with current design capacities can accrue considerable savings if an energy conservation program implemented by a microcomputer is installed.

The development of an energy conservation program for fractionating column control requires a thorough understanding of the design and operating characteristics of the column and the effects on its energy consumption due to changes in feedstock, throughput, composition of feed reflux rate, pressure variations, etc. To this end, several studies have been made, with promising results toward improved economic operation of fractionating towers. Studies done by Buckley[10] and others discuss how on-line steady state models can provide input information for various methods to achieve improved tower operations. The basis for this improvement, which can be achieved with the application of a local micro and possibly a host mini, are as follows.

- Adaptive tuning of feedback controllers.
- Feedforward and interactive compensation.
- More accurate determination of column constraints (flooding, feed rates, etc.).
- Energy and material balance control.
- The use of pressure and temperature measurements to deduce compositions.

Additional articles and studies indicate that attention should be paid to the operating performance of steam re-boilers; minimizing reflux ratios in order to minimize heat loss to cooling fluid; and control of heat input to the tower.

This section will briefly explain what each of these methods are and how they can assist in improving the performance (reduce energy use) in a fractionating column.

Adaptive Tuning of Controllers

Adaptive tuning of controllers is a concept by which an analysis is made of auxiliary variables, controllable or non-controllable, for the purpose of modifying principle control variables by changing the modes of control (gain, reset, derivative settings), so that desired system performance criteria may be more fully realized. E. H. Bristol defines adaptive control in four different ways.

1. An adaptive (or self-adapting) system is one which is capable of adjusting itself for acceptable operation over a range of external conditions.
2. A self-organizing system is one which is capable of changing its own structure.
3. A learning system is a system which accepts information and fixes it in some memory device.
4. A pattern recognization system is a system which is able to take on input pattern of data with a large number (usually infinite) of degrees of freedom and classify it as to membership among members of some set (usually finite) of abstract patterns such as letters of the alphabet.[11]

Though all definitions of adaptive control may be applied to fractionating columns, discussion will be limited to the self-adaptive methods.

The tuning of controllers using adaptive methods generally requires an operational mathematical model or equivalent steady state gain of the toner, usually non-linear, relating principle output requirements to input disturbances. This gain and the gains of the controllers on the toner define an overall loop gain. When an analysis of the results of a disturbance takes place by the adaptive control system, the tower loop gain is changed by manipulating the control mode setting on the controllers. The degree of manipulation of these control modes is a function of the severity of the disturbance variables and the effect on the principle functions of the tower as determined by the mathematical model or the gain constants. The adaptive system by its action attempts to minimize the error and duration.

Feedforward and Interactive Control

The concept of feedforward and interactive compensation was discussed earlier in this chapter. Studies have indicated[4,6,7] that feedforward interactive control of feed rate and feed composition is the primary area for the application of this concept. The result of feedforward compensation and interactive control does not require column models to the same degree as does the adaptive control concept. It can be applied as indicated in Figure 4-21.

The concept of a self-adapting system and the feedforward interactive philosphy may sometimes be confusing. A feedforward system could adapt the tower control to changes in feed rate and composition. There exists, however, one very important distinction between adaptive and feedforward control. The distinction is that the adaptive system can, in this case, calculate gain settings and reset them according to programmed instructions, whereas the feedforward system operates in a fixed gain controller mode resetting set points on the variables. Figures 4-25 and 4-26 show the fundamental difference.

The following advantages accrue as a result of feedforward control and/or adaptive control strategy.

1. Product compositions are held closer to requirements.
2. Minimization of stability problems.
3. Increased column capacity.
4. Reduced energy consumption.
5. Reduced propagation of disturbance through control loops.

Determination of Constraints on Column Operation

The purpose of accurate determination of constraints on column operation is to determine the condition where flooding and dumping occur in the column. The closer the column can operate to the flooding condition, the greater the capacity. The closer the column can run to dumping, the greater the column turn-down. Both conditions, when properly determined, offer a

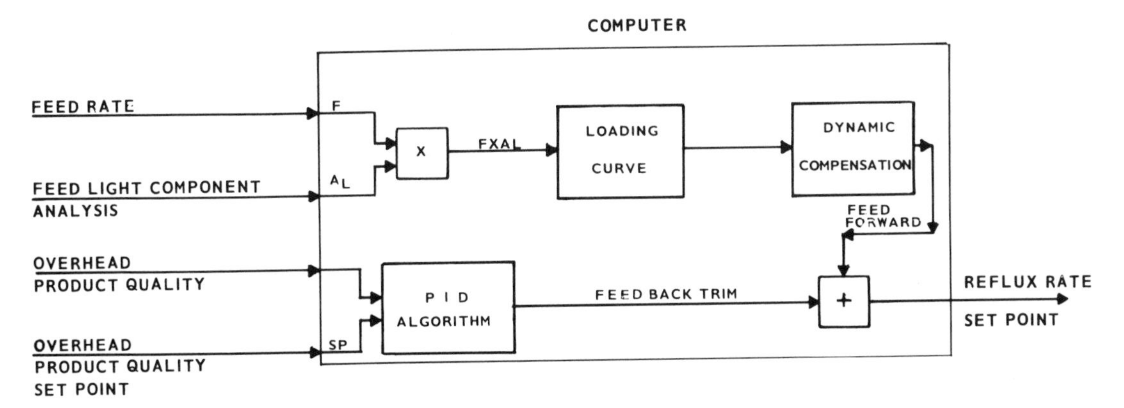

Figure 4-25. Typical feedforward computer control loop with feedback trim.

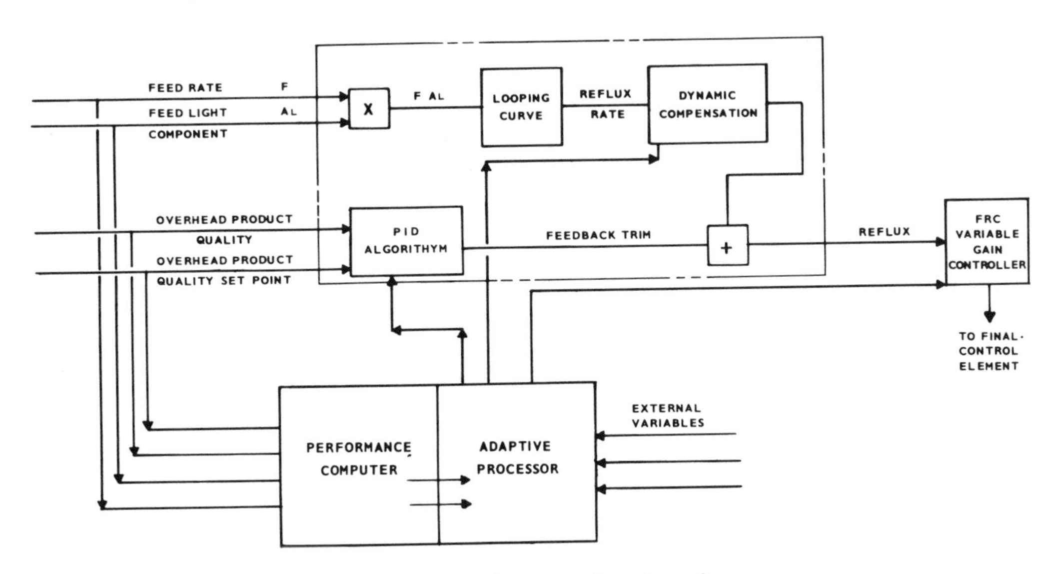

Figure 4-26. Adaptive control configuration.

degree of flexibility in the operation of the column that allows more efficient operation with varying constraints of feed and compositon.

Energy and Material Balance

Classical methods of column operation utilize either pressure or temperature to control column separation. Buckley[10] suggests that there is no technical or economic reason for pressure to be controlled; pressure should be allowed to float and find its own level or be manipulated to minimize steam consumption per pound of product. Essentially, he sees no reason to hold pressure constant. The advent of more reliable analyzers negates the dependence on pressure and temperature for determination of product composition. Analyses of product composition allow temperature and pressure to be manipulated for minimum steam consumption consistent with product quality requirements.

Reboilers

Operating performance of steam reboilers significantly affects the overall energy utilization because the variables which affect the transfer of energy often go undetected. The overall heat transfer coefficient is an excellent indication of operating performance of a reboiler. Operators generally increase the flow of steam to compensate for fouling. However, a calculation for heat transfer coefficient yields far more information as to the performance of the reboiler and is far more sensitive to changes in tube condition.

For example: Consider a heat exchanger reboiler with a heat input requirement given by

$$Q = UA\Delta T_m$$

where U = heat transfer coefficient (BTU/hr ft² of ft²)
A = Heat transfer surface area
ΔT_m = Heat transfer log mean temperature difference.

Let $\frac{1}{U} = \frac{1}{h_I} + \frac{1}{H_O}$ + other resistances.

In an article by A. E. Helzner,[12] the other resistances are defined as the total fouling resistance. When the reboiler is new, the resistances are zero and the overall heat transfer coefficient, U, is maximum under the given conditions. As fouling progresses, U decreases until the reboiler steam supply valve is wide open. By trending key variables over the operating cycle of a re-boiler, and making some minor calculations, decisions can be made as to the point when maintenance is indicated. These key variables are listed below.

- Steam flow.
- Steam pressure at boiler.
- Steam temperature in and out.
- Process temperature in and out.
- Control valve, ΔP.

Calculated variables indicating fouling are as follows.

- Log new temperature differences T_m (°F).
- Heat transfer coefficient U (BTU/hr ft², °F).
- Heat duty, Q (BTU/hr).

The following conditions indicate increased fouling.

- Increased steam valve open position.
- Increased T_m.
- Increased steam chest temperature.
- Reduction in U.
- Decreased control value, P.

All items discussed in these previous sections are not conducive to manual operator control. The concepts of process modeling and iterative search techniques to calculate minimums and maximums with respect to energy use and product production indicate the obvious need for some machine methods to perform their functions.

A typical energy consumption printout

for a fractionating column was shown in Figure 4-2. It is not the last word, but with the ever-increasing cost of energy and the requirements of tighter control, its use and improvement is assured.

CONCLUSION

This chapter should provide some answers to the question of how the chemical industry can integrate mini- and microcomputer technology into the control of fractionating columns. It is hoped that these will stir discussion and controversy, without which progress cannot occur.

The application potential for mini- and microcomputers—which lists among its uses supervisory set point analog control (SPC); direct digital control (DDC); communication interface with host computers; and advanced control techniques—is still in its infancy. The requirements for sound implementation with maximum usage of this potential necessitates utilization of the systems engineering approach. The conceptual, problem definition, problem solution and evaluation phases of the systems engineering method are important to a successful interaction between mini- and microcomputer fractionating column technologies. With the increasing cost of energy and feedstock projected over the next 20 years, efficient column operation is an absolute necessity. The need for advanced control strategy required for efficient column operation then becomes obvious. Adaptive tuning of controllers, feedforward and interactive compensation, determination of column constraints, energy and material balances and pressure and temperature parameters for composition identification are all prime applications for mini- and microcomputer technology.

Mini- and microcomputer technology is currently at the stage where electric control was 10 years ago.

Resistance to change ultimately succumbs to progress. In years past, United States industry had the luxury of making decisions for profit and technology enhancement, with foreign industry looking to us for knowledge. Increased competition from foreign industry plus worldwide energy scarcities require the application of better analytical thinking and more efficient operation of our industrial base. Many new mathematical tools are currently available to assist in this approach, specifically in the application to mini- and microcomputer systems. This increased competition will force the utilization of the new concepts discussed in this chapter.

REFERENCES

1. Buckley, P. S. *Techniques of Process Control,* John Wiley & Sons, New York, 1964.
2. Smith, Cecil, *Digital Computer Control,* Intext Educational Publishers, Scranton, Pa., 1972.
3. Fox, R. L., *Optimization Methods for Engineering Design,* Addison Wesley, Reading, Massachusetts, 1971.
4. Shinskey, F. G., *Process Control Systems,* McGraw-Hill, New York, 1967.
5. Liptak, B., *Instrument Engineers Handbook,* Chilton, Philadelphia, 1970.
6. Shinskey, F. G., *Distillation Control for Productivity and Energy Conservation,* McGraw-Hill, New York, 1977.
7. Buckley, P. S., *Multivariable Control in the Process Industries,* Proceedings of Conference, Multivariable Control Systems, Purdue University, April 1975.
8. Skrokov, M. R., "Benefits of Microprocessor Control," *Chemical Engineering,* October 1976.
9. Skrokov, M. R., Thor, M. G., and Weiss, M. D., "Energy Conservation Control in Olefins Plants by Mini and Micro Computers," Stone & Webster Engineering Corporation, New York. Unpublished paper.
10. "Distillation Column Design Using Multivariable Control, Part II," *Intech,* Instrument Society of America Pittsburgh, Pa., October 1978.
11. Bristol, E. H., *Adaptation in the Process Industries, Nonlinear and Adaptive Control Techniques,* Proceedings of Advanced Control Conference, Purdue University, April 1974.
12. Helzner, A. E., "Operating Performance of Steam Heated Reboilers," *Chemical Engineering,* February 1977, pp. 73–76.

5.
Sequence Control Systems

Morris Leitner
Senior Control Systems Engineer
Stone & Webster Engineering Corporation
New York, N.Y.
and

Maurice S. Cocheo
Control Systems Engineer
Stone & Webster Engineering Corporation
New York, N.Y.

INTRODUCTION—SPECIAL REQUIREMENTS FOR SEQUENCE CONTROL

There is little available in the literature pertaining to computer control of discontinuous type operations that exist in utility and other support areas of a chemical plant. Therefore, this chapter is dedicated to some engineering and design techniques necessary to successfully design a mini- or microcomputer based sequence control system for use in the chemical process industries (C.P.I.). The definition, analysis, specification, installation and documentation of typical sequence control systems applications are discussed, as are the functions performed by such systems in three application examples.

Most batch processes have such large numbers of simple operations (sequence steps) that the assistance of many operators is required on tasks that are not overly challenging. Automation relieves operators of these tedious jobs and makes them available for other productive activity. Fully automatic control of batch or intermittent-type operations also enables the coordination of these functions into the control requirements of continuous-type processes. Industrial design and expansion are continuing in the direction of more complex processes and vastly increased plant size. Process control computer systems are concurrently becoming more sophisticated, utilizing advanced control strategies and incorporating ever-larger sections of the plant within a given control scheme. Also being included in this increasing sophistication are the various downstream unit operations with many stepwise functions that are generally associated with batch or semi-continuous processes. Another reason for fully automatic control is the greater flexibility provided through automation in controlling the processing cycle. For example, it is much easier to change the operating cycle with a computer system than with an equivalent "programmer" or analog loop, and in either case, there is little need to further train an operator on how to manually implement the change.

In hazardous processes, automatic control also eliminates the possibility of accidents to personnel while reducing the possibility of costly damage to equipment. An automatically controlled batch-type process achieved with computer control also results in more stable plant operation, fewer operating errors, fewer product rejects and more uniform product quality.

The objective of a sequence control system is to automatically perform a variety of supervisory tasks, to respond quickly with predetermined actions for given input conditions and to allow complicated control of processes to take place in a safe, reliable manner with a minimum of operator supervision. Regardless of the reasons for automation, the sequence control system must be designed to provide good performance characteristics and to maintain this performance over a long period of time. The system must also be easy to install, maintain, trouble-shoot and service.

Requirements For Systems Design

Design Procedure

Before attempting to design the sequence control system, it is essential that the step-by-step operations and logic of the process be written in outline form. Time intervals between the steps must be included as well as flow, pressure, temperature, valve position and other time-dependent variables. Next, a definition of the process must be stated, and listings must be prepared of the necessary interlocks and safety devices, displays or pilot lights to indicate operations, accuracy of timing, proof of completed steps, power failure protection and emergency shut-down requirements. Modes of operation desired (such as full automatic, semi-automatic, step or manual control) must also be programmed into the system.

Logic Diagrams

These written outlines must finally be translated into the form of firm logic diagrams that will serve as the basic functional design document for the project. By this procedure, there is less possibility of omitting some essential requirement of systems design. Others familiar with and somewhat responsible for the operating requirements of the process, such as the contractor and the owner, electrical and operations engineers and the sequence control systems package manufacturer's engineers, can responsibly review this information and include any items which may have been overlooked.

Some diversified examples are cited to give a clear view of the steps required, so that the presentation of these principles may serve as a guide to the reader in formulating his own design.

Systems Manufacturer

Operating companies may prefer to design these systems completely in-house. However, when designed as part of a world-sized petrochemical plant (such as an ethylene plant), it may be advantageous to sub-contract such sequence control systems packages. In that case, it is important that these packages be clearly defined early in the specification phase to ensure that details previously discussed are covered. A section of this chapter is devoted to the preparation of a specification that addresses the responsibilities of sequence control systems package sub-contractors.

History of Sequence Control Systems Design

In the past, systems of this nature were designed using combinations of electromechanical relays, timers and cam programmers, in conjunction with analog control instrumentation. All components were jointly assembled and mounted in cabinets or panels. Rarely were these systems fully automatic, and significant operator interface was required to ensure that all steps were proceeding in proper sequence and that succeeding steps were indeed initiated.

If a change in product formulation was required, extensive manual adjustments were necessary in order to institute the changes. When out-of-tolerance conditions occurred, operator intervention was also generally required in order to avoid product waste and unnecessary process shutdowns.

As significant changes in process requirements occurred, the need for additional components became necessary. Finding more room on the panel or in the cabinet frequently was a problem and, of course, extensive wiring changes and check-out were sometimes required as a result.

As with all electromechanical systems, wear and tear, dirt and heat take their toll and make for unreliable system performance unless careful and regular maintenance is provided.

With the advent of solid state electronics, it became possible to improve the reliability of sequence control systems because of the replacement of mechanical relays with solid state relay systems. Solid state devices, however, did not provide any additional flexibility when process changes were indicated after the initial design of the system.

As digital computer hardware improved in reliability and reduced in size and cost, sequence control systems designers began to take advantage of the extensive flexibility these programmable instruments provided. Within the limits of the existing plant equipment, there was seemingly no boundary to the combinations of instructions that could be assembled into a system. Later expansion of the plant equipment further enhanced the versatility of these new systems.

Programmable Controllers

Figure 5–1 illustrates a small programmable controller which is intended for relatively simple sequence control systems. Programmable controllers are available in varied capacities from several manufacturers but are limited in scope when compared to minicomputer based systems. Another limitation of these devices is their inability to compile or assemble their own machine language program (object program) from the higher-level language or assembly programs (source programs). These small microprocessor based systems usually lack sufficient memory and input/output capability to do software compilation; therefore, they must rely on object programs generated and edited on larger systems. The large system downloads the developed software into the microprocessor based system.

Figure 5–2 shows a system in which a larger minicomputer based system is used as both a controller for large plant systems and as a compiler of its own object program. Minicomputer based systems are capable of being equipped with sufficient memory capacity and auxiliary peripherals to perform these functions. The development of small microprocessor based systems capable of performing functions equal to—or more sophisticated than—minicomputer based systems is rapidly approaching.

Applications Examples

The engineering experience described below is that of the computer control staff of a large contract engineering corporation engaged in the design and construction of chemical and petrochemical plants. Three typical applications for sequence control systems are briefly described.

Chemical Plant—Tank Batch Weight Control System

Three batch loading tanks are involved in this chemical process. The tanks are provided with weight-sensing load cells, high-level alarm switches, feedstock inlet and batch outlet control valves. One of these tanks is filled with particular chemical feeds for weight determination prior to being sent to the batch reactor. The other two tanks are filled, respectively, with predetermined weights of sludge residue transported from the batch reactor by conveyor and two other chemicals needed for dissolving the residue prior to disposal (see the process

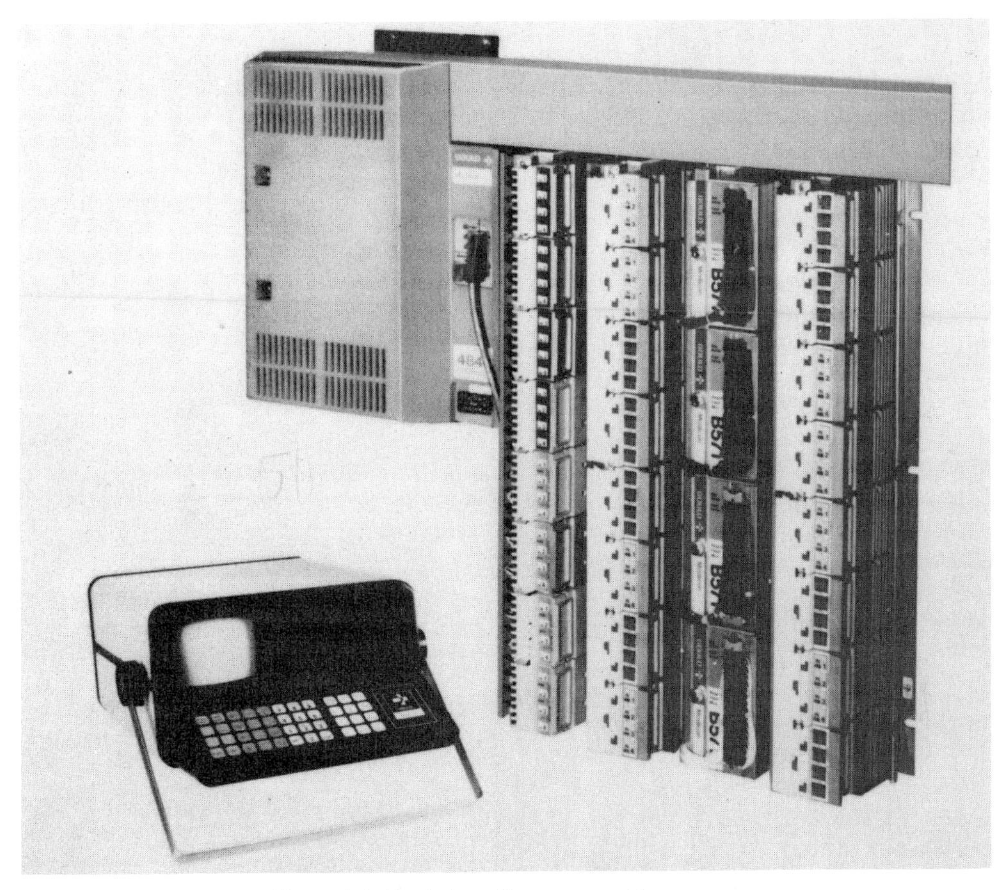

Figure 5-1. Typical small programmable computer.

control diagram for a tank batch weight system in Figure 5–8 [p.142]).

The purpose of the tank batch weight control system is to manually set and start, automatically monitor, stop and reset the filling and weighing operations of each tank, and then repeat the operations again, in accordance with scheduled time and batch setting requirements.

The system is designed for remote central control to operate without manual assistance from operating personnel. It is complete with the necessary accessories and includes load cells; all necessary relay outputs for actuation of solenoid-air operated valves; contact closure inputs for valve positions and tank levels; and remote control panel inserts (having the thumbwheel weight selectors necessary for setting batch requirements, push buttons for starting, emergency stop and reset, tank batch weight indicators and malfunction indicating lights).

Ethylene Plant—Gas Dehydrator Regeneration Control System

In the continuous process of ethylene production, feedstock such as naptha is cracked by heating in large furnaces. Dilution steam is introduced into the process to assist in vaporizing the feedstock and reducing the coke accumulation in the furnace tubes. Gas leaving the furnace passes

Figure 5-2. Large minicomputer based control system.

on through prefractionation and compressor stages and is discharged with much of the moisture removed. However, the gas continues to be saturated with water vapor which must not remain in the final product.

To remove this moisture, the gas is passed through a gas dehydrator sub-system—two large dehydrator vessels filled with desiccant and an auxiliary desiccant regeneration gas heating system (refer to the process control diagram in Figure 5–10 [p.149]). During normal operation, gas passes through one dehydrator to remove moisture, while the other dehydrator remains shut down as it undergoes regeneration. Each dehydrator is designed to have an on-stream gas drying cycle varying from 24 to 72 hours and a desiccant regeneration cycle of 24 hours or less.

The purpose of the gas dehydrator regeneration control system is to automatically or manually place the fresh (stand-by) dehydrator in service, and to shut down, remove from service, reactivate and return to stand-by mode the exhausted dehydrator, reset the master cycle program, and start the regeneration cycle again in accordance with scheduled time, cycle, high moisture or manual settings.

The system is furnished for remote central control, designed to operate without manual assistance from operating personnel. It is complete with necessary accessories and includes analog outputs for automatic temperature control; relay outputs for actuation of motor or solenoid-air operated valves; contact closure inputs for valve positions; pressure and temperature switches; name-plate light displays to indicate the status of the dehydrator reactivation cycle and the status of the valve positions; push buttons for manual operation of the valves; an adjustable cycle time programmer and multi-position key lock switch for automatic manual start, stop and reset of the control cycle. A name-plate light display indicates stages of operation: prereactivation permissives, depressurizing, purging, heating, soaking, cooling, repressurizing, reactivation complete and reset.

Ethylene Plant—Make-up Water And Condensate Polisher De-mineralizer Control System

A make-up water and condensate polisher de-mineralizing plant is a sub-system of a large ethylene refinery, providing de-mineralized, uncontaminated water to the steam boilers.

Steam is consumed continuously in parts of the process and is also used for heating and turbine-driven equipment. Condensate recovered from steam turbine condensers is continuously passed through a condensate de-mineralizer polishing system to assure that no contaminant is picked up from a leak in condenser tubes prior to return to the steam boilers.

The make-up water de-mineralizer plant consists of three common raw water sand filters, which feed filtered water to two independent de-mineralizer systems, each consisting of three cations, three anions and two mixed bed units; the necessary acid and caustic regeneration equipment; inlet; outlet; shut-off valves; and flow, level, temperature, conductivity, sodium, silica and pH instruments. (See the partial process control diagram in Figure 5–14 [p.154] showing anion, cation and mixed bed train).

The condensate polisher system is composed of two mixed bed units with associated regeneration equipment, valves and instrumentation similar to those indicated above.

The purpose of this de-mineralizer control system is to automatically shut down and remove from service an exhausted de-mineralizer or filter bed, regenerate it and return it to the on-stream mode, and to reset the master cycle, shut down and regenerate the next exhausted de-mineralizer or filter bed, in response to high silica, total flow, time settings or operator's request.

Two complete and totally independent minicomputer based control systems (*A* and *B*) are furnished, each designed for interconnection to and operation of its associated make-up water-treating de-mineralizer systems (*A* and *B*), mixed media filter system (*D*) and condensate polishing de-mineralizer system (*C*). Each computer system is dedicated for control of its associated make-up de-mineralizer system with means for manual selection of either computer (*A* or *B*) to operate the condensate polishing de-mineralizer and mixed media filter systems (*C* and *D*). Separate means are provided for manual selection of either computer (*A* or *B*) to operate either make-up de-mineralizer system (*A* or *B*).

THE SYSTEMS ENGINEERING APPROACH

Systems Analysis and Design

An overview of various parts of a computer control systems project as applied to a sequence control system is given in this section.

As in any project, the first requirement is to recognize that some kind of sequence control system is needed. Economic benefits that will accrue from these control systems must also be documented. Therefore, most projects will go through a cost justification stage prior to final application of the system.

Project Cost Justification

Figure 5–3 shows the important phases in evaluating what kind of sequence control system may be best for a given application. In some circumstances, it may be more practical to use one of the older systems. For example, in some plant locations, a lack of qualified personnel could be an overriding factor for going to a mechanical relay system. In other cases, the benefits to be obtained with a more sophisticated system may not justify the added cost.

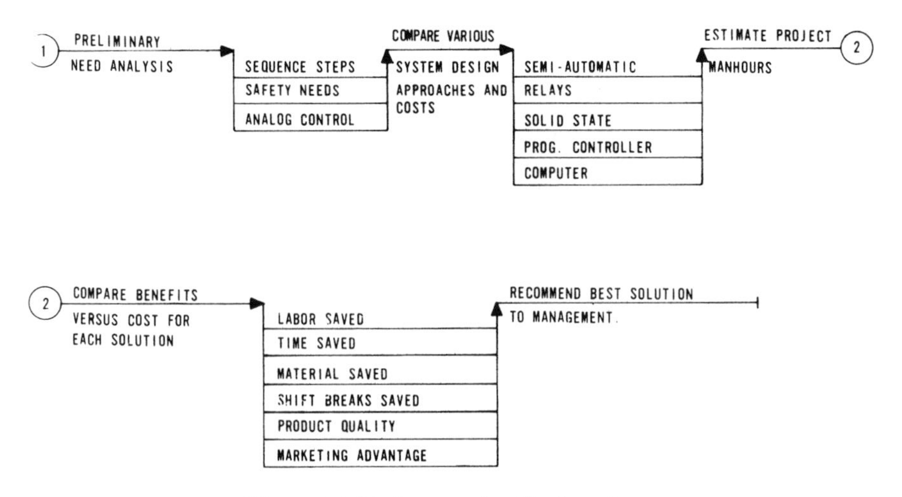

Figure 5-3. Project cost justification steps.

In proceeding through this determination phase, the engineer must perform a brief version of some of the steps in a complete computer sequence control systems project, in addition to evaluating alternate systems. After all the cost and man-hour figures have been accumulated for each system, they should be compared to the benefits and disadvantages they each entail.

In the application of computer-type sequence control systems to batch processes, there generally are many economic advantages to offset the higher cost of these systems. For example, improvements in product quality and reduction of raw material wastage are but two of these.

Improved product quality may sometimes reflect in the price of the product or, at the very least, the reduction or elimination of lost customers due to marginal products.

Another area in cost savings may be found in the reduction of some operating man-hours with these sophisticated control systems. Some labor and time savings are not always apparent at first examination. For example, in some batch plants, loading of the next batch may be delayed due to an impending labor shift change because the loading might be unsupervised at that time. With a computer in charge, the next loading can proceed immediately, saving both labor and production time. These lost production hours, over a period of time, can have significant contributions to company productivity and profit. In other words, it is wise for the engineer to dig for the benefits and not just consider the more obvious factors when he makes an economic analysis.

Project Execution Steps

In Figure 5–4, a diagram of typical project steps required to start and complete a computer sequence control system is illustrated. This chapter discusses those aspects of Figure 5–4 which are peculiar to sequence control systems: sequence analysis and presentation; hardware selection and layout; goal-oriented specifications for hardware/software procurement or computer systems sub-contract purposes; and checkout and installation.

In the beginning of plant control systems analysis, the process must be studied to determine the required sequence. The part each item of process equipment and machinery plays in plant operation must be clearly outlined, as well as how each plant section affects and is affected by other plant sections.

There are several tools that are useful in the planning, analysis and specification phases of designing sequence control systems. These tools serve as later input to

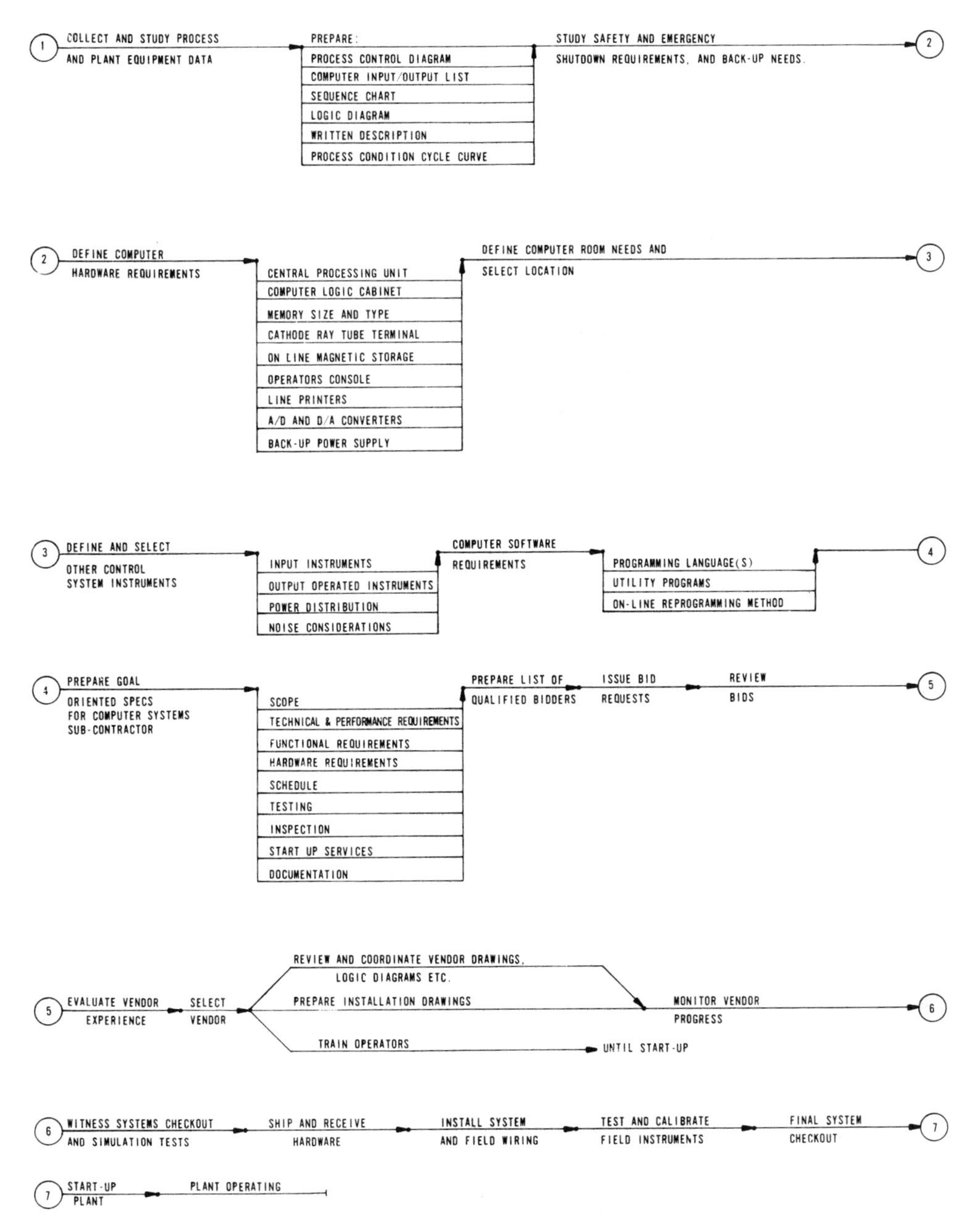

Figure 5-4. Steps in execution of a project.

other engineering disciplines within the design organization. They are listed below.

- Process control diagram (PCD).
- Process sequence chart.
- Logic diagram.
- Written description.
- Process condition cycle curve.

Process Control Diagram

Preparation of the PCD is generally the first analytical step taken. Figure 5–5 is a PCD of a simple batch mixing system. In this system, three fluids are individually allowed to flow into the mixing tank, where the weighing system senses completion of each incremental fluid stream. When the recipe for all three ingredients is satisfied, the blending motor is turned on for ten minutes for thorough mixing. The final step discharges the completed batch to the product storage tank.

Process Sequence Chart

The process sequence chart is a tabular form for describing each step in the process. Figure 5–6 illustrates a typical format which, in this case, shows the steps for the batch-fill and mixing system. The left-hand column enumerates the function performed at that step. Under each equipment column, the position of each valve or the operation of each item of equipment is indicated.

One advantage of the sequence chart is that it provides a clear overview of the normal sequence flow. It lacks, however, a method of showing the interlock conditions which must be satisfied before initiation of

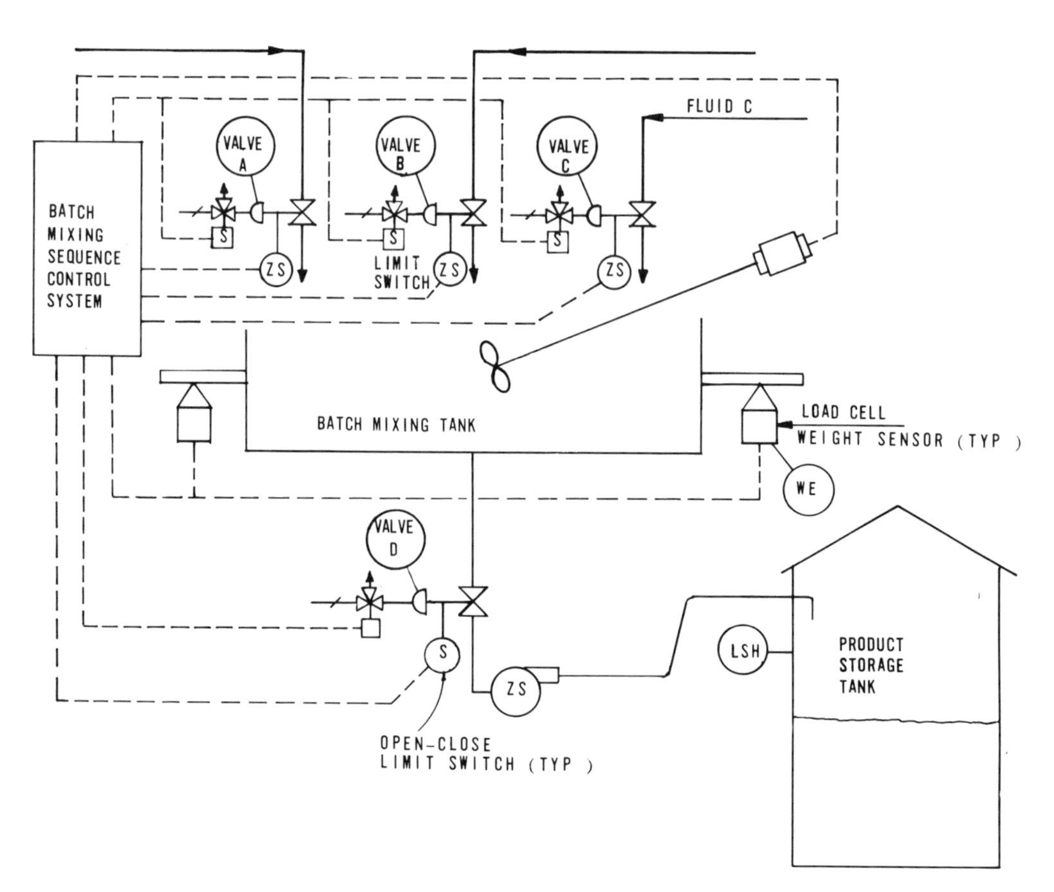

Figure 5-5. Process control diagram—batch mixing system.

EQUIPMENT OPERATION						
EQUIPMENT / SEQUENCE STEP	TIME	VALVE A	VALVE B	VALVE C	VALVE D	MIXING MOTOR
FILL FLUID A	EF	O	X	X	X	OFF
FILL FLUID B	EF	X	O	X	X	OFF
FILL FLUID C	EF	X	X	O	X	OFF
MIXING STAGE	*	X	X	X	X	ON
DISCH. STAGE	EF	X	X	X	O	OFF

O = OPEN
X = CLOSED
ON= POWER APPLIED
OFF= POWER OFF
EF = TIME IS APPROXIMATE ELAPSED FILL TIME
* = TEN MINUTES

Figure 5-6. Process sequence chart—batch mixing system.

the sequence of succeeding steps. For example, if the batch tank discharge valve failed to close fully after the last batch was emptied from the tank, further step initiation could ruin the next batch. This weakness may be overcome to some extent by adding notes to the sequence chart, but this adds complexity to the chart and clouds it with excessive detail. A better method of step-to-step procedure presentation is accomplished by use of a "logic diagram."

Logic Diagram

The logic diagram is a graphic presentation of the sequence and interlocking requirements of various plant systems. It records an understanding of control functions of individual instruments and equipment functions of the process, operating as a complete, integrated system. Logic diagrams are not intended to summarize or specify the hardware required. This may be implemented in many forms, such as electromechanical; solid state relay; mini- or micro-based computer, pneumatic or hydraulic systems; or combinations thereof. Hardware is shown in detail on specification sheets and on flow, elementary and instrument loop diagrams.

The logic diagram is not to be confused with computer-type logic diagrams, which serve as flowcharts for programming. These are usually prepared by the computer package manufacturer.

The logic diagram method is also an important medium as a development and communication tool. It is useful in the following phases of development and operation.

- Design and engineering.
- Installation.
- Start-up.
- Operation and maintenance.

As the plant to be sequence controlled grows larger and more complex, the logic diagram's value as a communication and development tool becomes more apparent. A number of engineering disciplines become involved in the design, and the control systems engineer will discover that the logic diagram engenders comments and questions from its recipients. These comments will bring to light additional factors, new information and differing views of operating philosophy, and may result in some changes in the logic.

During equipment check-out and installation, the logic diagram is an excellent reference document to ensure that equipment and software design will perform in accordance with the logic requirements. It is also useful in the verification of field instrument connections to the system.

As the plant reaches the start-up phase, the value of the logic diagram is shown again. At first it operates as a training tool for start-up and operating personnel. Then it functions as a trouble-shooting tool to diagnose problems encountered as attempts at start-up may fail.

When the plant is finally operating satisfactorily, the logic diagram becomes a basis for any future changes in plant logic if process changes become necessary. It also retains its value as a training aid for new operating and maintenance personnel and as a diagnostic aid whenever plant shutdown or upset occurs.

Logic Diagram Symbols

Figure 5–7a shows typical logic diagram symbols in general use by engineering contractor organizations. By a suitable combination of these symbols, all logic functions can be expressed.

Figures 5–7b and c show the logic diagrams corresponding to the sequence chart of Figure 5–6 for the batch mixing system. Examination of these two figures, which show the same sequence, will illustrate how much more information the logic diagram contains.

Note that the initial stage ensures that all equipment is properly set for an additional batch. Therefore, all valves must be closed, the reactor tank must be empty, the mixing motor must be off, and the product tank must be sufficiently empty to receive another batch.

When all these conditions are met, the logic interlock system allows the operator to initiate another batch by pressing the start push button. If another batch is not already in process, a command is initiated through the retentive memory symbol to open value A. When the set point for fluid A is reached, the reset of valve A's retentive memory is accomplished through the "or" gate, and the opening of valve B is initiated.

The process proceeds similarly through filling fluid B and C, at which time the mixing motor is run for 10 minutes until the "time delay" shuts off the motor and initiates the emptying operation by opening valve D and starting the product pump. As the weighing system senses zero product weight, these items are shut down and the batch cycle light is energized.

As the cycle progresses, panel lights, indicating which stage of the sequence is in operation, are utilized for operator information. If at any time the operator should become aware of an emergency condition, he can abort the cycle with the emergency stop button that enters the retentive memory reset through the "or" gate.

Note again that this type of logic diagram is a systems diagram and is not to be confused with a computer logic diagram. The systems logic diagram can be used for any method of implementation—electromechanical, relay, solid state, or computer based control; the computer logic diagram serves only as a flowchart for computer programming.

Written Description

The written description is sometimes used in lieu of, or in addition to, the sequence chart or logic diagram. It is an excellent form of logic/sequence description but requires the reader to visualize the many interlocks involved. The written description has been found most useful where it forms part of a specification to the computer systems sub-contractor, who is then required to generate the logic diagram for review and comments by the client and/or the engineering contractor.

In addition, the sub-contractor must pro-

SYMBOL	LOGIC FUNCTION	DESCRIPTION
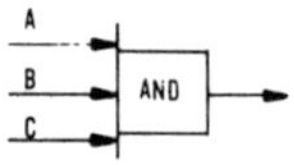 A, B, C → AND →	AND	ALL INPUTS A,B AND C ARE REQUIRED TO BE PRESENT BEFORE OUTPUT D CAN EXIST.
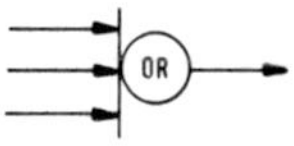 → OR →	OR	ANY OUTPUT A,B, OR C IS REQUIRED TO BE PRESENT BEFORE OUTPUT D CAN EXIST.
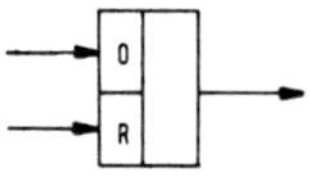 → O / → R →	RETENTIVE MEMORY (O = ON) (R = RESET)	MOMENTARY INPUT A CAUSES CONTINUOUS OUTPUT C. MOMENTARY INPUT B CANCELS OUTPUT C IF INPUT A IS NOT PRESENT.
 A → NOT → B	NOT	OUTPUT B EXISTS ONLY WHEN INPUT A DOES NOT EXIST.
 A → T.D. 10 MIN. →	TIME DELAY	CONTINUOUS INPUT OF A PRODUCES OUTPUT B AFTER DESIGNATED TIME. WHEN INPUT A IS REMOVED, OUTPUT B IS LOST AND TIME DELAY IS RESET
 LSH 237	INSTRUMENT OR ELECTRICAL SOURCE	THE BUBBLE SHOWN PROVIDES THE CONTACT CLOSURE OR OPENING,OR VOLTAGE ABSENCE OR PRESENCE TO INITIATE SUBSEQUENT STEP OR RECEIVES COMMAND TO PERFORM A REQUIRED FUNCTION.
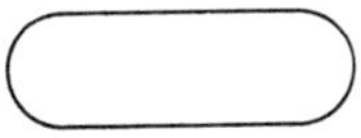 (rounded rectangle)	CONDITION	STATEMENT OF OPERATING CONDITION OR RESULTANT STATUS
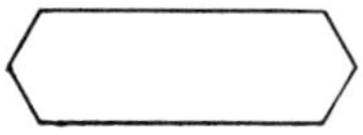 (hexagon)	MANUAL CONTROL ACTION	HS = HAND SWITCH PB = PUSHBUTTON
 2	NUMERICAL CONTINUATION REFERENCE ELSEWHERE ON LOGIC DIAGRAMS	
☒	INDICATING LIGHT	
△	ALARM	

Figure 5-7a. Logic diagram symbols.

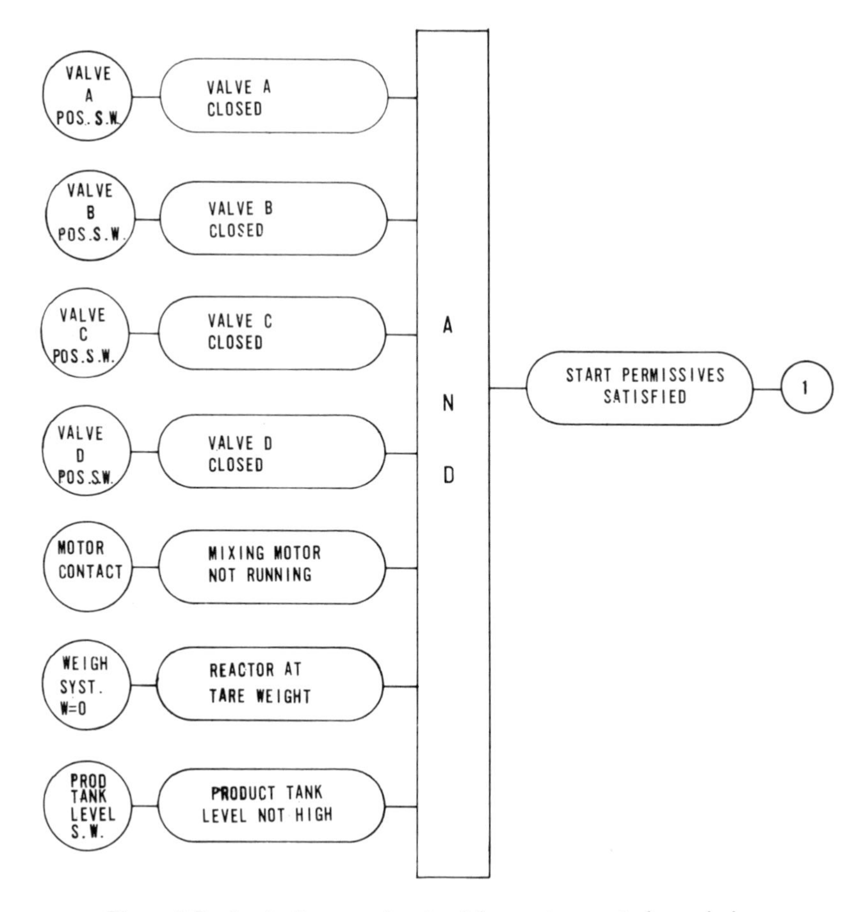

Figure 5-7b. Logic diagram—batch mixing system control permissives.

duce a written description of the final logic diagram. This step is taken for use in the final system operating manual, which is used by operating and maintenance personnel. It is also useful as an aid in the full understanding of the logic diagram, since the symbolism used may not be universally recognized.

An example of written description is given in the "Typical Applications" section of this chapter (p.140–146).

Process Condition Cycle Curve

The process condition cycle curve is a special document used to graphically illustrate a programmed analog variable that takes place within a sequence step or occurs over several steps. A typical process condition cycle curve is utilized in the application example of the gas dehydration regeneration system discussed later in this chapter.

TYPICAL APPLICATIONS

Tank Batch Weight Control System

The first actual system to be considered is a relatively simple tank weighing and batching system. Figure 5–8 shows three weight-batching tanks with input and output control valves, and related weight-sensing load cells. One of these tanks receives feedstock for weight determination prior to being sent to the batch reactor. The other two receive the residue from the batch reactor which is then dissolved by the addition of other chemicals and released to the next stage of the process.

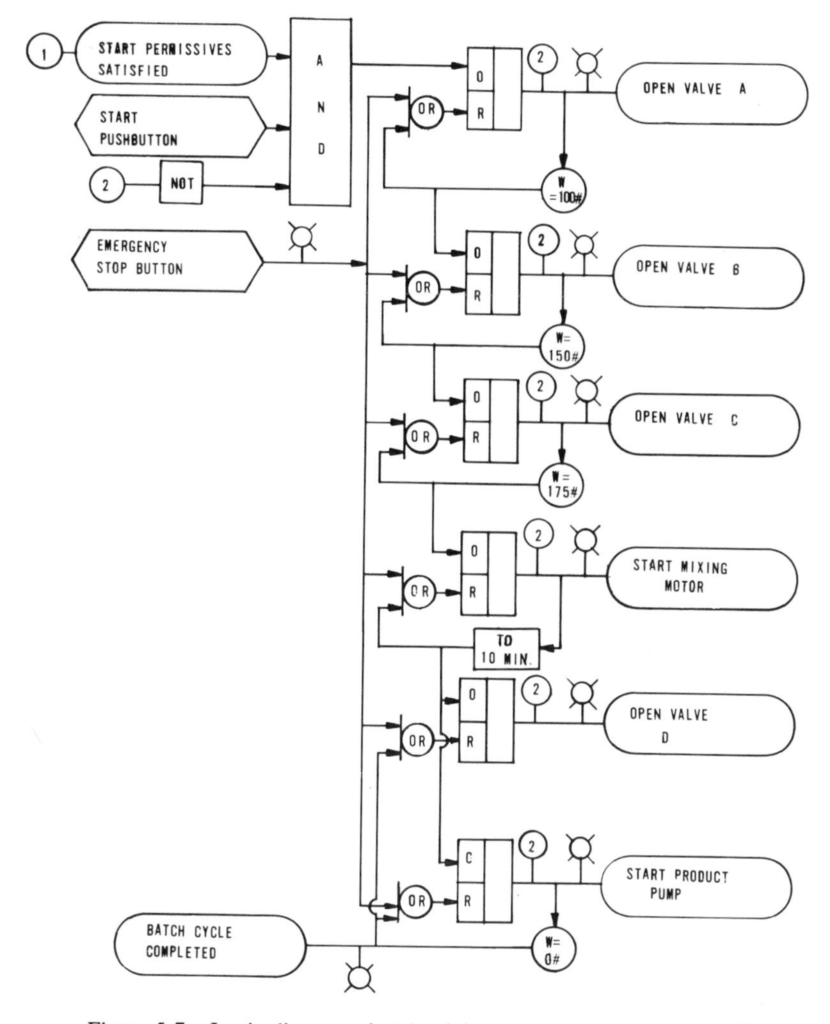

Figure 5-7c. Logic diagram—batch mixing system control start and fill.

A computer weight-batch technique utilizing tank weight load cell sensors is selected for three main reasons:

1. Weight determination requirements are 0.1% of set point. Weigh-belt or weigh-scale-type systems, because of insufficient accuracy, were not considered for this service.
2. Raw material losses.
3. Production time losses.

Basis of Design

Conditions of service. Three batch tanks including necessary fill and drain valves are provided as follows:

	INSTALLATION LOAD CELLS	TANK WEIGHT EMPTY	TANK WEIGHT, FILLED (MAX.)
Weigh tank I	Suspended—3 points	1500 lb	9,000 lb
Dissolving tank II	Floor mounted—4 legs	2500 lb	14,000 lb
Dissolving tank III	Floor mounted—4 legs	2500 lb	14,000 lb

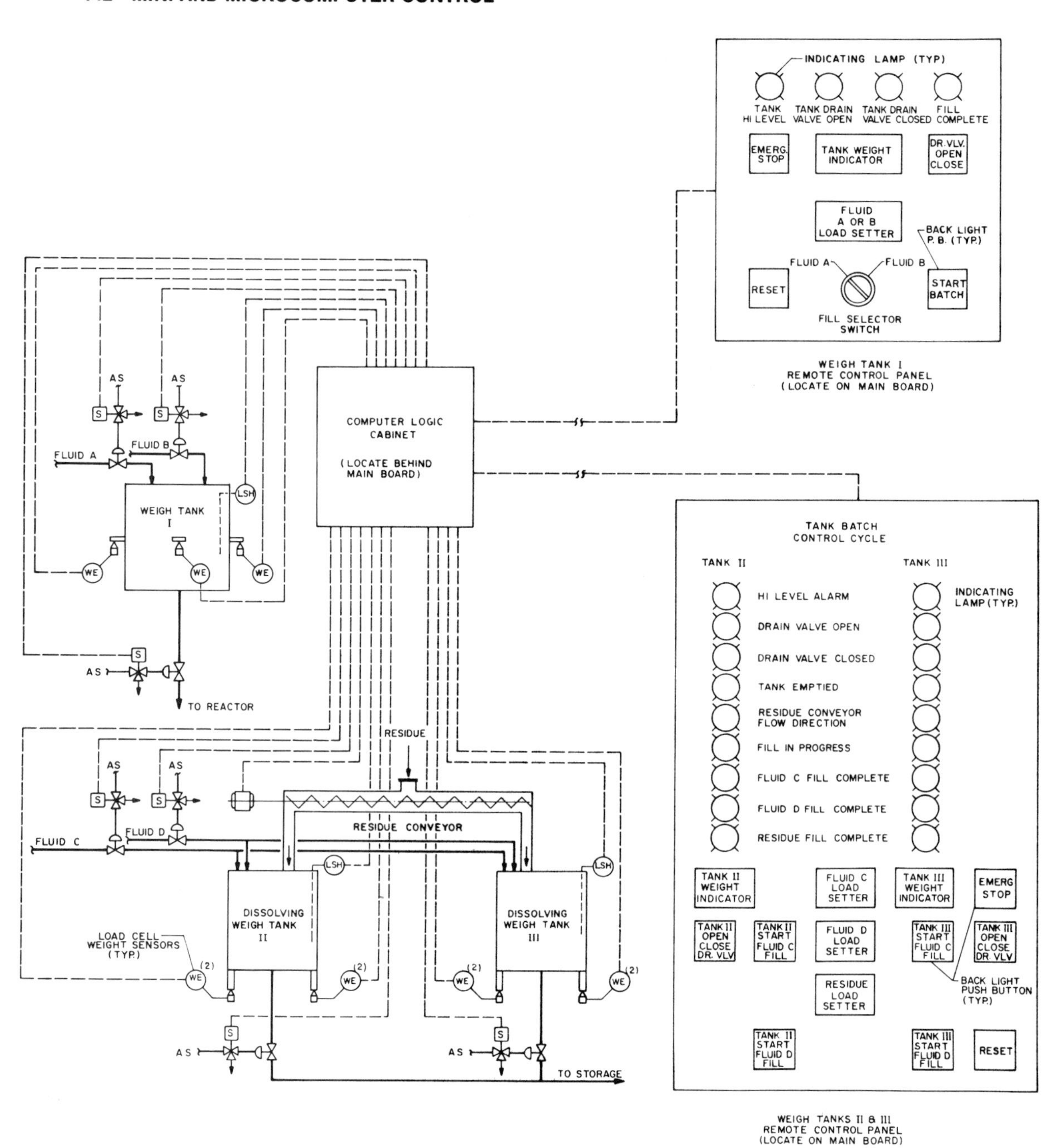

Figure 5-8. Process control diagram—tank batch weight system.

During normal operation, each tank is individually loaded, weighed and unloaded by a semi-automatic weight control system, in accordance with process batch requirements.

Functional requirements.

The purpose of the tank batch weight control system is to manually set and start, automatically monitor, stop and reset filling and weighing operations of three tanks, then repeat the operations again, in accordance with scheduled time and batch setting requirements. The system, furnished for central board control, is designed to operate with out manual assistance from operating personnel. It is complete with all necessary accessories and includes load cells; relay outputs for actuation of solenoid-air-operated valves; contact closure inputs for valve positions and tank levels; two remote control panel inserts (each having the thumbwheel weight selectors required for setting batch requirements; push buttons for starting, emergency stop and reset; tank batch weight indicator; and malfunction indicating lights).

Operation. The starting circuit for activating a tank batch cycle consists of several permissive interlocks. It requires that suitable conditions are established at each tank prior to starting for a successful batch loading operation.

1. Tank drain valve is closed (valve position switch).
2. Tank is empty (tank weight, zero contact).
3. Master cycle is reset. After manual start and completion or emergency stop of a tank batch cycle, the operator must press a separate push button on control panel insert to reset the cycle.

There are three indicating lamps (with identifying name-plates) on each control panel insert, one for each of the above conditions. When the conditions are satisfied, the lamps light up, advising the operator that all permissives have been met and the tank is ready for batch loading. If a condition is lost during these procedures, the indicating lamps will go out. The operator must then take corrective action to re-establish these permissives before proceeding to the next batch start sequence step. The operator must determine and fix the required weight settings for second stage (dribble flow) activation of each fill valve by means of individual adjustable switches located inside the control logic cabinet. Set points are specified under system descriptions for each tank. Operations are monitored and displayed continuously on control panels, as indicated in Figure 5–8. Note that tank weight indicators will continuously display actual weight of fluid in tanks at all times.

With the above satisfied, operations may then proceed manually, automatically and sequentially, as follows.

- Weight tank I is to be filled with a predetermined weight of either of two fluids.

In addition to the prerequisite conditions described above, the operator must manually select the fluid required by turning the fluid selector switch on the control panel insert to the proper position. With these conditions satisfied, loading of tank I may proceed sequentially, as follows:

1. The operator sets the batch requirement by turning the common thumbwheel weight selector on the control panel insert to the desired batch setting. Assume 5000 lb setting.
2. The operator presses the batch start push button on the control panel insert. This actuates the next sequence step.
3. Fully open the solenoid-air-operated fill valve. This permits fluid to flow into the tank at a maximum rate. The weight switch in the control logic cabinet, preset at 4900 lb for second stage

(dribble) activation, confirms that approach to the set point is near, and actuates the next sequence step.

4. Reset the solenoid-air-operated fill valve to the "dribble" position. The flow is thus decreased to allow the tank to fill to the set point at a slower rate, thus minimizing overrun when the valve is actuated to close. The weight switch in the control logic cabinet, confirming that the 5000 lb batch set point is reached, actuates the fill valve to close and the "fill complete" lamp on the control panel insert to light up, completing the loading cycle and alerting the operator that the tank is ready for draining.
5. If during these procedures the tank drain valve is opened and/or the tank level rises abnormally (sensed by the valve position switch and the level switch), lamps on the control panel insert indicating this will start flashing and the tank fill valve, if previously opened, will be actuated to close. In this manner, an operator can discern the source of trouble, take corrective actions and press the reset push button on the control panel insert to start a new batch cycle. The operator may at any time press the emergency stop button on the control panel insert to close the fill valve and stop the batch cycle.
6. When all of the above have been satisfactorily completed, the operator may proceed to manually open the tank drain valve (control switch on main control panel) and empty the tank in accordance with process batch requirements. The valve's open position switch actuates the "fill complete" lamp to go out and the "drain valve open" lamp to light up.

- The two dissolving tanks (II and III) are to be alternately filled with predetermined weights of fluids *C* and *D* and residue.

In addition to the prerequisite conditions described above, the operator must manually select and set the operating mode required for activation of the residue conveyor by turning the control switch on the main control panel to "automatic cycle" or "manual" start position and by turning the adjacent control switch to tank II's or III's initial start flow direction. With these conditions satisfied, loading of tanks II and III may proceed sequentially, as follows:

1. The residue conveyor is assumed to be set on "automatic" and tank II is assumed to be set as the first tank to be filled with residue on initial start.
2. The operator sets batch requirements for filling either tank by turning common thumbwheel weight selectors on the control panel insert to the desired batch settings:

   Fluid *C* 4000 lb
   Fluid *D* 4000 lb
   Residue 2500 lb

3. The operator presses tank II's fluid batch start push button on the control panel insert. This actuates the next sequence step.
4. Fully open the solenoid-air-operated fill valve. This permits fluid *C* to flow into the tank at a maximum rate. The weight switch in the control logic cabinet, preset at 950 lb for second stage (dribble) activation, confirms that approach to the set point is near, and actuates the next sequence step.
5. Reset the solenoid-air-operated fill valve to the "dribble" position. Fluid *C* flow is thus decreased to allow the tank to fill to its set point at a slower rate and minimize overrun when the valve is actuated to close. The weight switch in the control logic cabinet, confirming that the 1000 lb batch set point is reached, simultaneously actuates the fill valve to close and the "fluid *C* fill complete" lamp on the control panel insert to light up. This alerts the

operator to proceed to the next sequence step.

6. The operator presses the tank II fluid *D* batch start push button on the control panel insert, actuating the next step.
7. Fully open the solenoid-air-operated fill valve. This permits fluid *D* to flow into the tank at a maximum rate. The weight switch in the control logic cabinet, set at 3900 lb for second stage (dribble) activation, confirms that approach to the set point is near, and actuates the next sequence step.
8. Reset the solenoid-air-operated fill valve to the "dribble" position. Fluid *D* flow is thus decreased to allow the tank to fill to its set point at a slower rate and to minimize overrun when the valve is actuated to close. The weight switch in the control logic cabinet, confirming that the 4000 lb batch set point is reached, simultaneously actuates the fill valve to close; the "fluid *D* fill complete" lamp on the control panel insert to light up; and the next sequence step.
9. Start the residue conveyor flow to tank II. The conveyor motor drive circuit contact actuates the "residue flow to tank II" lamp on the control panel insert to light up. Alerted by "tank II fluid *C* fill complete" and "tank II fluid *D* fill complete" lamps (steps 5 and 8 above), the operator presses separate tank III start push buttons on the control panel insert to actuate the tank III fluids *C* and *D* fill valves in the same manner as specified in steps 2 through 8. During this period, the weight switch in the control logic cabinet, confirming that the tank II 2500 lb residue batch set point is reached, simultaneously actuates the residue conveyor to stop; reverse-set itself; the "tank II residue fill complete" lamp on the control panel insert to light up, indicating completion of the tank II loading cycle; and, provided the tank II fluids *C* and *D* fill operations are satisfactorily completed, automatically start the residue conveyor flow to tank III.
10. If during these procedures the tank drain valves are opened and/or the tank level rises abnormally (sensed by the value position switches and the level switches), lamps on the control panel insert indicating this will start flashing, the tank fill valves (if previously open) will be actuated to close and the residue conveyor (if previously running) is tripped. In this manner, an operator can discern the source of trouble, take the necessary corrective actions and press the reset push button on the control panel insert to start a new batch cycle. The operator may at any time press the emergency stop button on the control panel insert to close the fill valves, trip residue conveyor and stop the batch cycle.
11. Prior to the start of (or during) the tank III filling cycle, with the tank II fill satisfactorily completed, the operator may proceed to manually open the tank II drain valve (control switch on main control panel) and empty tank II in accordance with the process batch requirements. The valve open position switch actuates the "fill complete" lamps to go out and the "drain valve open" lamp to light up. When tank II is drained, the zero weight switch in the control logic cabinet actuates the "tank II empty" lamp on the control panel insert to light up, indicating the tank is ready for next filling cycle. With the tank III fill satisfactorily completed, the operator may proceed to empty tank III by actuation of the drain valve (control switch main panel) in same manner as outlined for tank II above. Manual and/or automatic filling and manual draining

of tanks II and III will thus be cycled continuously until the operator shuts down the system.

Gas Dehydrator Regeneration Control System

This sytem involves a gas dehydrator regeneration system which is a sub-system of an ethylene refinery.

In this plant, a feedstock such as naptha or kerosene gas-oil is cracked in large furnaces. The feedstock first makes two passes through the economizer section of the furnace, prior to passing through the radiant section, where it decomposes into ethylene and various by-products. Dilution steam is introduced in the second economizer pass. This steam serves two functions:

1. It helps to vaporize the feedstock.
2. It assists in the reduction of coking of the radiant tube walls. (This is the more important function.)

The cracked feedstock is quickly quenched and the heavier by-products removed in a prefractionation stage. The residual gas, which contains the ethylene, and various by-products, such as ethane, propylene, butane light oils, etc., now go to a four-stage compressor with knock-out drums and cooling heat exchangers that remove much of the moisture and most of the heavier by-products not eliminated by the prefractionation stage.

We are now left with a process gas saturated with water vapor that must not remain in the final product. The gas is now passed through a large dehydrator, which absorbs the moisture until the desiccant is saturated and must be regenerated.

The sub-system of interest consists of two of these dehydrators. Only one dehydrator is on stream at any given time. The second unit is either in stand-by status or in the process of regeneration of the accumulated moisture in the desiccant bed.

Moisture removal is not simple in this case, since air cannot be used for drying. The introduction of oxygen into a hydrocarbon gas might create an explosive hazard. A fuel gas, which is a plant by-product, is used as a drying medium.

A PCD of the system is shown in Figure 5–9a (associated computer control panel indications and devices are shown in Figure 5–9b). The process of taking dehydrator *A* off-stream and replacing it with dehydrator *B* is initiated in one of three modes. The first mode is time based, wherein regeneration is automatic; the second is based on moisture analysis; operator manual control is the third mode.

Once off-stream, the spent dehydrator is put through a number of sequential steps, all of which are under computer control. (See the sequence chart in Figure 5–10 for an illustration of a typical cycle.)

As the computer steps through the program, it not only follows a time sequence, but it also verifies the completion of each step. Input for the verification is sensor based and is made by various pressure, temperature, flow and limit switches.

During certain steps in the regeneration phase, temperature control of the regeneration gas—and thereby the desiccant bed—is performed by computer set point manipulation of the regeneration gas analog temperature controller. This control technique is utilized just after the dehydrator is slowly depressurized to avoid upset to the desiccant bed. Cool fuel gas is then introduced to the spent desiccant under program control.

Some fuel gas is diverted through the gas heater and blended with unheated gas. Gradually, the desiccant bed is heated with the resultant vaporation of moisture into the hot fuel gas. The desiccant is maintained at the highest temperature for a soaking period, after which the temperature is gradually reduced by program control of the temperature set point. The dehydrator is then slowly repressurized and placed on stand-by for the next cycle.

Reference to Figure 5–11 shows the first few steps in the logic diagram correspond-

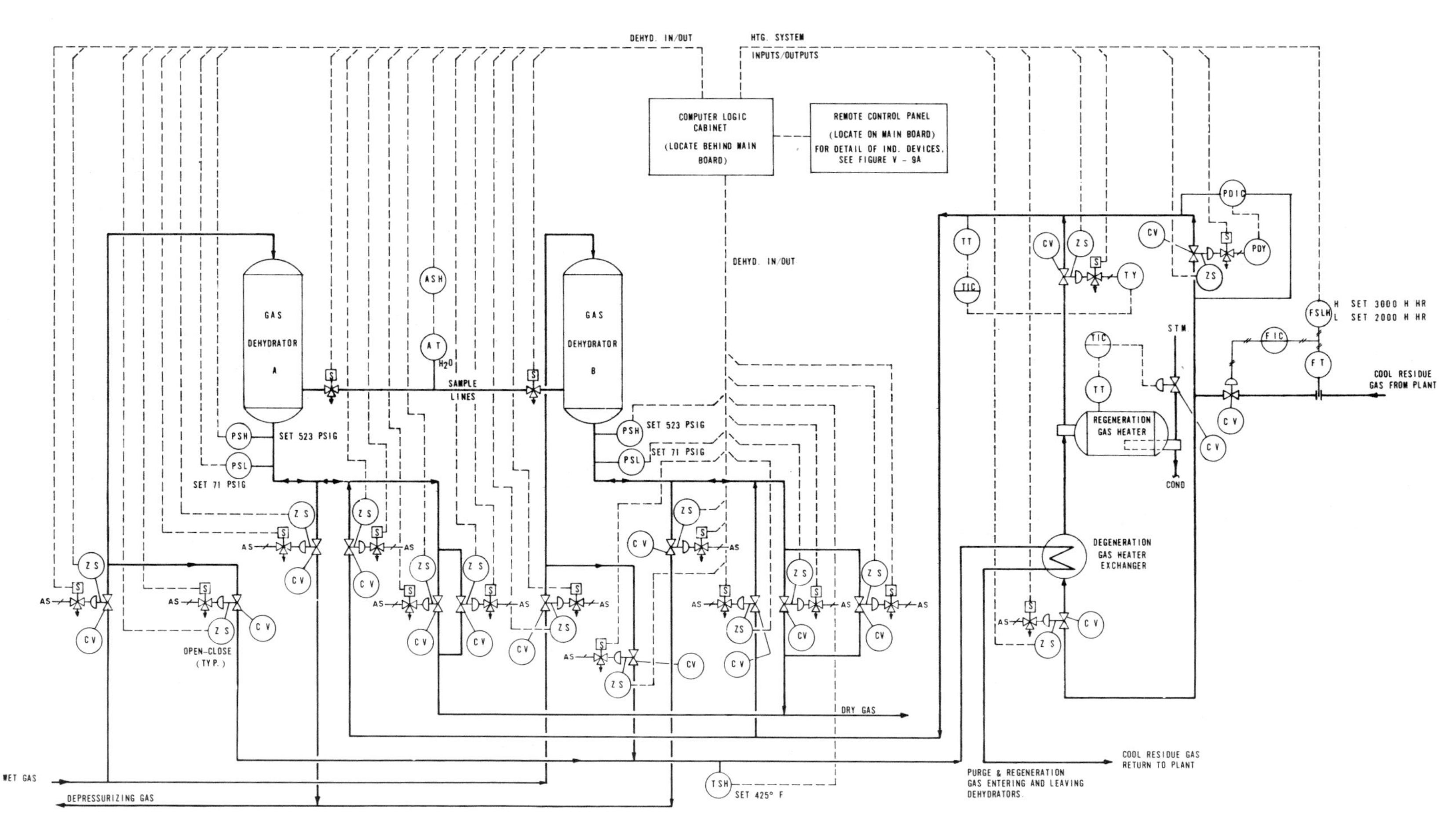

Figure 5-9a. Process control diagram—gas dehydrator regeneration system.

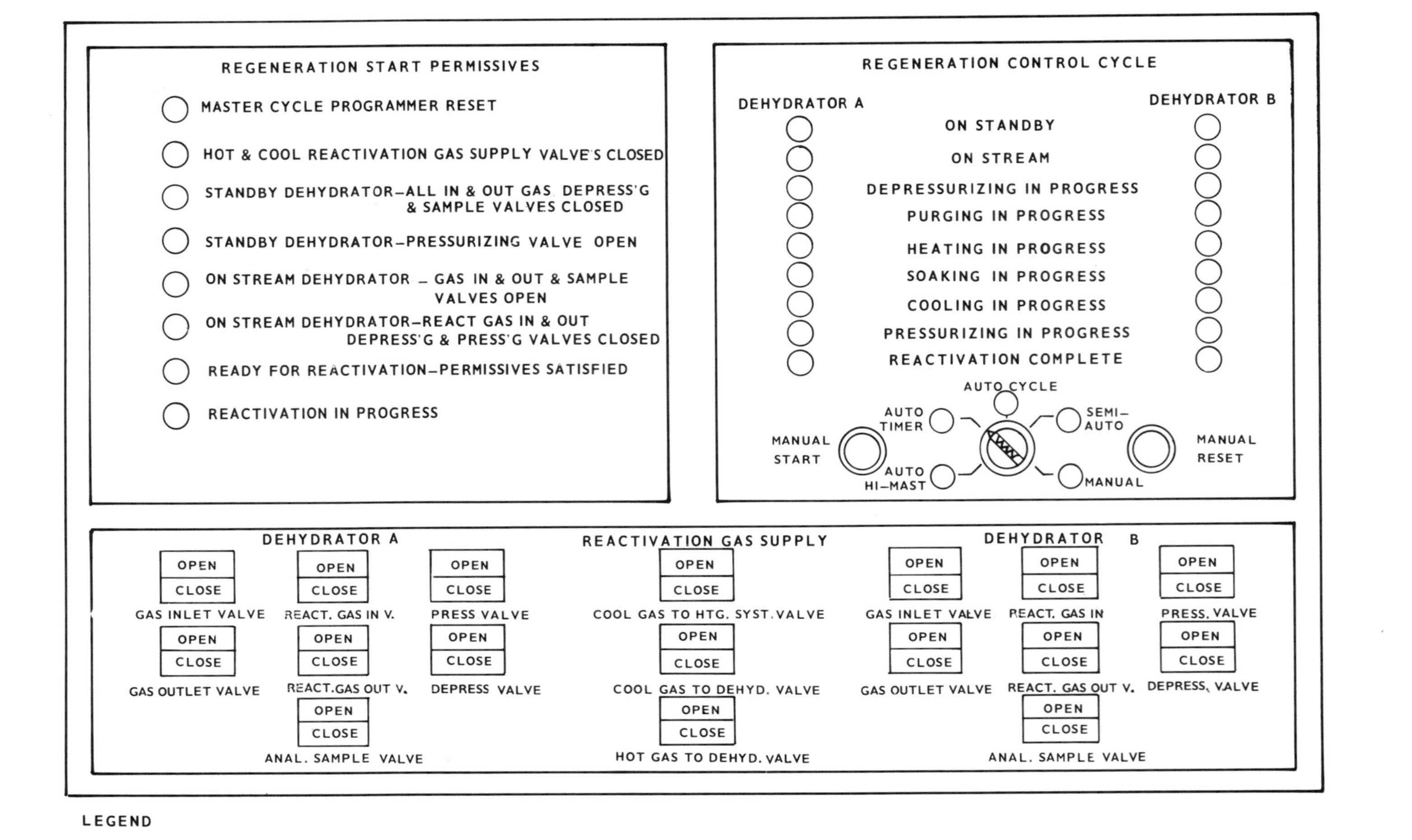

Figure 5-9b. Remote control panel indications and devices for gas regeneration control system.

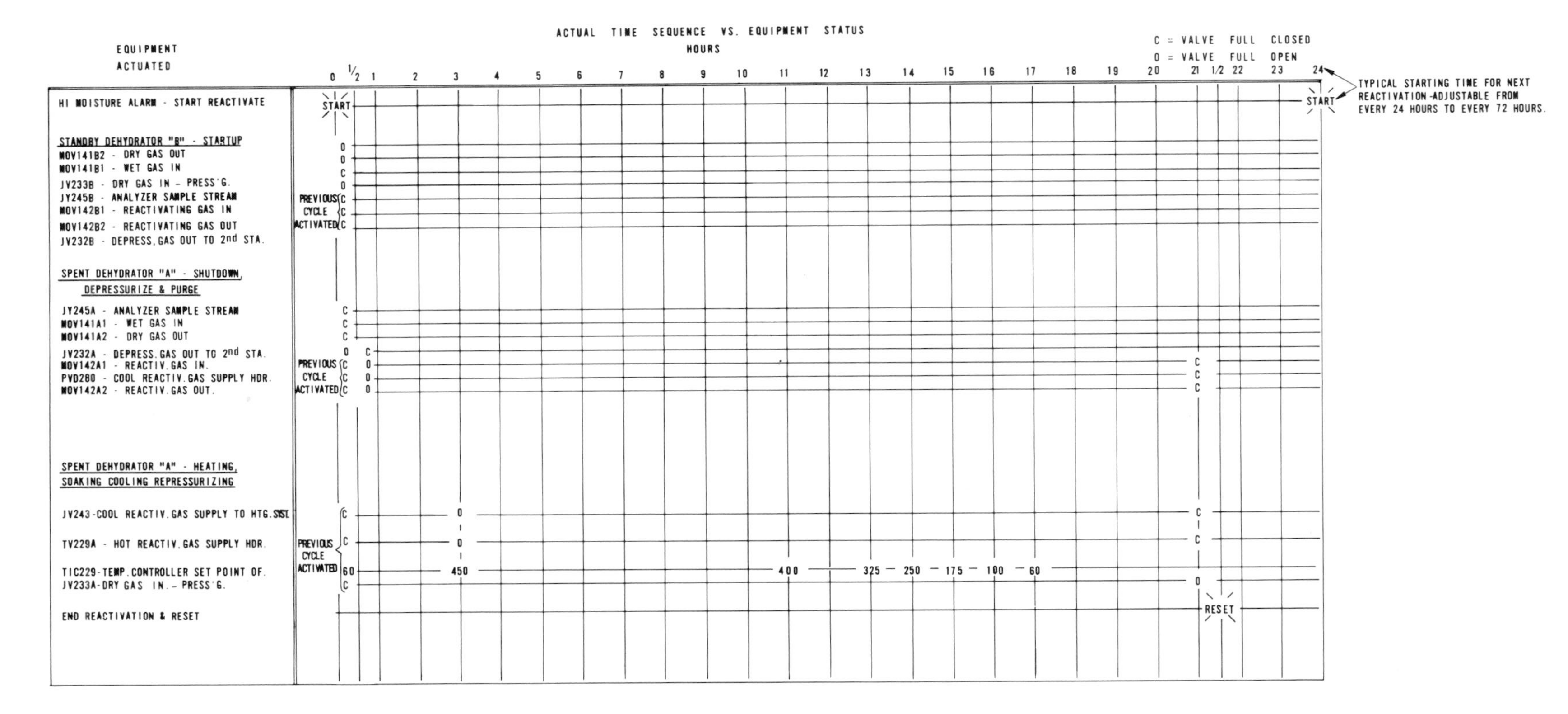

Figure 5-10. Sequence chart—gas dehydrator regeneration control system.

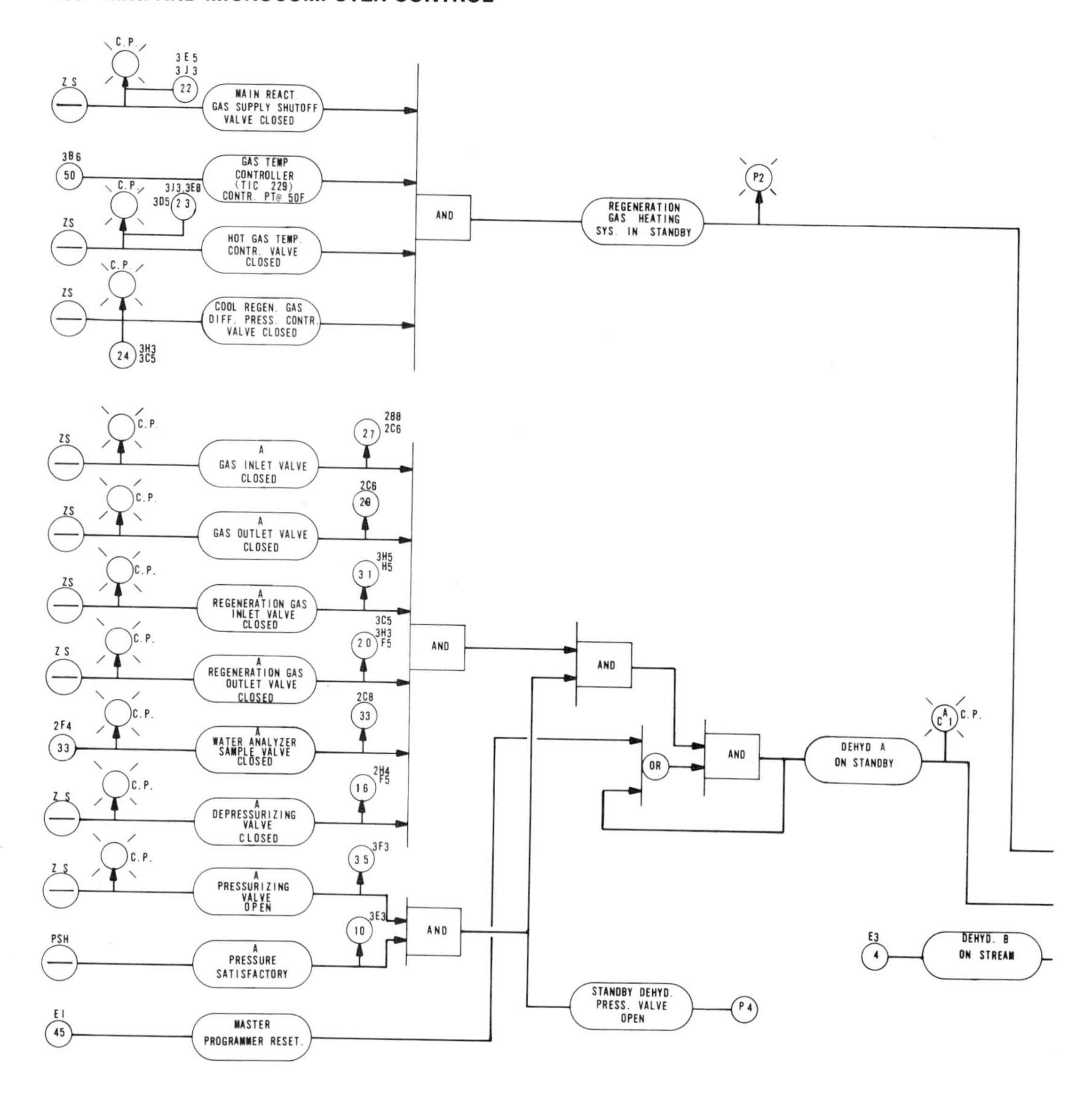

Figure 5-11. Logic diagram—gas dehydrator regeneration control system.

ing to the beginning of the sequence chart in Figure 5–10.

Figure 5–12 graphs the imposed temperature control steps of the regeneration fuel gas and the anticipated temperature of the desiccant bed as it heats, soaks and cools in response to the heating/drying cycle. This is an example of a process condition cycle curve.

Water De-mineralizing Plant Controls

In Figure 5–13, a block diagram for a water de-mineralizing plant is illustrated. This plant is a portion of the utility area of a large ethylene refinery, and it provides cleaned and de-mineralized water to the steam boilers. (The PCD is shown in Figure 5–14.)

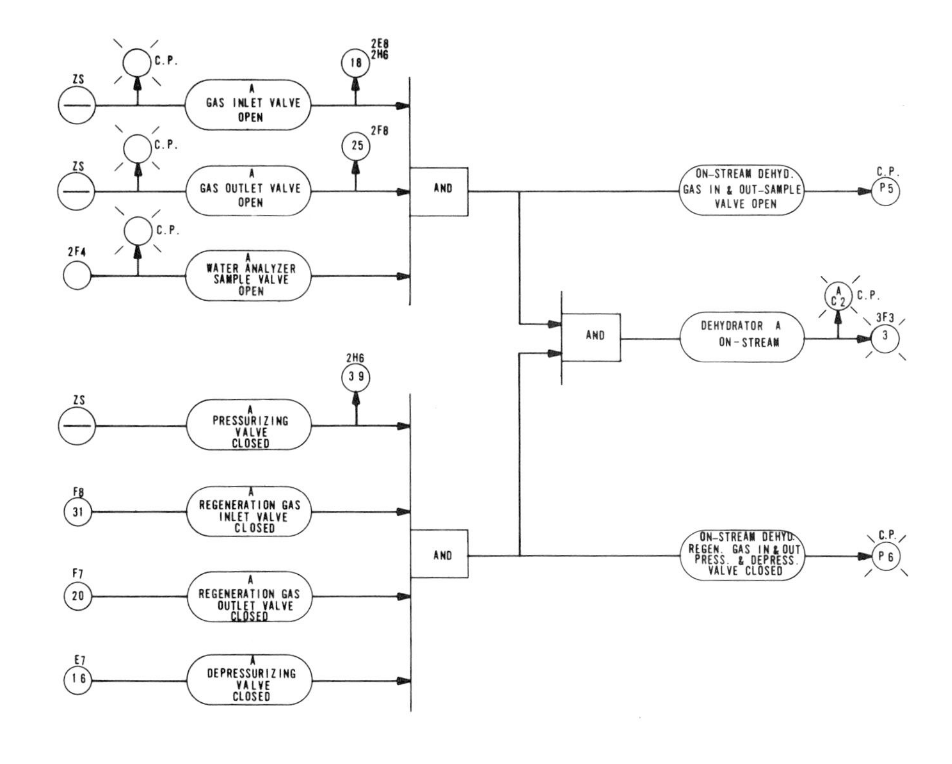

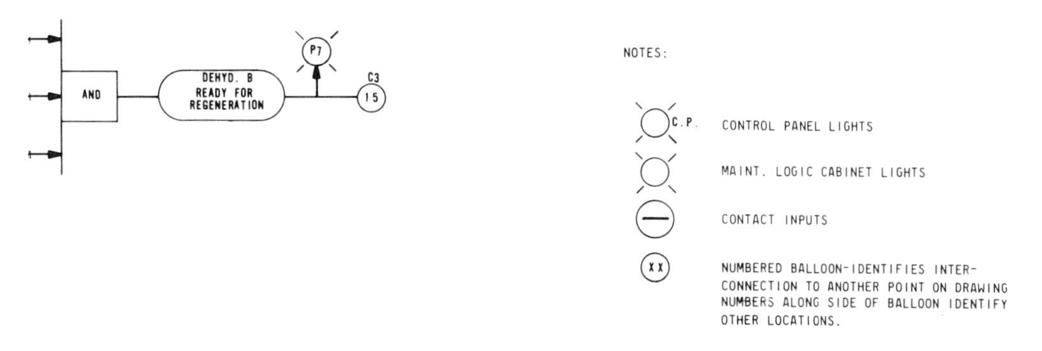

Figure 5-11. (continued)

Some of the steam generated is used for large turbine drives on process gas compressors, some is used as dilution steam to the feedstock entering the ethylene cracking furnaces, and the balance is used for other turbine-driven equipment and various heating uses throughout the plant. Condensate collected from steam turbine surface condensers in the plant is continuously passed through a condensate de-mineralizer polishing system.

A utility plant of this type consists of a number of filters and water-treating tanks that contain filter media and chemical beds which require periodic cleansing and regeneration. Therefore, the route of raw water and condensate entering the system is subject to multiple paths as various filters and

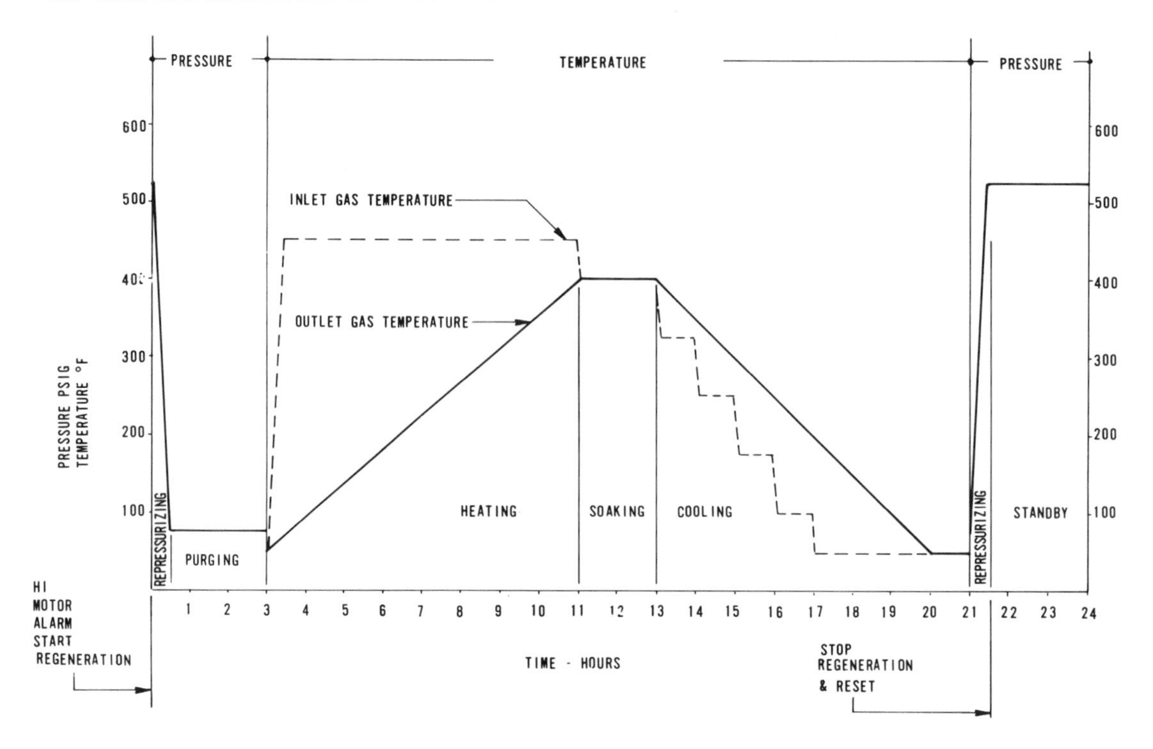

Figure 5-12. Process cycle curve—gas dehydrator regeneration system.

treating beds are cycled on- and off-stream.

The anion, cation and mixed bed sections are chosen to illustrate the typical procedure each section of a de-mineralization plant goes through as it normally operates, and then is regenerated. The function of this portion of the plant is to remove the mineral ions by reacting them with suitable treatment chemicals.

The anion and cation tanks remove most of the positively and negatively charged ions. Those which remain are removed by passing through a second stage of de-mineralization in the mixed bed tanks. As their name implies, these mixed bed tanks have both anion and cation reactive chemicals. This can be viewed as a final polishing operation.

As the chemical beds are used, they eventually become spent and require regeneration. Procedures within the sequence program now take the spent beds off-stream and bring stand-by beds on-stream. At any instant, there are two beds each of cation and anion treatment on-stream, with one mixed bed unit. The stand-by units will cycle on-stream as programmed or as conditions dictate.

Off-stream equipment goes through a series of steps as the filters and treating beds are cleaned and chemically regenerated. Each of these procedures involves a variety of flow paths through each item of equipment. These paths are arranged by the proper manipulation of open/close control valves, typically indicated in the combination sequence chart/written descriptions of Figure 5–15.

Adding to the complexity of this cyclic plant are two factors. The first involves analog control of the level in the de-mineralized water storage tank. As in the last case, this can be done by SPC or DDC. In this case, the latter method is chosen by client preference. This control is accomplished by varying the rate of water flow through the de-mineralization section.

The second factor is caused by the analytical measurement of water quality, such as conductivity, pH and dissolved silica and sodium, by individual analyzer systems for each of these variables.

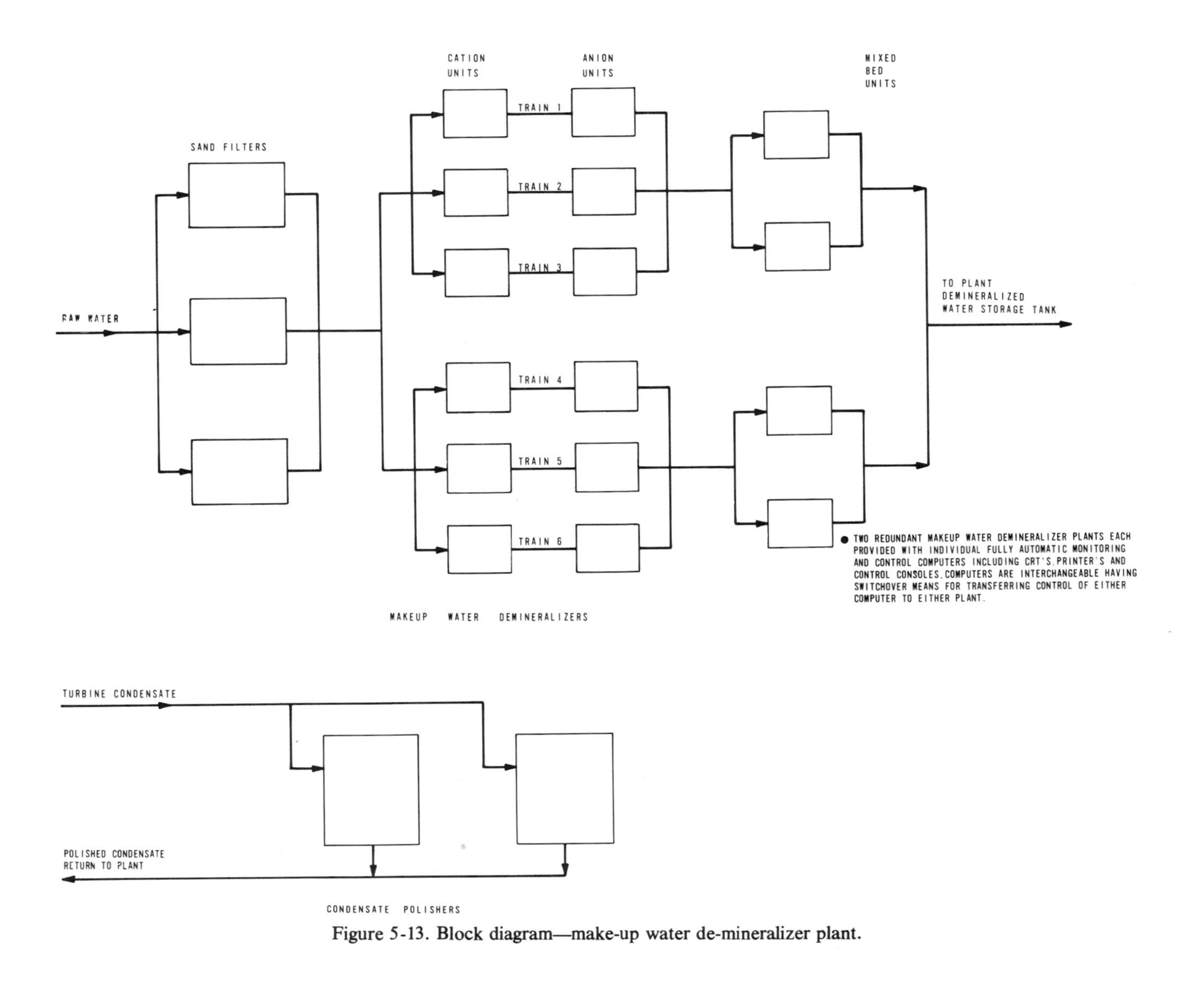

Figure 5-13. Block diagram—make-up water de-mineralizer plant.

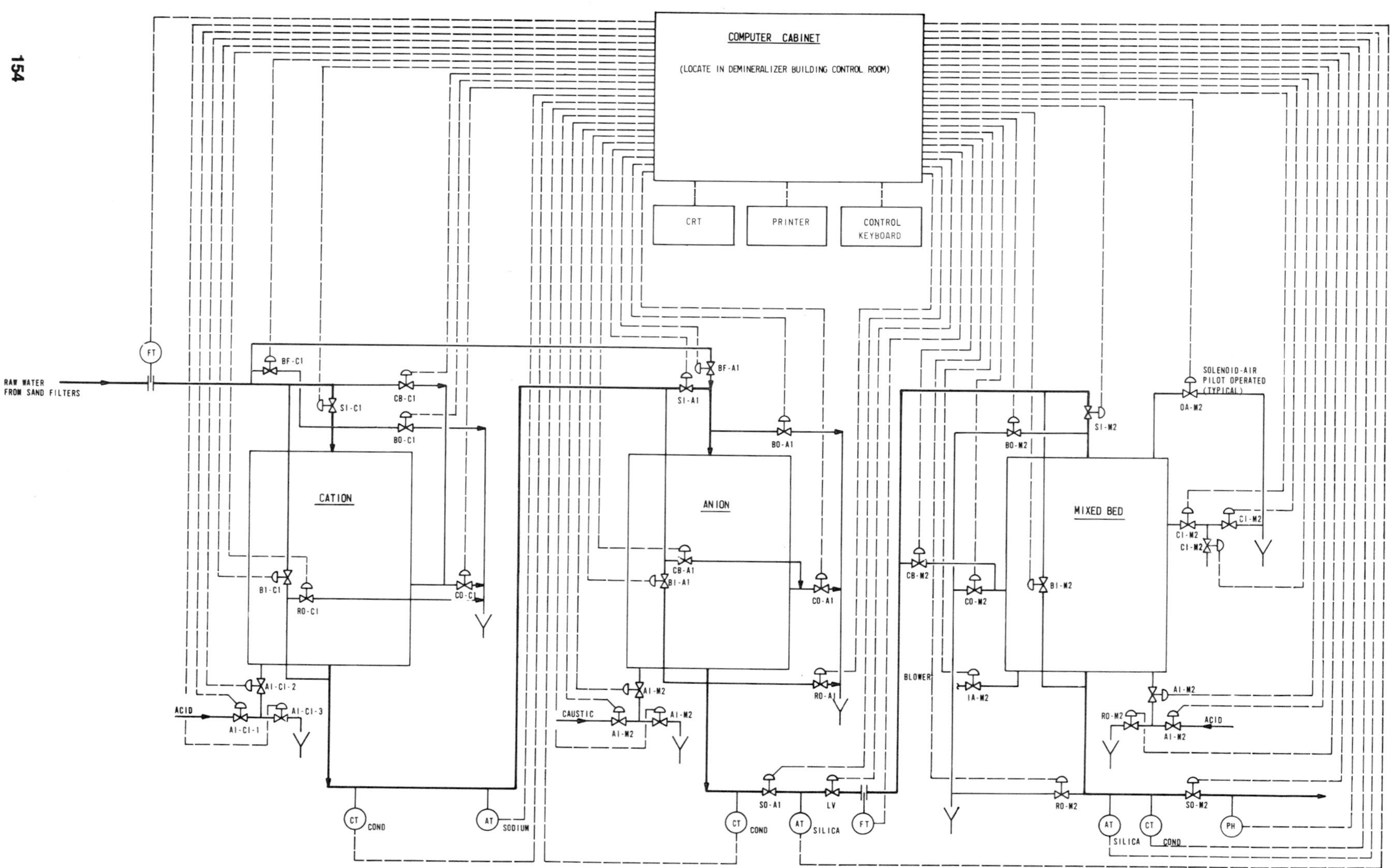

Figure 5-14. Process control diagram—make-up water de-mineralizer.

	VALVE IDENTIFICATION	FLOW GPM	TIME MINUTES	SI-CI	BI-CI	BO-CI	CB-CI	CO-CI	BF-CI	RO-CI	AI-CI-1	AI-CI-2	AI-CI-3	SI-AI	SO-AI	BI-AI	BO-AI	CB-AI	CO-AI	BF-AI	RO-AI	CI-AI-1	CI-AI-2	CI-AI-3	SV-AI	
SERV.	SERVICE	260		0	X	X	X	X	X	X	X	X	0	0	0	X	X	X	X	X	X	X	X	0	0	A
	CLOSE ALL VALVES	0	3	X	X	X	X	X	X	X	X	X	0	X	X	X	X	X	X	X	X	X	X	0	X	
PREHEAT	PREHEAT ANION CLOSE ALL VALVES ON CATION	94	15	X	X	X	X	X	X	X	X	X	0	X	X	X	X	X	0_1	0_1	X	0_1	0_1	X_1	X	E
A DRAW C DRAW	ACID DRAWER FOR CATIONS CAUSTIC DRAWER FOR ANIONS	357 / 100	30 / 30	X	X	X	X	0_1	0_1	X	0_1	0_1	X_1	X	X	X	X	X	0	0	X	0	0	X	X	B F
C RINSE A RINSE	CATION SLOW RINSE ANION SLOW RINSE	354 / 100	20 / 20	X	X	X	X	0	0	X	0	0	X	X	X	X	X	X	0	0	X	0	0	X	X	C G
	CLOSE ALL VALVES	0	3	X	X	X	X	X	X	X	X	X	0	X	X	X	X	X	X	X	X	X	X	0	X	
FSTRIN	FAST RINSE CATION ANION ALL VALVES CLOSED	520	15	0	X	X	X	X	X	0	X	X	0	X	X	X	X	X	X	X	X	X	X	0	X	
SERV	RETURN CATION TO SERVICE ANION ALL VALVES CLOSED			0	X	X	X	X	X	X	X	X	0	X	X	X	X	X	X	X	X	X	X	0	X	
SERV DSH_2O	CATION IN SERVICE DISPLACE RAW H_2O IN ANION	520	5	0	X	X	X	X	X	X	X	X	0	0_1	X	X	X	X	0_1	X	X	X	X	0	X	
SERV FSTRIN	CATION IN SERVICE ANION FAST RINSE	520	25	0	X	X	X	X	X	X	X	X	0	0	X	X	X	X	X	X	0_1	X	X	0	X	D
STD BY	PLACE CATION IN STAND BY PLACE ANION IN STAND BY			X	X	X	X	X	X	X	X	X	0	X	X	X	X	X	X	X	X	X	X	0	X	
SRVRIN	SERVICE RINSE	520	15	0	X	X	X	X	X	X	X	X	0	0_1	X	X	X	X	X	X	0_1	X	X	0	X	H

NOTES:

A. Monitor flow, low flow to alarm at 100 gpm.
Monitor conductivity, alarm high conductivity at 0.7 micromhos.
Monitor silica, alarm high silica at 0.02 mg/liter.
Monitor sodium, alarm high sodium at 200 ppm.

B. Monitor flow, allow 2 minutes to establish acid dilution water flow and filtered water for blocking flow, alarm low flow at 230 gpm and alarm high flow at 260 gpm, interlock acid pump with low flow alarm - pump to start only after 230 gpm has been exceeded.
Monitor acid strength, alarm high acid strength at 3%, alarm low acid strength at 1%, on high acid strength alarm - shut off pump, on low acid strength alarm - restart pump.

C. Monitor low - rinse flow rate, alarm low flow at 330 gpm for cation.

D. Monitor flow, high flow to alarm at 550 gpm and low flow to alarm at 470 gpm. Monitor conductivity - return unit to service immediately on conductivity dropping within acceptable limit (below 0.5 micromho). If conductivity does not drop to acceptable limit in 15 minutes - alarm high conductivity and place unit in hold for regeneration position.

E. Monitor flow, allow 2 minutes to establish caustic dilution water flow, alarm low flow at 85 gpm and alarm high flow at 105 gpm.
Monitor temperature, alarm high temperature at 120°F., and low temperature at 90°F. At high temperature alarm close st - ct valve.

F. Monitor flow, allow 2 minutes to establish caustic dilution water and filtered water flow for blocking flow, alarm low flow at 60 gpm and alarm high flow at 80 gpm, interlock caustic pump with low flow alarm - pump to start only after 60 gpm has been exceeded.
Monitor temperature, alarm high temperature at 120°F., and low temperature at 90°F., on high temperature alarm close st - ct valve.
Monitor caustic strength, alarm high caustic strength at 5%, alarm low caustic strength at 3%, on high caustic strength alarm - shut off pump, on low caustic strength alarm - restart pump.

G. Monitor slow - rinse flow rate, alarm low flow limit at 85 gpm for anion.

H. Monitor flow, high flow to alarm at 550 gpm and low flow to alarm at 100 gpm. Within 1 minute of completion of rinse check conductivity - if not within acceptable limit (below 0.7 micromho) alarm high conductivity and initiate extended rinse for 15 minutes, if conductivity does not drop within acceptable limits during this time period alarm for re regeneration. If conductivity drops to acceptable limits at any time during extended rinse, terminate extended rinse.

1. Operation to be performed first.
2. Operation to be performed second.
3. All operations not numbered can be performed simultaneously.
4. *3 valve block & bleed - each group of valves indicated requires 2 computer output for actuation and 2 computer inputs for position.
5. Time, flow rates, quality monitoring parameters (conductivity, silica, and sodium etc., where applicable) are subject to adjustment/revision during start up.

Figure 5-15. Sequence chart—make-up water de-mineralizer system.

As water quality degrades toward unacceptable limits, it sometimes becomes necessary to change the basic plant cycle in order to bring various off-spec items of equipment into regeneration sooner. In addition, equipment readiness is analytically checked at the end of the regeneration cycle. These analytical complications require careful programming to properly integrate the out-of-tolerance conditions into the overall control system.

Figure 5–16 illustrates a small portion of the computer logic corresponding to the earlier steps shown in the sequence chart of Figure 5–15. This logic flow diagram was prepared by the computer systems sub-contractor according to the specifications and sequence charts supplied by the engineering contractor.

SYSTEMS MANUFACTURER

The best approach for fabrication of the sequence control systems package is to have it designed and built by a qualified specialist manufacturer with many years of demonstrated experience and total responsibility in the execution of similar projects. There are many advantages to having the package designed by such a specialist.

- A specialist manufacturer will utilize the latest proven techniques of design, with undivided attention given to proper selection, procurement and assembly of hardware components and preparation of software.
- Systems can be fully demonstrated and tested under simulated operating conditions; major design, equipment and construction faults can be corrected before shipment.
- A fully tested and reliable system will be shipped to the job site, ready to operate.

Investigating acceptable fabrication shops, obtaining proposals and placing an order for a prepackaged sequence control system is not very different from any other materials procurement management procedure. It is more complicated than placing an order for a flow transmitter or a control valve because of the diversified aspects involved. However, the number of qualified manufacturers or "systems houses" in this field has grown over the past few years and a good choice of firms is available.

With this approach, it is incumbent on the contract engineer to provide a sequence control system specification which clearly and substantially defines the scope of the work; the basis of the design; the conditions of service; and all technical, performance, functional, equipment and schedule requirements. It must also include contractors' guidance sketches; logic diagrams; and descriptions of associated contractor furnished equipment, material and services. The following manufacturer's responsibilities must also be included in the scope of the specification:

Design; manufacture; assembly; internal interconnecting cabling; supply of documentation, including logic diagrams, system description, flowcharting, programming, seller's testing, purchaser's witness testing, on-time shipment and "de-bugging" of the computer system (computer and associated components); training of purchaser's personnel in programming; use and maintenance of the system coincident with purchaser's system factory testing; spare parts list; provisions for supervision of installation, interconnection and start-up, if required; reprogramming in accordance with purchaser's return of marked-up logic diagrams and comments.

SYSTEM SPECIFICATION

A systems house furnished sequence control system package combines mechanical and electrical hardware and software programmed to automatically provide monitoring and control of a unit. However, every sequence control application is unique, and the systems house cannot be expected to define all process needs, operator interfaces or management requirements. To as-

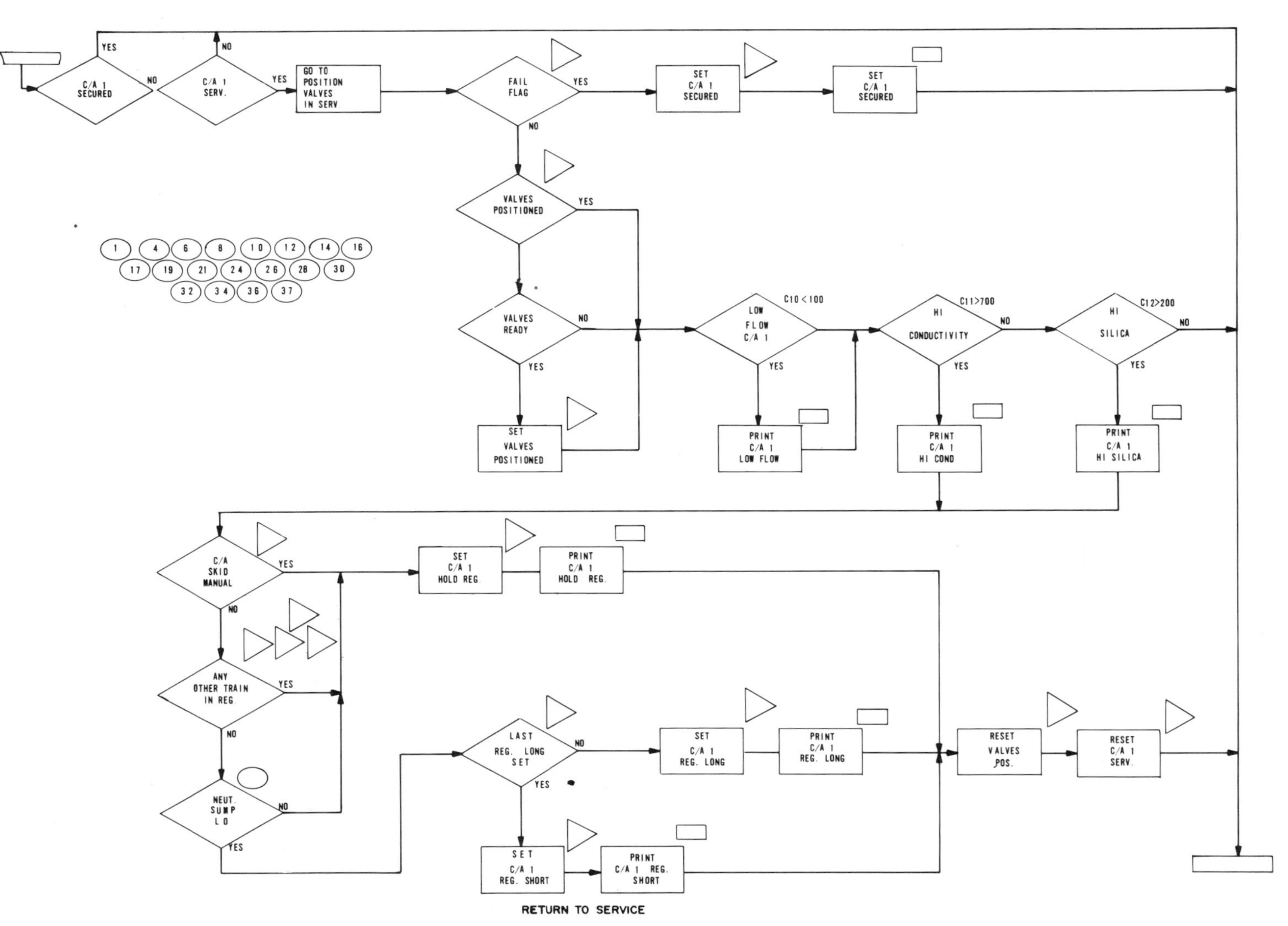

Figure 5-16. Computer logic flow diagram—make-up water de-mineralizer system.

sure successful execution of a project, the design engineer must provide a procurement specification which will achieve a complete and technically correct specification delineating what is to be furnished according to the requirements of the buyer.

The organization and content of a typical sequence control system procurement specification which substantially fulfills this need is illustrated below.

Scope

The seller shall furnish and deliver a gas dehydrator regeneration control system, factory calibrated and tested, to demonstrate that it meets all requirements of this specification.

Basis of design

This system will be used to control two gas dehydrators. During normal operation, gas passes through one dehydrator for drying service while the other dehydrator remains shut down on stand-by or undergoing regeneration. Each dehydrator is designed to have an on-stream drying cycle varying from 24 to 72 hours and a regeneration cycle of 24 hours or less.

Regeneration gas system. A regeneration gas heating system is provided, consisting of gas heater, exchanger, cooler and separator. The regeneration gas supply at 50°F, 75 psig, is passed through the regeneration gas heating system, where it is heated to and maintained at 450°F for regeneration purposes. During normal operation, with no demand for regeneration, the gas heating system is maintained in a hot condition of readiness; i.e., the steam supply to the regeneration gas heater is maintained at a minimum rate and the residue gas valves in lines to and from the heating system remain closed, ready to be opened on demand for regeneration.

Functional requirements—The purpose of the gas dehydrator regeneration control system is to automatically or manually place the fresh (stand-by) dehydrator in service; to shut down, remove from service, reactivate and return to stand-by mode the exhausted dehydrator; reset the master cycle programmer; and start the regeneration cycle again in accordance with scheduled time, cycle, high moisture or manual settings.

The system is usually furnished for remote central control, designed to operate without manual assistance from operating personnel. It is complete with necessary accessories and includes analog outputs for automatic temperature control, relay or solid state logic outputs for actuation of motor or solenoid-air operated valves, contact closure inputs for valve positions, pressure and temperature switches, name-plate light displays to indicate status of dehydrator reactivation cycle and status of valve positions, pushbuttons for manual operation of valves, adjustable cycle time programmers and multi-position key lock switches for automatic manual start, stop and reset of the control cycle. Nameplate light display are used to indicate stages of operation: pre-activation permissives, depressurizing, purging, heating, soaking, cooling, repressurizing, reactivation complete and reset. The seller usually guarantees that the equipment and all parts thereof are suitable for the service described in the specification or that they can be reasonably inferred therefrom.

Regeneration Cycle—The sequence and timing of automatic and/or manual actuations required for regeneration is shown on the attached Figures A and C (descriptions are based on dehydrator "A" as the "Spent" unit and dehydrator "B" as the "Standby" unit):

Pre regeneration requirements. The starting circuit for regeneration of a spent dehydrator consists of a series of permissive interlocks. It usually requires that suitable conditions are established, prior to starting, for a successful regeneration.

The following requirements are necessary for the start of a successful regeneration cycle:

- The operator determines and selects the operating mode required for reactivation, by inserting a key in the multiposition key lock switch on the control panel insert, and turning it to either "Automatic-High Moisture Start," "Automatic Timer Start," "Automatic Cycle Start," "Semi-Automatic," or "Manual" position.
- Prior to this, the operator must establish the required operating time settings on the starting and heating period timers by an adjustment of various setting devices located in the control logic cabinet.
- Note that the "Automatic Cycle Start" mode is a fixed setting designed to start the reactivation cycle immediately upon reset at end of the previous cycle. (The starting period timer adjustment is not required for this control mode.)
- The master cycle programmer is reset. When the selector switch described above is set on either of the three "Automatic" positions indicated, reset occurs automatically; when set on "Semi-automatic," the operator must press a separate push button to reset; when set on "Manual," no reset is required.
- Hot and cool reactivation gas supply valves must be closed.
- Fresh dehydrator inlet and outlet gas, depressurizing and sample valves must be closed.
- The fresh dehydrator pressurizing valve must be open and the pressure be satisfactory.
- The spent dehydrator gas inlet and outlet valves and sample valve must be open.
- The spent dehydrator reactivation gas inlet and outlet depressurizing valves must be closed.

When all the preceding conditions have been satisfied, an indicating lamp on the control panel insert lights up, advising the operator that all permissives have been met and that the spent dehydrator is "ready for regeneration."

If a condition is lost during these procedures, the indicating lamp will go out and the "ready for reactivation" lamp will also go out, if it was previously on. In this manner, an operator can discern the source of any inadequate setting and take the necessary corrective actions quickly to re-establish the "ready for regeneration" condition.

When the above has been satisfactorily completed, and provided the multi-position key lock switch described above is in either of the three "Automatic" positions indicated, the reactivation cycle will start automatically. If the selector switch is set in the "Semi-automatic" or "Manual" positions, the operator must press a separate push button on the control panel insert to start the reactivation cycle. In either case, a "regeneration in progress" lamp will come on, indicating that the cycle timer is running or that operations are being conducted manually, and the "ready for regeneration" lamp will go out. Operations will then proceed in automatic (or manual) sequence as follows.

1. Fresh dehydrator (*B*)—place in service.
    - Open the gas outlet valve. The fully open valve position switch actuates the next sequence step.
    - Open the gas inlet valve. This places the fresh dehydrator in service parallel with the spent dehydrator. The fully open valve position switch actuates the next two sequence steps, and the indicating lamp on the control panel insert lights up, advising the operator that the "fresh unit is in service."
    - Simultaneously close the pressurizing and water analyzer sample valves on the spent dehydrator and open the solenoid-

operated water analyzer sample valve on the stand-by dehydrator.

2. Spent dehydrator (*A*)—shut down.
   - Close the gas inlet valve. The fully closed valve position switch actuates the next sequence step.
   - Close the gas outlet valve. This shuts down the spent dehydrator and prepares it for regeneration. The fully closed valve position switch actuates the indicating lamp on the control panel insert to light up, advising the operator that the "spent unit is shut down," and the next sequence step begins.

Depressurization. Open the depressurizing valve. The fully open valve position switch actuates the indicating lamp on the control panel insert to light up, advising the operator that "depressurizing is in progress." The shut down, spent dehydrator pressure (at 523 psig) is thus ramped down gradually, over a 30 minute period, to the 75 psig operating condition of the second stage discharge drum. The pressure switch on the spent dehydrator, set at 75 psig to confirm that the depressurizing is complete, actuates the depressurizing valve to close. The fully closed valve position switch actuates the next sequence step.

Purging.

- Open the generation gas inlet valve of the spent dehydrator. The fully open valve position switch simultaneously actuates the next sequence step, lights up the indicating lamp on the control panel insert advising the operator that the "purge is in progress" and extinguishes the "depressurizing in progress" lamp.
- Switch the cool regeneration gas supply differential pressure control valve from shut-off to differential pressure controller service and open the spent dehydrator gas outlet valve. The cool purging gas flow is thus directed up through the spent dehydrator desiccant bed to the regeneration gas separator system for a period of 2½ hours to remove any absorbed high-boiling hydrocarbons. During this period, the flow switch, set to confirm that the flow is satisfactory (not less than 30,000 lb/hour), plus the 2½ hour elapsed timer switch, actuate the next step.

Heating and soaking.

- Open the regeneration gas supply heater to the heating system's shut-off valve. The fully open valve position switch simultaneously actuates the next sequence step, lights up the indicating lamp on the control console—advising the operator that "heating is in progress"— and extinguishes the "purge in progress" lamp.
- Switch the hot regeneration gas temperature control valve from shut-off to temperature controller service. The existing 50°F control point of the temperature controller, set at the start of the cycle, causes this valve to remain closed. The fully closed position switch actuates the next sequence step.
- Simultaneously actuate the milliamp signal generator and 8 hour timer to output a 16 milliamp signal to the reactivation gas inlet temperature controller, to *increase* the set point to 450°F; hold this condition for 8 hours. Note that the scale range is 0° to 600°F for a 4 to 20 milliamp input.

The controller will maintain its temperature by automatic regulation of the hot regeneration gas control valve. Hot regeneration gas flow is thus directed upflow through the dehydrator desiccant bed and out to the reactivation gas separator system. The desiccant bed is thus heated gradually over an 8 hour period to approximately 425°F.

At end of this period, the 8 hour timer switch plus the dehydrator regeneration gas outlet temperature switch (set to confirm that 425°F has been reached) actuate the milliamp signal generator to output a 14.7 milliamp signal to the controller to decrease its control setting to approximately 400°F for 2 hours of soaking. Also, an indicating lamp on the control panel insert lights up, advising the operator that "soaking is in progress," and the "heating in progress" lamp goes out. The 2 hour timer switch actuates the next sequence step.

Cooling. The indicating lamp on the control panel insert will light up advising the operator that "cooling is in progress," extinguish the "soaking in progress" lamp and actuate the milliamp signal generator to output signals to the temperature controller to *decrease* its control setting in graduated, timed steps as follows.

REQUIRED OUTPUT SIGNAL (MILLIAMPS)	TEMPERATURE CONTROLLER SET POINT (°F)	ELAPSED TIME (HOURS)
12.7	325	1
10.7	250	1
8.7	175	1
6.7	100	1
5.3	50	4

The controller will maintain these temperatures by automatic regulation of the hot regenerating gas control valve. The dehydrator desiccant bed is thus cooled down gradually to approximately 50°F over an 8 hour period and reactivated for standby service. At the end of this period, the 8 hour timer switch, plus the dehydrator regeneration gas outlet temperature switch (set to confirm that 50°F has been reached), actuate the next sequence step.

Pressurizing and reset.

- Close the regeneration gas inlet valve of the reactivated dehydrator. The fully closed valve position switch simultaneously deactivates the low-flow alarm on the purchaser's main control board, actuates the indicating lamp on the control panel insert to light up—advising the operator that "pressurizing is in progress"—extinguishes the "cooling in progress" lamp and actuates the next sequence step.
- Simultaneously close the residue gas supply to the heating system shut-off valve, switch the solenoid-air-operated hot regeneration gas temperature control and cool gas differential pressure control valves to the shut-off condition and close the regeneration gas outlet valve of the reactivated dehydrator. The regeneration gas flow is thus shut down, and the heating system is maintained in readiness for the next regeneration cycle. The fully closed valve position contacts on these valves will actuate the next sequence step.
- Open the pressurizing valve located in the bypass around the closed gas outlet valve. (Dry gas from the on-stream dehydrator is thus employed to gradually repressurize the reactivated dehydrator to 523 psig for stand-by service.) The pressure switch, set to confirm that dehydrator repressurization is satisfactory, permits reset of the control system starting mechanism for the next scheduled regeneration cycle; lights up the indicating lamp on the control panel insert, advising the operator that "reactivation is complete"; and extinguishes the "pressurizing in progress" lamp.

Equipment requirements—The overall scope and general arrangement of the gas dehydrator regeneration control system are as follows:

- The system shall be minicomputer based, principally employing solid state electronic-type logic, memory, processing and control components of the highest quality and highest degree of reliability for the various control and monitoring functions. Digital input/output functions shall be electrically isolated by means of optical isolators with AC output limited to 115 volts AC, 1.5 amps. The integrity of the system shall not be affected by electrical noise, spikes and surges normally expected in a petrochemical plant.

- The sequence control system must be sufficiently subdivided to permit in-service check-out and maintenance without impairing the reliability of the overall control system. The computer cabinets shall include status-indicating lamps for all important solid state logic components and relays operated during regeneration start-up and shut-down; for monitoring availability of control power at all sub-systems; and for monitoring the continuity of important relay coils, all as required to facilitate system check-out. Means for stepping the logic through its sequences shall be included.
- Each sequence action should be successively satisfied before the next step or action can be effected. Interlocks shall not operate in reverse. (If, due to malfunction, equipment actions are not satisfactorily completed in a predetermined time period, the control system will identify the faulted valve and/or pump by means of flashing or other appropriate display on the control console, actuate the common annunciator on the remote panel and stop the control system in a "safe" condition—as determined by field devices—in its last sequence step. In this manner, an operator can determine the trouble and take the necessary corrective action quickly. When the operator corrects the condition, or if it self-corrects, the operator must acknowledge and manually actuate the "trouble corrected" reset at the control console to permit the control system to resume operation and proceed to its next sequence step.)

- A watchdog timer should be provided to indicate computer failure. Any detectable digital input/output failure should be indicated by a lamp on the appropriate input/output card.
- The computer provides complete flexibility for reprogramming applications software.
- Computer programming, memory capacity and speed, including input/output auxiliaries, must be adequate to accomplish the complete sequence control of the system. Additional input/output capacity may be provided as an option.
- Elapsed time setting for starting successive automatic regeneration cycles are fully adjustable from 24 hours to 72 hours (or other time base).
- Elapsed time settings of individual sequence steps within the automatic regeneration cycle must thus be fully adjustable.
- A milliamp signal generator should be provided for output to the regeneration gas temperature controller set point. Capability is provided to output a maximum of 10 discrete signals, sequenced as required over the 4 to 20 milliamp range.
- Means should be provided on the control console to permit transfer from one operational mode to another, namely, "Automatic-Hi Moisture Start," "Automatic Timer Start," "Automatic Cycle Start," "Semi-automatic" or "Manual," without requiring operator adjustments and without changing the status of the sequence cycle existing at the time of such a transfer
- When the control system is in the manual mode, the proper actuation sequence steps required for operator guidance must be successively indicated by an appropriate display on the control console.
- Multiplying relays or solid state logic shall be provided where necessary for duplication of limit, auxiliary or other automatic switch contacts where it is not practical to supply such additional switches.
- Contacts closed in the alarm position shall be suitable for 125 volt DC annunciator and miscellaneous controls

reasonably required for this type and size of installation.

- The input/output circuits of the control system is usually designed to be noise tolerant. The need for shielded wiring, suppression devices or buffering relays shall be eliminated. It is usually permissible to route all field wiring in conduit or trays together with power and control wiring, 120 volts, 60 Hz, without any degrading influence on the control system functions.
- All equipment shall conform to National Electric Code requirements for general purpose areas, in addition to conforming to the requirements indicated in this specification.

Computer cabinet. A control cabinet of minimum size is usually furnished. The cabinet is usually a single unit construction, sized as necessary to contain the programmer unit, processor and all control logic components, power converters, circuit protective devices, controls for solenoid valve actuations, valve position indicating lights, accessories and termination facilities for all purchaser's cables, all factory assembled, wired and terminated as described below.

Standard termination blocks arranged for top entry shall be provided to receive the purchaser's miscellaneous incoming cables from the seller's remote control panel insert and the purchaser's equipment interlocks.

The cabinet should be a rugged, self-supporting structure with suitable means for handling. It shall have adequate ventilation, openings, front and rear access doors, top cable entries, terminal boards (including 10% spares), fittings for prefabricated cables and 10% spare space for adding control logic components.

In the event of an air conditioning failure, and during initial plant starting, the system is required to operate satisfactorily at plant altitude without air conditioning or forced cooling, with ambient temperatures between 35° and 120°F, and, for short periods (24 hours), up to 120°F; at such time, it should also operate satisfactorily at a 90% relative humidity.

The seller should state the expected heat release value in btu's/hour from the system cabinet.

Remote control panel insert. A remote control console of minimum size (20 inches high by 24 inches wide, maximum) shall be provided. It usually includes the necessary control and monitoring devices for dehydrator transfer, depressurizing, purging, heating, soaking, cooling and repressurizing logic sub-systems and power monitoring lights.

Power supplies. Power distribution within each cabinet is arranged such that the loss of power or system fault at an individual sub-system does not impair the reliability of other sub-systems or create an unsafe condition. Loss of power at any sub-system shall activate the purchaser's annunciator.

A contact closure to indicate complete loss of power shall be furnished for the purchaser's use.

All power supplies are sized to operate within their self-cooled rating at 100% load.

The seller is responsible for providing a system which is properly protected from voltage surges that are normally experienced in a petrochemical plant. Should the operation of the system or equipment be affected by any surge, the seller does usually at his own expense, make the necessary corrections to the system or equipment as required.

6.
Computer Controlled Analyzers

Marvin D. Weiss

Senior Control Systems Engineer
Stone & Webster Engineering Corporation
New York, N.Y.

INTRODUCTION

The microcomputer can provide considerable gain when used in connection with a process control analyzer. This chapter will provide descriptions of analyzers available for continuous, on-line control, and outline how the addition of microcomputers can aid the collection of information, the operation of the analyzer and the control of associated chemical processes. We shall describe a typical ethylene manufacturing plant, for the purpose of demonstrating how advanced control, by the aid of microcomputer elements, can improve performance of this typical ethylene plant.

Before the development of on-line analyzers, samples of the chemical condition of process streams had to be collected and brought to the laboratory. The time constant of the analysis was then said to be the time it took a boy on a bicycle to get the sample to the laboratory. Improvements in process control had to await the report on the analysis and subsequent action by the operator to correct on-line composition difficulties.

In the 1940's, during World War II, online analyzers were developed in connection with the Synthetic Rubber Program, but the industrial development of on-line analyzers as a business, and the availability for purchase of on-line analyzers as a commodity, dated from 1955. At that time, there was a concentrated effort to move analyzers from the laboratory to the line. Laboratory instrument manufacturers attempted to adapt their instruments so that they could be operated continuously online. These attempts, for the most part, were unsuccessful. But the operating companies were successful in developing continuous, on-line instruments, using more reliable components and a more intimate knowledge of the grueling operating conditions present in the plant. In this manner, the continuous infrared analyzer, the continuous moisture analyzer, the continuous refractometer and, finally, the continuous process gas chromatograph were developed. The gas chromatograph proved to be an instantaneous success for two reasons. First, it was economically attractive; one analyzer could be used for up to 12 components on 6 streams, the equivalent of 72 single stream infrared analyzers. Second, it was resistant to the usual dirt and scale in process lines; the chromatographic column filtered out the interferences, allowing the pure samples to proceed through the columns to be analyzed. Definitions arose to differentiate between the laboratory instruments and methods, and between the continuous process analyzers and their methods of operations.

DEFINITIONS

The terms we must define differentiate between an *analytical instrument,* an *instrumental method of analysis* and *analysis instrumentation* (this last term was invented by the Instrument Society of America's Division on Analysis Instrumentation to describe the instruments developed specifically as "continuous process analyzers.")

An *analytical instrument* is defined as a "laboratory" instrument that uses some physical, chemical or physicochemical method to determine the presence and concentration of a single molecular component in a gas, liquid or solid sample. This laboratory instrument is designed for intermittent operation by a skilled chemist (or technician) using a procedure such as the following.

The chemist obtains a sample, places it in a proper container which fits into the instrument. After inserting the sample, the chemist twists appropriate knobs and operates appropriate switches until a deflection is obtained upon a meter. The chemist reads the needle deflection of the meter, and by use of a table and/or complicated calculation, he determines the concentration of the component or components of interest in the sample. The laboratory instrument is operated in an air-conditioned laboratory, in a dust-free environment, with controlled temperature and humidity.

Before each test or series of tests, the instrument must be turned on and allowed to warm up; it must be calibrated with standard samples; its zero and linearity must be checked; and a calibration curve must be plotted. Only then is it ready for use on the sample. The sample itself, which has been obtained from a plant stream or from a product or raw material storage tank, has been placed in a sample "bomb" and carefully transported by the legendary "boy on the bicycle" to the laboratory, where it is carefully maintained at proper temperature and pressure, so that parts of it would not evaporate while it awaits its turn for insertion into the instrument for analysis.

After the test, or at the end of the day's operations, the analytical instrument is turned off and stored in an environmentally controlled location. Its electrodes or sample cells are cleaned and stored in an appropriate solution; its mirrors and lenses are cleaned; and the instrument is then restored to perfect working order for the next day's operations.

Instrumental methods of analysis are analytical methods devised to use the analytical instruments in the obtaining of analytical informaton from a specific sample. For example, included in the procedure for an instrumental method of analysis for the analysis of a hydrocarbon sample from an ethylene furnace will be most of the steps given below.

1. *Obtaining the sample.* The following procedure is the usual method used for obtaining the sample from the process stream. For example, a 100 milliliter sample may consist of a collection of 10 milliliter samples collected each hour for 10 hours. Once, each hour, an operator goes to the sample bomb in the plant and opens a connecting valve for a predetermined period that will allow 10 milliliters of sample to enter the sample bomb. The bomb may be cooled with water, dry ice or liquid nitrogen to maintain the sample in liquid phase. The procedure is designed so that a sample representative of the full shift's production is obtained.

2. *Preserving the sample.* A procedure has to be outlined for maintaining the thermodynamic condition of the sample so that the liquid that reaches the laboratory has the same composition as the sample removed from the process stream. Continued treatment with the cooling medium used in Step 1 may be required.

3. *Physical separation.* Upon the sample's arrival in the laboratory, physical methods may be employed to separate undesirable parts of the sample from the desirable parts. For example, filtration may be employed to remove dust, dirt, tar and scale; distillation may be employed to obtain desired temperature cuts of the sample for analysis; and solvents may be used to extract the active components from interfering components in the sample.

4. *Chemical separation.* Chemical methods may be required, such as:

- *Precipitation*. Adding a reactant to precipitate out unwanted or interfering components in the stream. Or, in the case of acetylide, adding a silver salt to precipitate out the silver acetylide concentration in the original sample (an obsolete method now that a flame ionization detector can detect parts per million concentrations of acetylene in hydrocarbons, without precipitation).
- *Catalytic oxidation*. For example, carbon monoxide can be converted to carbon dioxide, so that it may be determined as such.
- *Salting out*. Adding salt to a solution to make the desired ingredient come out of solution.
- *Drying*. Adding a desiccant to the sample to remove the water so that the analysis can be obtained on a "dry basis," etc.

5. *Titration*. The sample is placed in solution and titrated with a reagent of known concentrations, so the amount of desired component can be calculated by equivalence at the end point (or an undesirable component can be titrated out of the solution, to keep it from interfering with an analysis method).

6. *End point detection*. Detection of the end point of a titration may itself employ an analytical instrument such as a conductivity or pH meter, photometry, spectrometry, radio- or microwave frequency absorption, dielectric constant change or detection of radioactive emission or absorption.

The term *analysis instrumentation* was invented at a meeting in 1955, which the author attended, when the IMA (Instrumental Methods of Analysis) Division of the Instrument Society of America changed its name to AID (Analysis Instrumentation Division). The new term was invented to describe the new instruments being developed for *continuous, on-line* analysis of chemical streams in process plants. These instruments were designed to continuously perform all the required operations for an analysis selected from those listed above for laboratory procedures. The analysis instrument thus was a system consisting of some or all of the following parts.

- Sampling system.
- Pretreatment unit.
- Injection unit.
- Analysis unit.
- Read-out unit.
- Memory unit.
- Alarm unit.
- Control unit.

Sampling system. A sketch of a typical sampling system is shown in Figure 6-1. The functions of the sampling system include obtaining a sample from the process, continuously, and conditioning the sample to proper temperature and pressure.

Pretreatment unit. The pretreatment unit performs all the required operations on the sample equivalent to those contained in Steps 2, 3, 4, 5 and 6 in the laboratory analysis: a filter may be included to remove dust, dirt and scale; a coalescer may be used to remove entrapped water mist; etc. Figure 6-2 illustrates a pyrolysis gas sample conditioner.

Injection unit. In order to have a fresh sample continuously near the analyzer, a bypass loop is usually provided through which the sample from the process is brought back to the process at a point of lower pressure. In that loop is included an injection valve which periodically injects a sample into the analyzer. Some analyzers (such as infrared instruments) take continuous samples. In this case, flow through the injection loop is continuous, diverting a smaller flow from the main bypass loop through the analyzer sample cell.

Analysis unit. Details of the analysis unit will be described below under the respective analysis technique.

Read-out unit. The read-out unit replaces the reading of a meter by the chemist in the laboratory and the associated determination of the meaning of that reading. This may be as simple as a properly calibrated

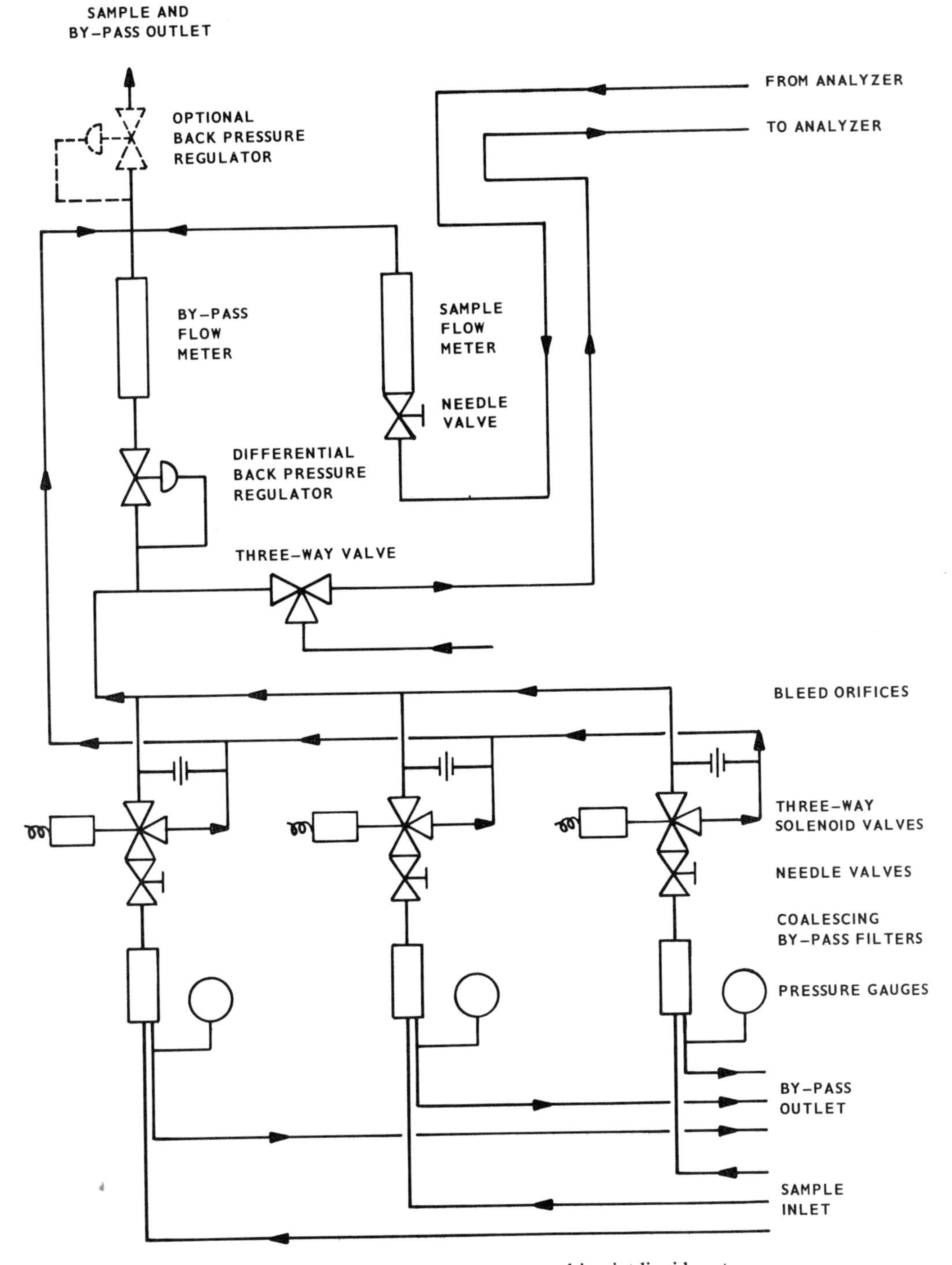

Figure 6-1. Typical sampling system—multi-point liquid system.

recorder which records the carbon dioxide concentration obtained as an electrical signal from an infrared analyzer. Or it may be a minicomputer which extracts a matrix of information from a mass spectrometer and solves a series of simultaneous algebraic equations to obtain the concentration of methane, ethane and propane in a mixed gas stream.

Memory unit. How does one keep track

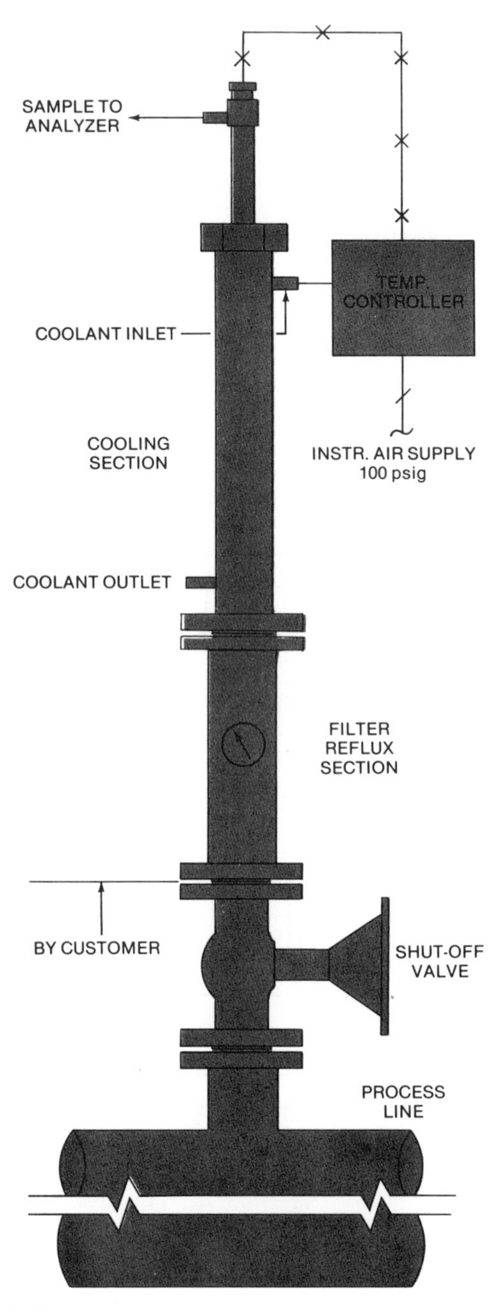

Figure 6-2. Pyrolysis gas sample conditioner. (*Courtesy of Fluid Data, Inc.*)

of the analyses? Figure 6-3 illustrates the bar graph display of a three-component, three-stream chromatograph, in which the data is preserved by pen markings on a recorder chart. Figure 6-4 shows the same data recalled from storage in a computer memory. These are the extremes of the memory unit design.

Alarm unit. The alarm unit is optional, but it serves to alert the operator when unusual conditions arise. For example, an alarm unit is mandatory on the carbon monoxide analysis on the feed to an acetylene hydrogenation reactor, since the presence of too much carbon monoxide may cause disastrous effects in the reactor.

Control unit. There are two types of control units: one control unit controls the analyzer itself; the other type provides a control signal for controlling the process. The first type occurs in every analyzer. In the past, an electronic stepping unit was used to control the analyzer. It is now being designed as a microprocessor. The second type, also in the form of a microprocessor, will be required for the advanced control techniques to be described later in this chapter.

Analysis instrumentation has become a multi-million dollar business because of the economic gains to be made by the incorporation of such instruments into refineries

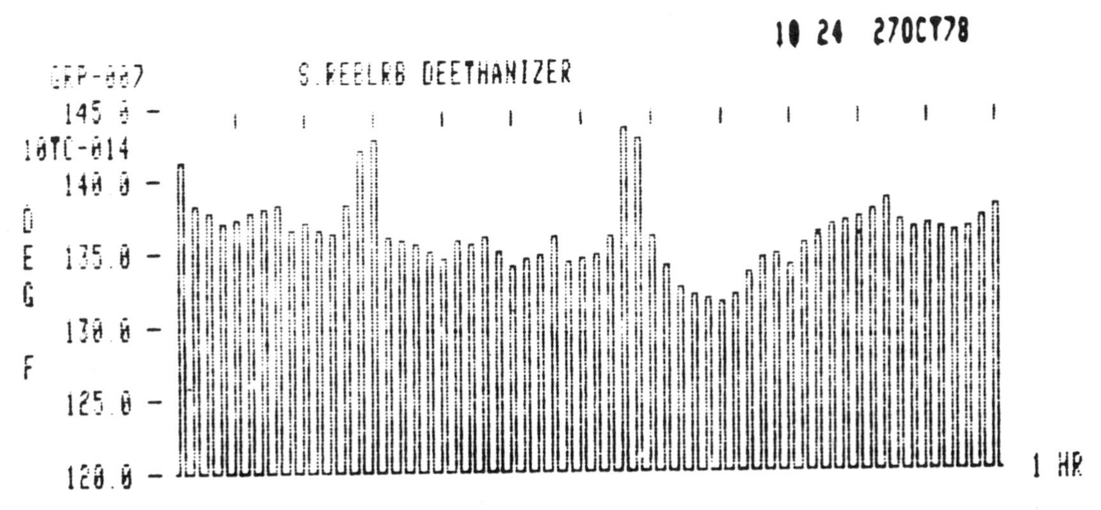

Figure 6-3. Bar graph display.

	18TC-818	18FC-815	18TC-814	18FC-812	18FRC-821	18IC-814	18PC-812	18FC-822
	DEGF	GPM	DEGF	KLB-HR A	KLB-HR B	INCHES	PSIG	GPM
	C2--FEED	REFLUX	TRAY 9	68LB STM	68LB STM	KBLR A.B	DEETH TP	DEPRO FD
SP/Y	98.8	675.8	136.8	78.8	188.8	52.2	388.8	665.8
PV/X	186.2	558.3	138.8	76.1	189.8	52.2	338.2	665.2
OUT %	54.9	94.9	78.4	48.6	63.2	74.8	54.8	72.4
	M	COM	COM	CAS	CAS	L(A)	A	A*

TRAY 9 CAS TO-68LB STM REBLERS

Figure 6-4. Computer display.

and chemical process plants. Connected directly to the process, such instrumentation provides up to date information on the nature of the process stream. A control system can function more efficiently if it operates on the chemical nature of that process rather than on indirectly related temperatures, pressures and flows.

ELECTROMAGNETIC INSTRUMENTATION

Each part of the electromagnetic spectrum shown in Figure 6-5 can be associated with an instrument using that part of the spectrum. In each case, a *source* is required to generate the energy and a sample cell is needed to carry the sample between the source and the detector. A *detector* is required to convert the resultant specific energy to an electrical signal that can be used for measurement or control. A table of typical sources and detectors for useful regions of this spectrum is contained in Figure 6-5.

Analytical Uses of the Electromagnetic Spectrum

Each part of the frequency spectrum has its unique effect upon matter, and thus can be used to provide a useful analytical instrument. Below, typical instruments in each region are described.

The DC Region

In the DC (direct current) region, the field of voltammetry measures the DC current flow between electrodes at different voltage potentials. Each ion has a threshold potential at which it starts to move in a given solution. The concentration of ions affects the current flowing. Hence, a qualitative measurement based on voltage can be coupled with a quantitative measurement based on current flow. One such instrument is the "polarographic" oxygen analyzer (see Figure 6-6).

The partial pressure of oxygen outside the electrode causes oxygen to diffuse through the membrane into the polarographic solution. The electrodes in the solution are maintained at the proper voltage for oxygen. The resultant current flow reflects the concentration of oxygen in the surrounding medium. This electrode is applicable both to liquid and gaseous environments. It is also used to measure oxygen in the blood, using a microelectrode.

Potentiometry

Instruments based on DC measurements include measurements of the oxidation-reduction potential of a solution, useful for detecting the end points of redox catalyzed processes, such as vinyl chloride polymer-

RANGE	FREQUENCIES	SOURCE	DETECTOR
DC	0Hz	Electrodes	Ammeter
AC	50 to 60 Hz	Electrodes	AC ammeter
Sonic	50 to 6000 Hz	Speaker cone	Microphone
Ultrasonic	6000 to 60,000 Hz	Quartz crystal	Piezoeletric crystal
Radio-frequency	50,000 Hz to 160 KHZ	Radio-frequency antenna	Tuned radio-frequency circuit
Microwave	1000 to 300,000 MHZ	Klystron tube	Tuned resonant cavity
Infrared	10^{11} to 10^{14} HZ	Glowing filament	bolometer
Visible	10^{14} to 10^{15} Hz	Light bulb	photocell
Ultraviolet	10^{15} to 10^{17} Hz	UV lamp	UV photo tube
X-ray	10^{16} to 10^{19} Hz	X-ray tube	Photographic plate
Gamma ray	10^{17} to 10^{22}	Radioactive	Scintillation counter
Cosmic ray	10^{22} (and up)	Outer space	Detected by generated emissions in lower regions

Figure 6-5. Electromagnetic sources and detectors.

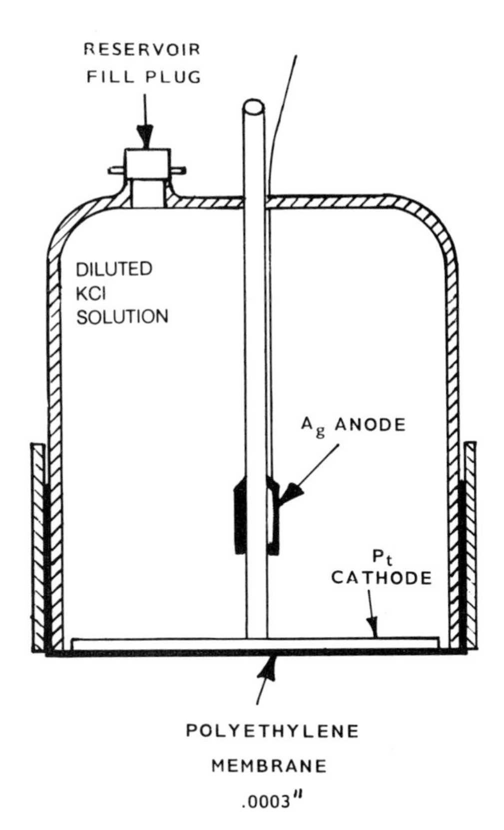

Figure 6-6. Polarograpic oxygen analyzer.

ization. When the potential is the pH, it is a voltage which equals the negative logarithm of the hydrogen ion concentration. Hence, the pH electrode is a potentiometric electrode sensitive to the hydrogen ion concentration of the solution. Recently, *specific ion electrodes* have been developed which are sensitive to specific ions such as calcium, sodium, potassium, sulfate and phosphate.

The AC Region

If an AC (alternating) current is passed between electrodes immersed in a solution containing ions, the AC current flowing is proportional to the concentration of ions in solution. This property makes possible the detection of trace quantities of ions in pure water, such as in boiler feed water. The extensive use of this method has led to the characterization of the quality of water by its *conductivity.*

Sonic Instruments

Sonic and ultrasonic instruments are used to measure the flow of liquids without insertion of an obstructing element in the flow stream. By alternately using two quartz crystals as generator and receiver of sonic signals, and by locating these crystals at an angle to the stream flow, the difference between upstream and downstream velocities can be electronically converted into a measure of the flow of the stream.

Ultrasonics are used increasingly in biomedical instruments for visualizing soft tissues inside the body that X-rays pass through.

Radio Frequency Spectroscopy

When a nucleus (of an atom) is subjected to a magnetic field, its magnetic moment reacts with the field to produce a radio frequency region signal. Radio frequency spectroscopy is used as the detector for NMR (nuclear magnetic resonance), which will be discussed in detail later in this chapter.

Microwave Frequency Spectroscopy

The rotational spectra of gaseous molecules falls in the microwave region. Analyzers using the microwave region have been devised for the detection of moisture in solids. This is useful because the microwave signals will pass (for example) completely through a paper box, measuring the moisture in the solid product inside the box. This method has been used in the measurement of moisture in starch, benzene, aromatic hydrocarbons and other organic materials.

Infrared Spectroscopy

All organic materials have characteristic spectra in the infrared region. The infrared spectrum is one of the most specific properties of a molecule known. Infrared reflects the vibrational frequencies of the molecule; the number of carbon atoms; the geometrical arrangement of the carbon atoms; and the position of other attached atoms, all of which affect the infrared spectrum at specific frequencies. An ingenious instrument that makes use of the specificity of the infrared spectrum of each molecule, yet does not require obtaining the detailed spectrum of the material, was developed in parallel lines during World War II, by Ludwig Luft of I. G. Farben Industries, and in the U.S. at Johns Hopkins, by Professor Pfund. It is called the non-dispersive infrared analyzer.

Figure 6-7 shows the schematic arrangement of the non-dispersive infrared analyzer. There are two parallel optical paths through which infrared energy generated by hot filaments alternately passes. The alternation is provided by a rotating disk with a hole in it, which alternately opens each path to the infrared radiation. One path contains the continuously flowing sample stream; the other contains a sealed sample, consisting of all the components of the process stream except the ingredient to be detected. After the infrared energy passes through each of the two sample cells, it passes into one of the two detector cells, each of which contains the gas to be detected.

Each organic material has a characteristic spectrum in the infrared region. Each bond in the compound makes its own contribution to the spectrum, so that the entire spectral pattern provides a "fingerprint" of the infrared compound. Classical infrared analyzers were dispersive; i.e., they looked at one frequency at a time and plotted the absorption of the compound on an intensity/frequency graph. The non-dispersive analyzer uses the material itself to define the entire spectrum pattern it absorbs.

Thus, in the reference cell, none of the parts of the spectrum characteristic of the infrared absorber to be detected are absorbed. In the sample cell, the amount of this spectrum absorbed is a function of the concentration of the desired ingredient in the process stream.

In the twin detector cells, separated by a sensitive diaphragm, the gas to be detected is heated by the infrared radiation absorbed, and being heated in an enclosed chamber increases its pressure, moving the diaphragm between the cells. The diaphragm between the cells is metal, and is one of the plates of a capacitor microphone. The oscillation amplitude of this microphone is dependent upon the concentration of the desired component in the sample stream. A signal produced by the detector and amplifier can be used to record this concentration, and also to feed a signal into the appropriate control system.

This ingenious technique, of forcing a material to be its own detector, made the infrared analyzer the first workhorse of analysis instrumentation. By using this strategem, interference by dirt, dust or oth-

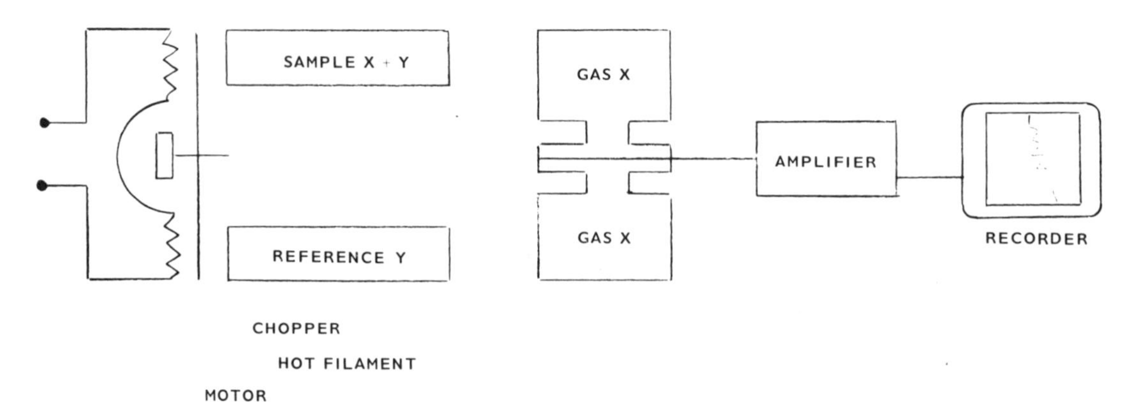

Figure 6-7. Non-dispersive infrared analyzer.

er infrared absorbing materials in the stream was minimized.

Ultraviolet Process Analyzers

Since ultraviolet (UV) radiation does not produce heat, it is sometimes called "cold light." It does however, force some electrons to leave their valence shells, thus causing the atoms to ionize. Ionized atoms of an organic nature often polymerize. Other ions will react with each other, and perhaps even explode. Hence, the artifice used to detect an infrared, non-dispersive signal cannot be used for UV signals. But the non-dispersive principle still works: Figure 6-8 illustrates the typical implementation of a UV non-dispersive analyzer. A UV source projects energy through a sample cell, after an optical filter removes the visible and other undesirable components of the radiation. A quartz sample cell allows the UV radiation to pass through the sample, and a UV phototube, which operates on the ionizing power of the UV energy, generates a current which can be amplified, recorded and calibrated to represent the concentration of the UV absorbing material in the sample. Some materials, such as mercury, absorb UV so well that one part per billion of the toxic vapor in air can be easily detected by this simple principle.

For materials such as benzene, and similar aromatic hydrocarbons, the signal to noise ratio is not as great, and the *dual beam* instrument shown in Figure 6-9 must be used. The UV radiation from the source is collimated by the lens, filtered to only the desired UV frequencies and passed through a splitting mirror, which transmits a fixed portion of the incident energy. The other portion is reflected by the mirror to a second phototube. The main portion passes through the sample cell to a mirror and measuring phototube. The phototubes

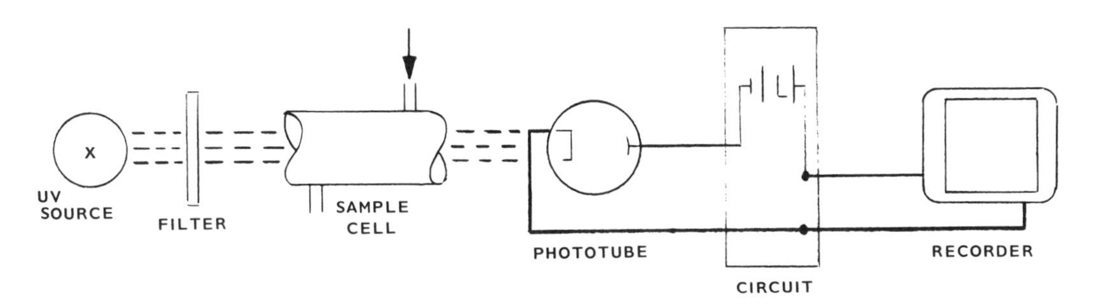

Figure 6-8. Single beam ultraviolet instrument.

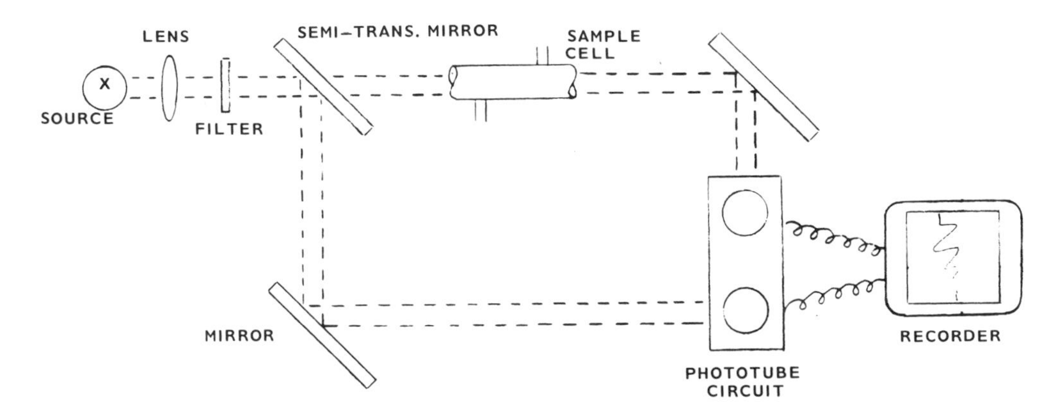

Figure 6-9. Double beam ultraviolet instrument.

are connected into a balanced bridge circuit whose output is the difference between the signals. This difference is proportional to the UV absorber in the sample stream. UV absorbers occur in many critical applications. Butadiene, vinyl chloride, benzene and toluene are all detected by their UV absorption. Sulfur dioxide, in some emissions of stationary power plants, is sometimes detected by UV analyzers.

Optical Transmissometers

With the requirements of the Clean Air Act for clean smokestack emission, microcircuits are coming into major use in the development of *optical transmissometers* for monitoring the *opacity* of smoke coming out of a smokestack.

Figure 6-10 shows the schematics of such an instrument. A single light source, filtered to a wavelength of 0.55 microns, is separated into a measuring beam and a reference beam. The measuring beam is projected across the stack to a reflector unit, which returns the beam back across the path to the detector. The reference beam passes through a preset path, including some quiescent gas from the stack, free of smoke.

The beams are chopped alternately, and both are directed to a single photocell, where successive impulses appear, and are made to match by moving a variable optical density disk in the reference beam. The motion of the optical disk is connected to the recorder pen and indicates the relative transmittance of the sample.

PROCESS CHROMATOGRAPH

The one instrument that has done more to introduce continuous analysis into process plants than any other is the process chromatograph. The electromagnetic analyzer can be programmed to read only a single component in a stream, and the composition of that component must be carefully limited so that the logarithmic signal generated remains appreciably limited within the linear range. The process chromatograph, however, can be used to measure many components in the same stream, and can be programmed to measure many streams, each with different concentrations of the same components.

By selection of appropriate columns and detectors, the process chromatograph can be used to measure parts per million concentrations of impurities in product gases up to sizable concentrations of controlled compositions in process gas mixtures.

The process chromatograph is limited to the analysis of gases or liquids which can be readily vaporized in the instrument; hence, it is often referred to a *gas* chromatograph.

Figure 6-11 is a sketch of the parts of a gas chromatograph. The sampling system is somewhat more complex than that of a simple infrared analyzer, since many streams may be cycled to the analyzer, some of which may drop out liquid components if they are allowed to cool or drop too rapidly in pressure. Heat-traced sample lines are fundamental in such sample systems, as are rapid recycling lines from which a small side-stream is taken to the analyzer.

At the analyzer, the side-stream itself continually passes through a loop of a sample injection valve. During the injection phase, a measured sample of the stream is injected into a chromatographic column. The sample is then carried through the column by a carrier gas, usually helium.

The *chromatographic column* consists of 3 to 10 feet of stainless steel tubing, ¼ inch in diameter, containing a porous solid on the surface of which a liquid has been absorbed. On passing through the column, the sample is dissolved and then evaporated into the column space many thousands of times. This is analogous to a distillation column operating with thousands of plates. The components of the mixed sample are thus separated according to their different affinities for the liquid separating agent and they emerge from the end of the column separated in time.

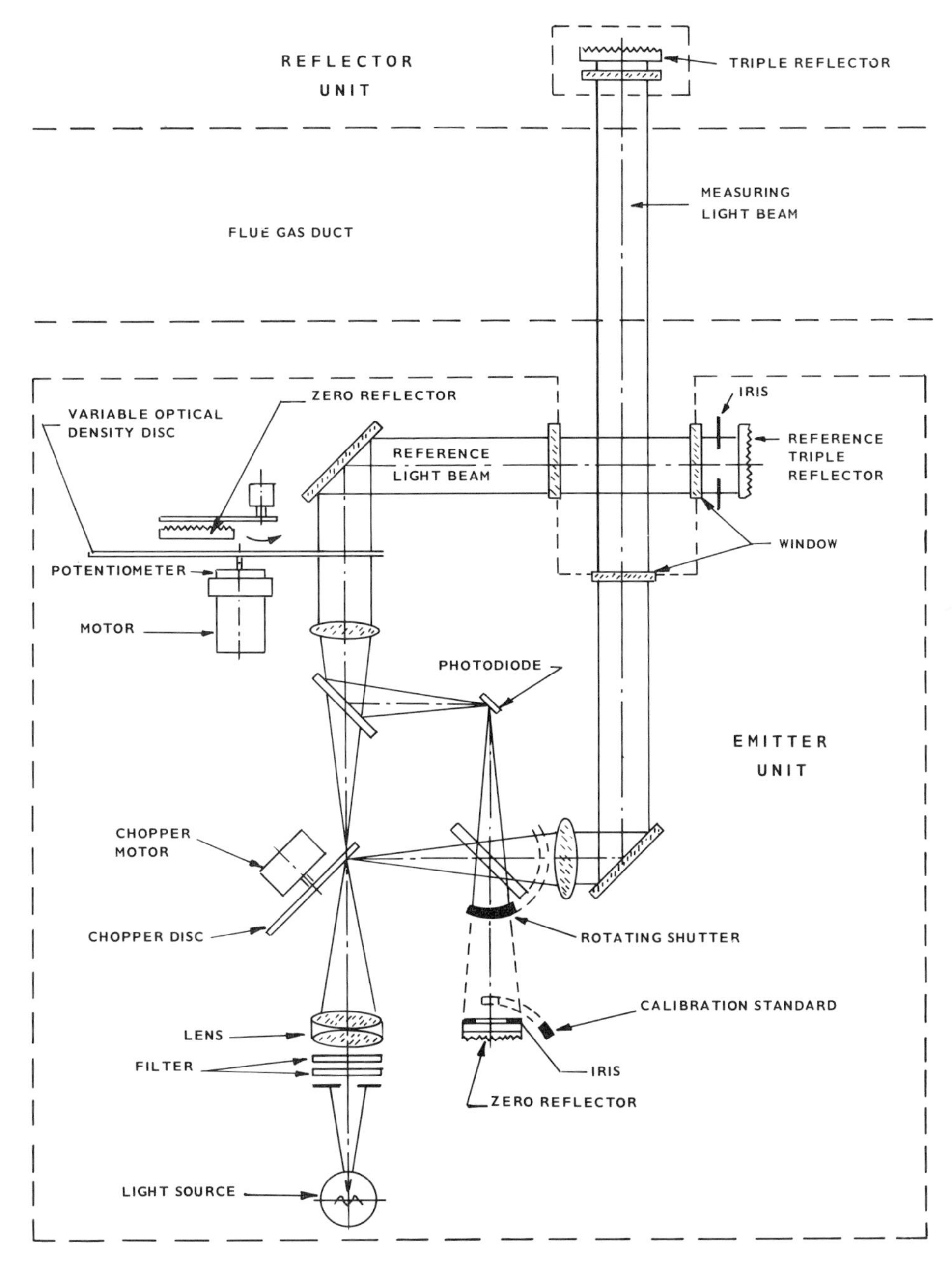

Figure 6-10. Optical transmissometer.

Upon emerging from the column, the sample encounters a *detector* which senses the concentration of each ingredient in the mixture. The location of a component at a particular moment in time defines its nature, while the intensity of the detector signal defines its concentration; thus we have both a qualitative separation and a quantitative detection in the gas chromatographic column.

In the hands of a skilled column technologist who is equipped with such accessories

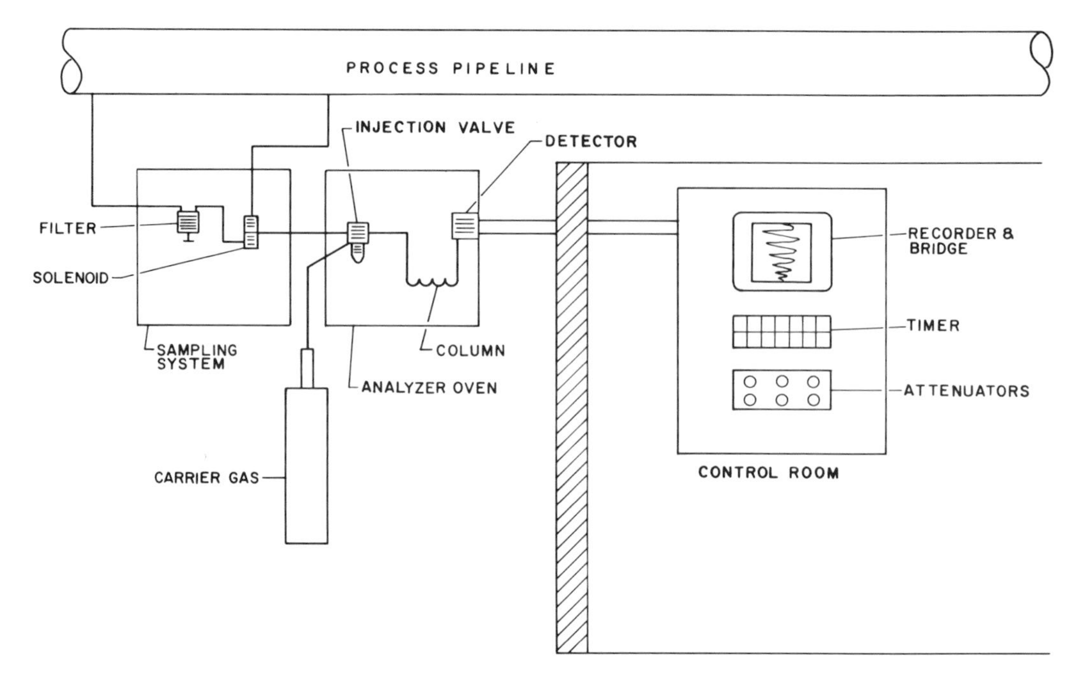

Figure 6-11. Process gas chromatograph.

as column switching valves, dual and triple column systems, dual detectors and back-flushing capability (to get rid of the heavy materials which would otherwise never emerge from the column), all kinds of miraculous analyses can be performed.

Computer Operation

The addition of a computer system to program the gas chromatograph, and to efficiently use the output of the one or more detectors, adds icing to the cake. For example, instead of recording the height of a chromatographic peak as the composition of an ingredient, the computer integrates the area under each curve; multiplies it by its calibration factor (previously determined by the computer); and sums up all the areas under the curve to normalize the total. Then, dividing each area by this total, the computer calculates the accurate calibrated value for each component, corrected not only for the individual calibration of that component, but for the present state of the instrument and for the detector itself (see Figure 6-12).

The computer is programmed to detect the area under peaks which aren't even separated; to detect areas over shifting base lines; and to detect the shifting base lines themselves. Alarms are available when an unexpected change in concentration is detected; when calibration shows an unlikely drift in detector sensitivity; and for a myriad of other undesirable conditions which may arise.

The following is a list of alarms available in a typical computerized gas chromatograph.

- Instrument too far out of calibration.
- A shift in elution time of a peak.
- Inordinate base line drift.
- Malfunction of an analyzer.
- Power failure and automatic restart.
- Watchdog timer unit.
- Limit check for each component.
- No information transfer to computer.
- Alarm when an expected peak fails to appear.

A single computer can operate and monitor all of the chromatographs in a typical plant. The additional hardware required consists of an interface unit at each analyz-

	PEAK HEIGHT		AREA		ACTUAL %
BDE	5.60	39.27%	15.96	41.55%	42.00
Butene 1	4.96	34.78%	13.02	33.90%	34.00
Butene 2	3.70	25.94%	9.43	24.55%	25.00
	14.26		38.41		

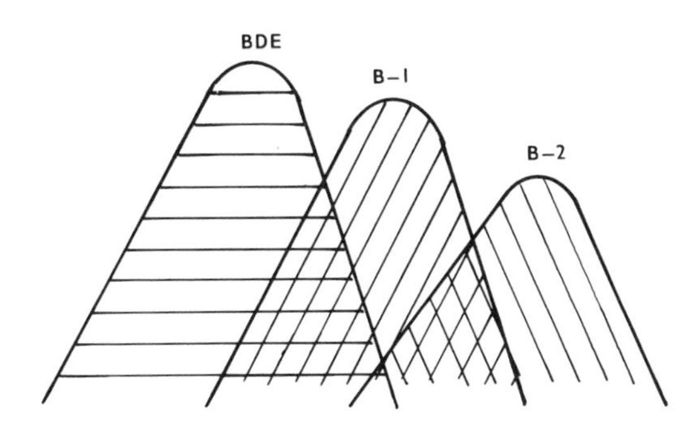

Figure 6-12. Peak height versus area calibration.

er, and an interface unit at the computer which scans all the analyzer interface units and converts and stores the signals for use as required by the computer. (A general block diagram of the system is shown in Figure 6-13.)

MASS SPECTROMETER

At the same time that the process chromatograph was being perfected in the U.S. in the 1960's, various instrument development groups in the process industries were trying to perfect the mass spectrometer as a continuous process monitor. Mass spectrometers had been used in the Atomic Energy Program to measure the separation of uranium isotopes, and it was hoped a similar instrument could be used in process streams. But when used in process streams, the very fine capillary required to lead the process sample into the high vacuum of the ionization chamber became obstructed by the dirt and dust in the process sample, and the initial development of the mass spectrometer for use in process environment was abandoned.

In 1978, several continuous process mass spectrometers were introduced to industry, and stories of their successful use appeared in the literature. One of the keys to the modern use of the mass spectrometer is the microcomputer, but other features of the new machines are that a very high vacuum is not required, that durable ionization chambers are now available and that techniques like quadrupole mass spectrometry are well proven.

How the Mass Spectrometer Works

In a mass spectrometer, charged particles of different atoms are separated by their mass to charge ratios (m/e). Separation has to take place in a high vacuum to eliminate collision with other molecules. A small sample (see Figure 6-14) is admitted through an inlet leak, and ionized under reduced pressure by an electron beam. The charged particles pass through a magnetic field, where the path of each is determined by its m/e ratio. Collectors are spaced along the target area to collect specific ions and pass the charges to electrometers, where

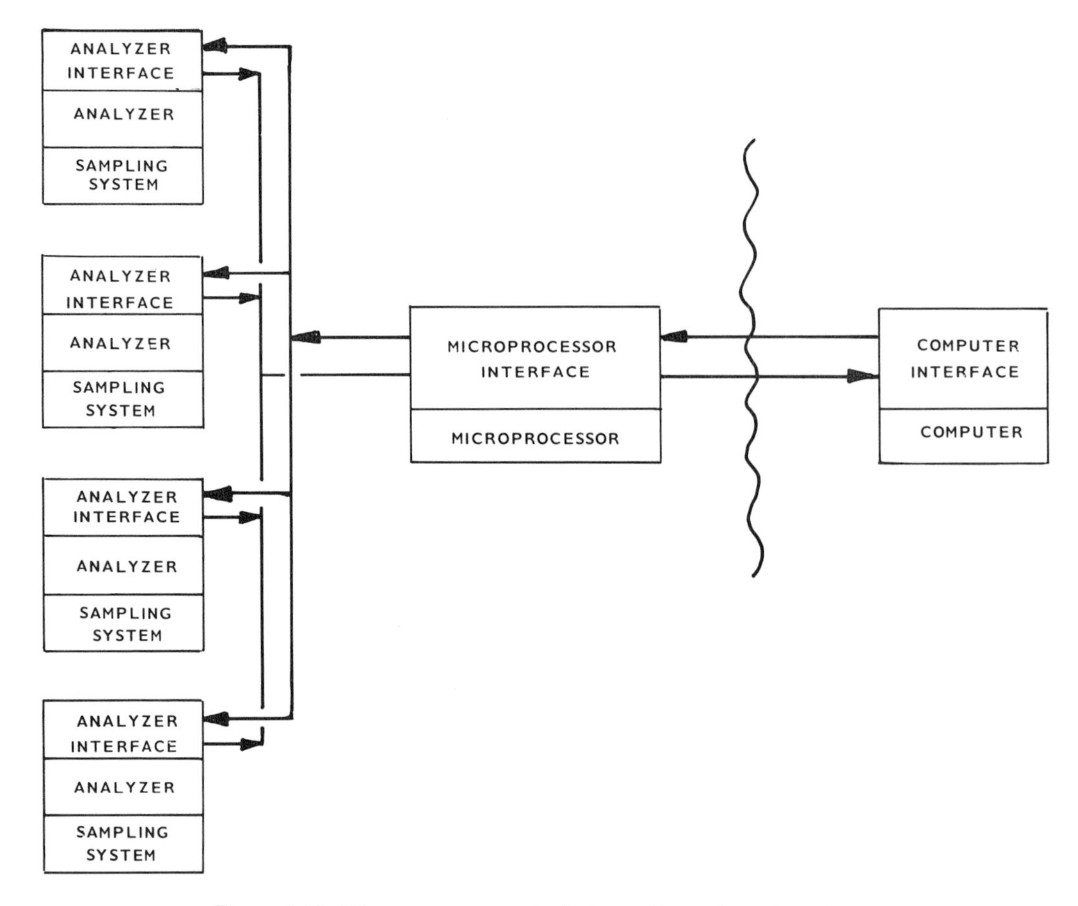

Figure 6-13. Microprocessor controlled gas chromatograph system.

they are amplified into voltages proportional to the composition of the resolved species. A patented closed-loop control system compensates for any change in sensitivity by forcing the sum of the readings to be equal to 100%. The output signals are available as analog outputs or as digital display.

The use of a microcomputer controller makes the system self-calibrating and able to operate without human attention other than that required for the selection of the desired panel display.

Ion Source Operation

To provide continuous operation, a rugged, dual filament, non-magnetic ion source is provided (similar to the filament developed for orbiting satellite instruments), as shown in Figure 6-15. Filaments of tungsten-rhodium are 5 mil in diameter, with estimated lifetimes of more than 20,000 hours. Electrons emitted from the filament are focused on the gas sample input area, ionizing the particles accelerated toward the analyzer region with an energy of about 500 electron-volts.

Sampling System

The new mass spectrometers can use one of two primary inlet systems built into their sampling systems.

1. The capillary bypass (Figure 6-16) inlet system, which requires a sample of 1 cm^3/sec. (The sample need not be pressurized; flow is maintained by a pump within the analyzer; and systems are available with one, two or four inlets, controlled by a manual, automatic, or digital programming sequencer.)

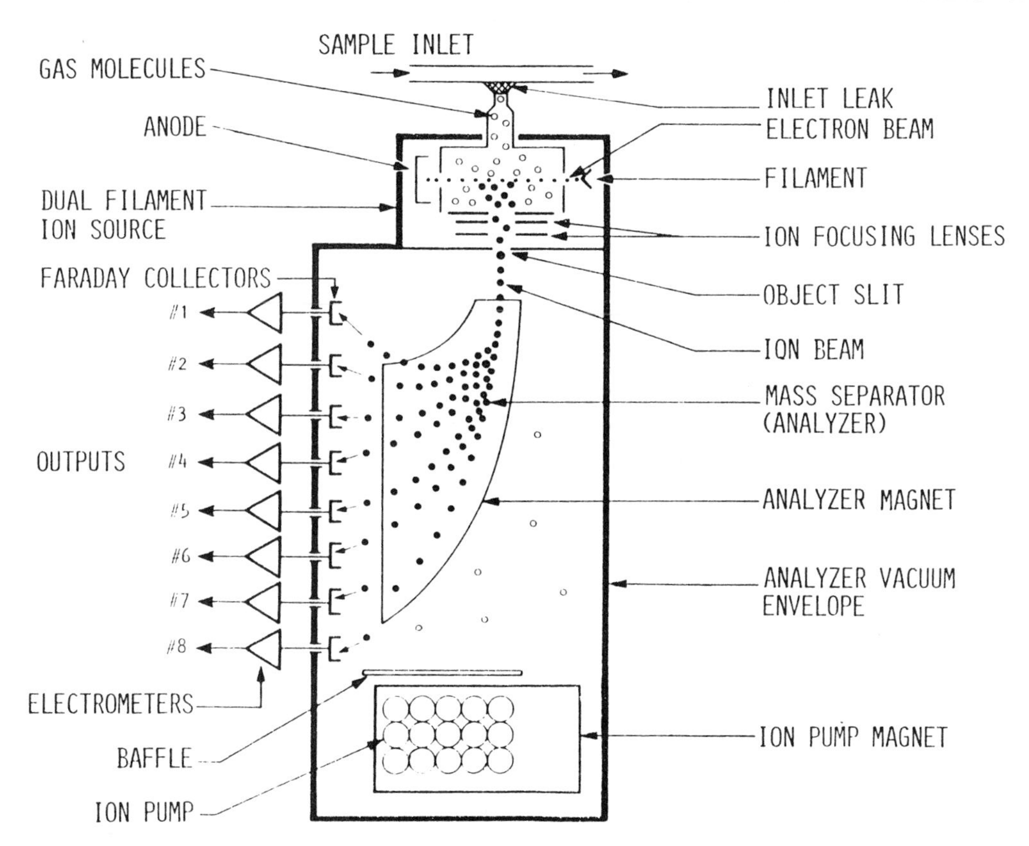

Figure 6-14. Mass spectrometer operation. (*Courtesy of Perkin-Elmer Corporation.*) The inlet sample is bombarded by an electron beam to form ions. The magnetic field causes the ions to follow individual paths based on their respective m/e ratios. Targets are located at appropriate positions to collect specifications, which are converted by the electrometer into corresponding voltages. The summing network forms part of a closed-loop control system which automatically compensates for any change in the system amplification factor. Each individual ion species provides both an analog output and a digital display, manually selectable in the Model MGA-1200 Instrument depicted here.

2. The flow-by inlet system (Figure 6-17), which requires a sample of sufficient pressure to flow by the inlet and return back to the process at a lower pressure point. (The sample is sucked into the analyzer by an internal pump, and proper sequencing is maintained by optional equipment such as manual selection, automatic sequencing or a computer addressable selector switch.)

Applications

The mass spectrometer is recommended for gases with m/e ranges from 2 to 135, in concentration ranges of 0 to 1, 0 to 2, 0 to 10, 0 to 20 and 0 to 100%. Gases such as nitrogen, oxygen and carbon dioxide are readily measured (see the list of gases in Figure 6-18). In a test run made by Dow Chemical, carbon balances were maintained over nine months between 99.5 and 100.5%, and deviated on only four occasions from between 99.7 and 100.3%. The previous use of a gas chromatograph and oxygen analyzer for the same application provided a range of 98 to 102%, with three calibrations per month required for the chromatograph, and weekly calibrations performed routinely for the oxygen analyzers. The mass spectrometer thus appears to be much more stable and easier to maintain than an equivalent gas chromatograph system.

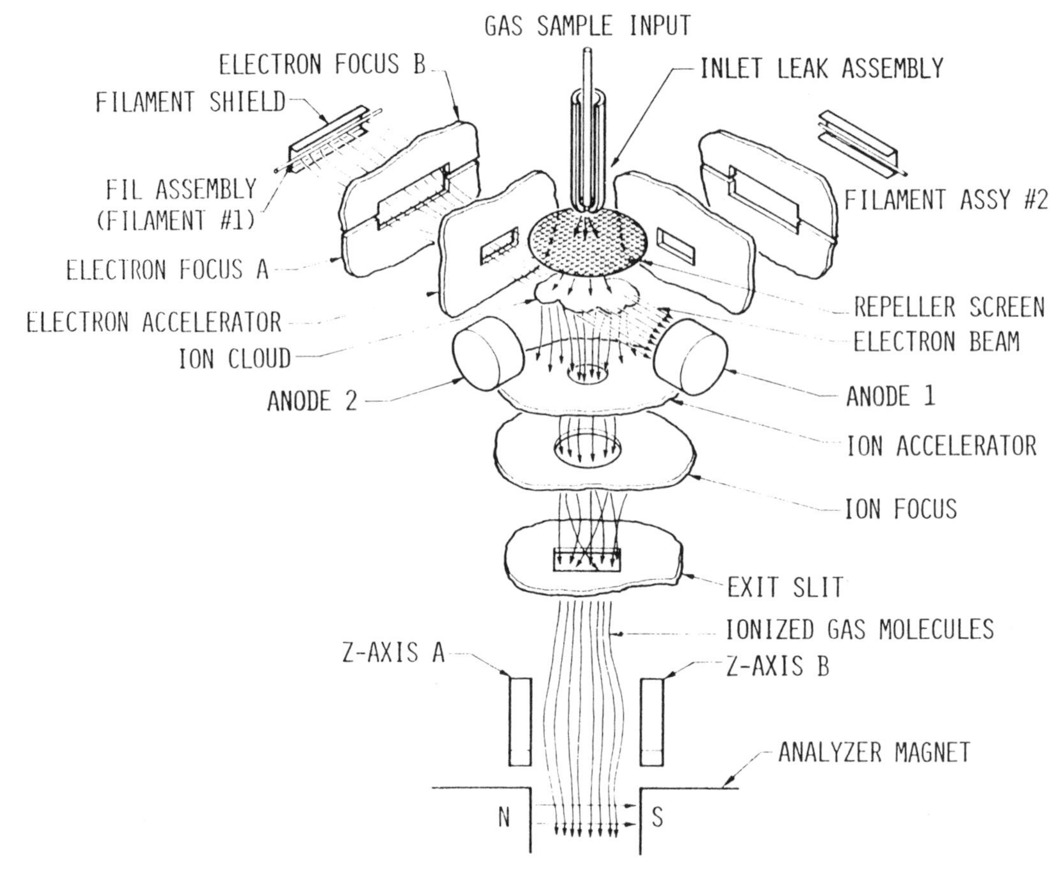

Figure 6-15. Ion source Operation. (*Courtesy of Perkin-Elmer Corporation.*) Electrons are emitted from the filament and focused on the gas sample input area. Ions thus formed are accelerated toward the analyzer region with about 500 electron-volts. They are focused by an electronic lens into the magnetic field, where the ionized molecules are separated by their m/e ratios.

The Microcomputer and Mass Spectrometry

In order to exploit the high speed of the mass spectrometer, a process computer is required. With the advent of the microprocessor as an optimizing controller for fractionation towers, the use of the high-speed mass spectrometer—analyzing the feed, overhead and bottoms streams—would provide the means for feedforward control, which some sources say will improve tower efficiencies by 5 to 20%. Appreciable energy savings are also available if high-speed analysis is coupled with material and heat balance control, rather than the old-fashioned, fixed temperature/fixed pressure control.

Other Applications for Mass Spectrometry

The mass spectrometer can do in one analyzer what is now performed by combined gas chromatograph, oxygen and infrared analyzers in ethylene cracking furnace applications, ethylene oxide production, vinyl chloride processes, natural gas production, digestive sludge processes and in-situ gasification control. In the latter process, the yield of combustibles can be maximized by controlling the amount of oxygen in the fire front. Too much oxygen reduces the yield; too little oxygen allows the fire to go out. The quick response of the mass spectrometer provides the capability of control. In large ethylene plants, a number of former

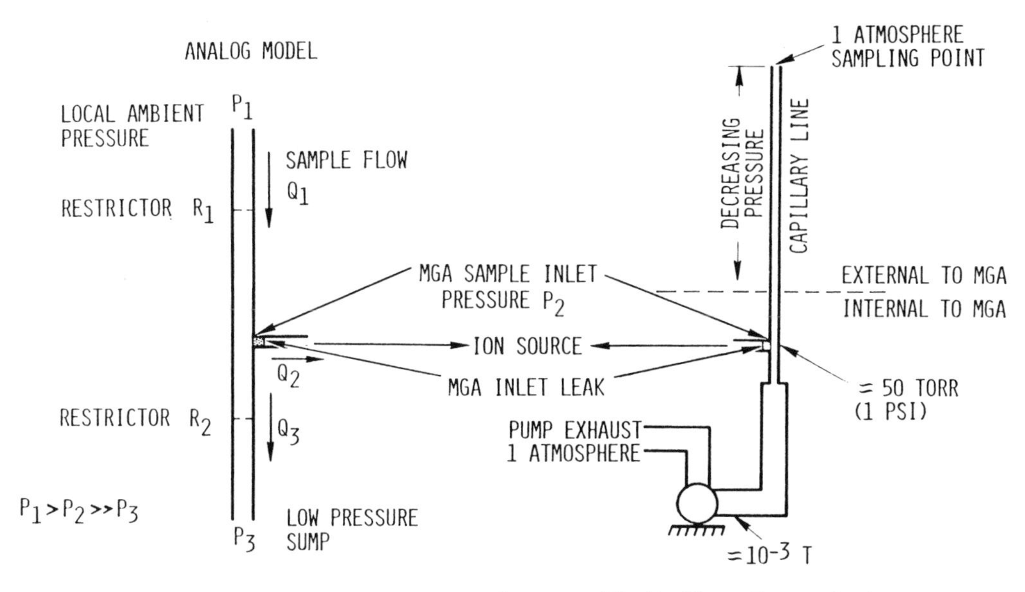

Figure 6-16. Capillary inlet system. (*Courtesy of Perkin-Elmer Corporation.*)

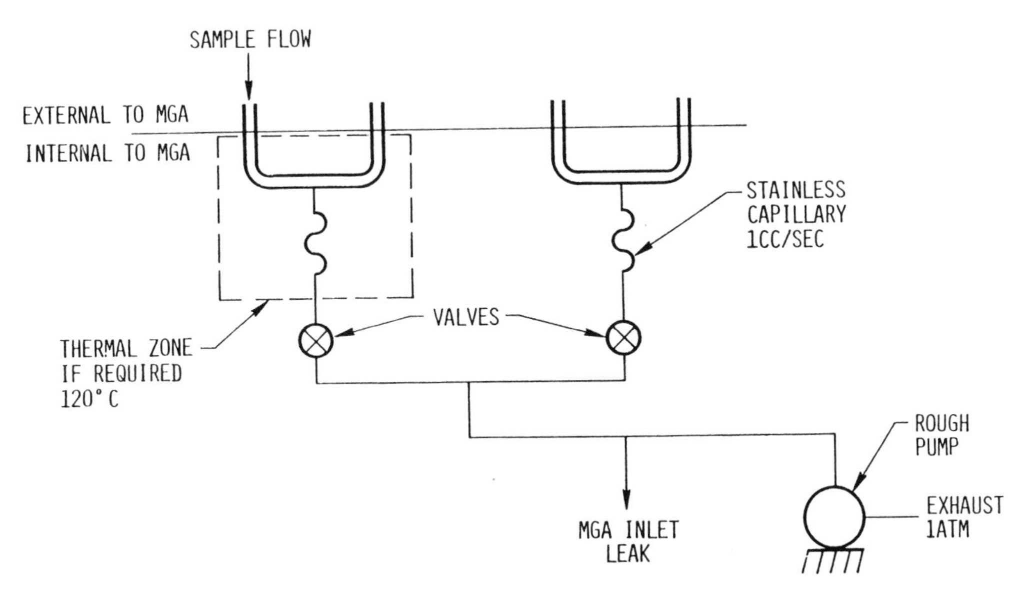

Figure 6-17. Flow-by inlet system. (*Courtesy of Perkin-Elmer Corporation.*) Pressurized samples provide the head for the master flow of the flow-by inlet system. Three-way valves provide continuous purge of transport lines during non-measurement periods, and a single sample is switched to the analyzer flow-by from which an inlet leak is obtained through a sintered metal disc. A programmer unit is available to automatically switch the three-way valves as needed.

Hydrogen	Methane	Acetylene
Oxygen	Ethane	Ethylene dichloride
Water	Propane	Vinyl chloride
Nitrogen	Dimethylether	Carbon dioxide
Ethylene	Helium	Propylene
Neon	Ethylene oxide	Argon
Propylene oxide	Sulfur dioxide	Krypton
Nitrous oxide	Sulfur hexaflouride	Xenon
Nitrogen dioxide	Hydrogen cyanide	Carbon monoxide
Methyl alcohol	Chlorine	Ethyl alcohol
Hydrogen chloride	Acetone	Hydrogen sulfide

Figure 6-18. Typical compounds measured.

gas chromatograph functions at the furnace and at downstream units (up to eight at present) can be combined on one mass spectrometer system.

Gas Chromatograph Versus Mass Spectrometer

The new mass spectrometer cannot replace the primary role of the gas chromatograph in trace analysis. Much of the monitoring of plant streams is in loss control or, for example, the monitoring of by-product streams for ethylene, in the ethylene plants. The process mass spectrometer discussed here does not have the sensitivity to monitor parts per million losses in side-streams.

However, when it comes to high-speed control, the mass spectrometer can provide a new high-speed facility to replace the 3 to 20 minute analysis time required by the gas chromatograph and even the 30 to 120 second analysis time available in the high-speed model of the gas chromatograph. This is especially valuable in conjunction with the current proposals to use microcomputers as distributed local controllers of fractionating columns and their unit operations. The mass spectrometer can provide high-speed access to the compositions of key streams needed for the material balance control, feedforward control and column optimization which the microcomputer now makes possible.

NUCLEAR MAGNETIC RESONANCE

To understand nuclear magnetic resonance (NMR), imagine a nucleus as a spinning sphere with a magnetic moment, placed in a uniform magnetic field. The field exerts a torque upon the sphere which causes it to precess around the direction of the fixed magnetic field, absorbing the amount of energy required to raise the nuclear spin to the next quantum level. This absorbed energy is on the order of 10 MHz for a field of a few thousand gauss.

There are two fundamental methods of measuring this small absorption of electromagnetic radiation at the resonant frequency. In one method, the sample is placed in the field of a radio-frequency coil, whose Q is changed by the absorption of the sample. The second method utilizes nuclear induction, in which the absorption by the sample causes an electromagnetic force to be induced in a coil placed at right angles to that supplying the magnetic field.

The NMR spectrum can be scanned either by making a small change in the applied magnetic field, or by changing the frequency of the radio-frequency oscillator. The former is common practice.

NMR Spectrometer

Figure 6-19 illustrates the operation of an NMR spectrometer. The sample is placed between the pole faces of a DC electromagnet whose field can be varied from 0 to 23,000 gauss. A low-power crystal controlled oscillator provides a fixed field in a direction perpendicular to the magnetic field. A few turns of wire around the sample tube will pick up the resonant frequency signal from the sample. The detector coil is perpendicular to the other two fields, to minimize pick-up from these fields. Ab-

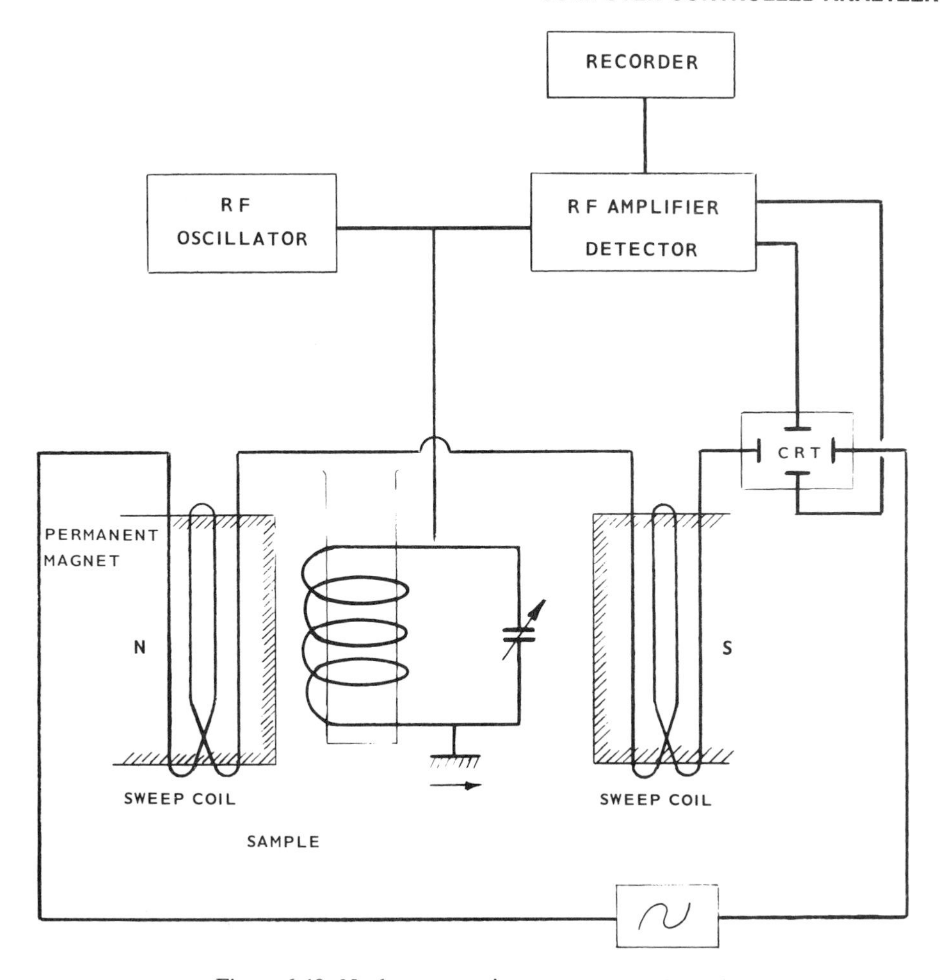

Figure 6-19. Nuclear magnetic resonance spectrometer.

sorption of energy causes the radio-frequency voltage across the receiver coil to drop. This drop in voltage is amplified and detected by a tuned radio-frequency amplifier. The resultant signal is placed on the vertical plates of an oscilloscope and displayed as intensity as a function of frequency, which is the NMR spectrum.

NMR is used mainly for the proton, the hydrogen ion. The NMR spectrum of a proton changes with the atoms attached to the hydrogen. Thus, a proton in water shows a different characteristic absorption than a proton attached to a primary saturated carbon, or a proton attached to a secondary carbon (e.g., the spectrum for ethyl alcohol shows differences between the three types of protons, depending upon the carbon or hydrogen to which each is bonded).

THE USE OF COMPUTERIZED ANALYZERS IN PROCESS CONTROL OF AN OLEFINS PLANT

In order to illustrate how computerized analyzers function in process control, we shall describe the analyzer functions in a typical olefins plant, and then show how advanced control functions can be made available by the use of microcomputer modules attached to the analyzers to improve control of the plant. In an olefins plant (Figure 6-20), the feedstock is passed through a furnace, where it is thermally cracked to ethylene, propylene, methane

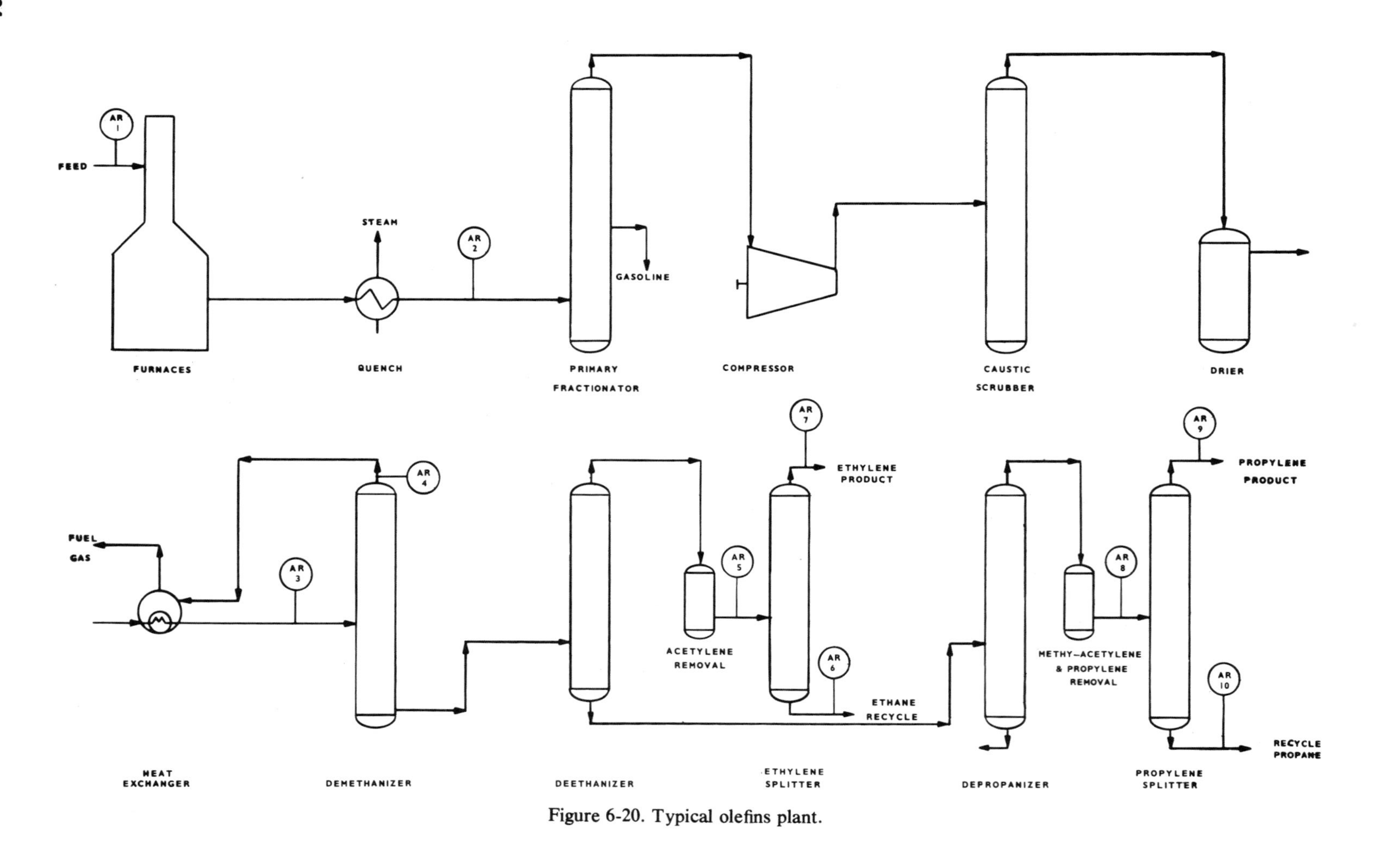

Figure 6-20. Typical olefins plant.

and other hydrocarbons, depending on the feedstock, the cracking temperatures and pressures (severity) and the rate of passage of the feedstock through the furnace. Typical feedstocks are ethane, naphtha, gas-oil and other hydrocarbon fractions from a refinery or gas separation plant.

The purpose of the rest of the plant is to upgrade the ethylene concentration from about 30% at the furnace to 99.95% in the product stream. (See table below for other specifications of ethylene.)

TYPICAL ETHYLENE SPECIFICATIONS

Ethylene	99.95% weight
Methane	less than 500 ppm mol. %
Ethane	less than 500 ppm mol. %
Propylene (and heavier)	less than 100 ppm mol. %
Acetylene	less than 5 ppm mol. %
Carbon dioxide	less than 10 ppm mol. %
Total sulfur	less than 5 ppm mol. %
Hydrogen sulfide	less than 1 ppm mol. %
Water	less than 15 ppm mol. %
Oxygen	less than 5 ppm mol. %
Hydrogen	less than 1 ppm mol. %
Carbon monoxide	less than 5 ppm mol. %

To meet these specifications, the typical olefins plant process includes the operations shown on Figure 6-20, which can be described as follows.

After leaving the furnace, the hot gas is quenched with steam, then passed into a refinery-type column called the primary fractionator. Here, the heavy oils are removed and the light ends passed into the de-methanizer. Prior to passage into the de-methanizer, the hot gas from the fractionator is cooled by many heat exchanges with cooler gases from the downstream plant. It is also allowed to expand in a turbine, which cools it further. It is washed with caustic to remove the carbon dioxide in it, then dried to remove the water. In the de-methanizer, the methane and hydrogen are removed to be used as fuel gas for the furnace. The bottoms from the de-methanizer go to the de-ethanizer, where ethane and ethylene are removed in the overhead and propane and propylene are removed in the bottoms. The overhead goes to the ethylene column, where pure ethylene is obtained, and ethane is recycled to the furnace. The bottoms of the de-ethanizer go to the de-propanizer for the removal of the C3's. The C3's are processed through the propylene splitter, where propylene product is produced and propane is recycled to the furnaces. The heavy ends from the bottom of the de-propanizer may be further processed in a de-butanizer, to produce C4 products.

CONTROL MODES FOR COMPUTERIZED ANALYZERS

The following control functions can be accomplished by adding computer capability to process analyzers.

- Composition monitoring.
- Material balance control.
- Loss control.
- Feedback control.
- Feedforward control.
- Adaptive control.
- Efficiency monitoring.
- Optimization techniques.

Composition Monitoring

In a typical olefins plant, the following compositions are monitored (numbers correspond to the numbers on Figure 6-20).

1. Furnace feed: methane, ethane, propane and C_4 (and heavier).
2. Primary fractionator feed: same as above plus ethylene and propylene.
3. De-methanizer feed: carbon dioxide, carbon monoxide, moisture analyzer.
4. De-methanizer overhead: ethylene and ethane.
5. Acetylene removal effluent: acetylene.
6. Recycled ethane: ethylene, propylene and propane.
7. Ethylene product: methane, acetylene, ethane, carbon monoxide and carbon dioxide.

8. Propylene splitter feed: methyl acetylene and propadiene.
9. Propylene product: ethylene, ethane, methyl acetylene, propadiene butadiene and total C_4 (and heavier).
10. Recycled propane: ethylene and propylene.

These data are logged for the benefit of the operator who can adjust his plant if any of the impurities of the stream approach the high side. Automatic alarms are provided when any of the compositions approach the point where they should be brought to the operators' attention. Different levels of alarms and shut-downs are provided, depending upon how critical the infracton may be.

In the composition monitoring mode, the plant is drifting from the most efficient operation to the least efficient operation, and so long as the product and the critical streams remain within specification, no action is taken to improve conditions.

Material Balance Control

With the increased costs of energy these days, it takes little improvement to obtain payout from the next step, *continuous material balance control.*

Material balance control of a unit in the plant (e.g., a column) is obtained by controlling the material removed from the column based on the material fed into the column. With mere environmental measurements of temperatures and pressures, material balance control would not be feasible. If the analysis and the rate of flow of all the streams entering and leaving a column are known, it is possible to continually balance the draw-off to match the rate of feed of each component entering the column, and to avoid oscillations and accumulations of material in the column which would waste energy and thus increase the unit cost of the products moving through the column.

For example, in an ethylene tower, in which we are extracting ethylene from ethane, if our specification reads 100 parts per million ethane in the product, and we are obtaining 10 parts per million, we are using excessive energy to overpurify the product, and by increasing the reflux rate to do this, the energy required to boil up the large additional reflux is further amplified.

The *simple material balance control* of an ethylene column consists of adding analyzers to the feed, overhead and bottoms. The ethylene and ethane concentrations in the feed are sent to the computer, as are the ethane concentration in the overhead and the ethylene concentration in the bottoms. The computer then performs the following tasks.

1. It multiplies feed rate by concentration of ethylene in feed.
2. It multiplies feed rate by concentration of ethane in feed.
3. Using a model of the column (loading curve), the computer estimates the optimum rate of overhead flow, bottoms flow, reflux rate, heat rate to the re-boiler and cooling fluid rate to the overhead condenser.
4. With a dynamic model of the column, the computer gradually changes the rates of the above (Task 3) at a rate consistent with the time constants of the column itself.

Estimates of 5 to 20% of the energy requirements of the column have been saved by the use of material balance control. Savings may amount to 5 million BTU/hour or about $100,000/year. The addition of a microcomputer doing continuous material balance control at a capital cost equal to or less than a one year payout appears to be highly favorable.

Continuous Loss Control

Another refinement can be added to the material balance controller without very much increase in capital cost. Continuous loss control monitors the losses in the effluent streams from a column and continuously minimizes those losses. A critical survey of the analyzers in a standard ethylene plant will soon reveal that many analyzers are monitoring the streams for losses; e.g., we are looking at ethane in the ethylene column's overhead, and ethylene in the column's bottoms, both of which constitute losses. Closing the control loop merely means using a feedback signal from the loss measurement to reset the appropriate control loop in the previously defined hierarchy. That is, we might increase the heat to the re-boiler slightly if we find too much ethylene loss in the bottoms of the column. A model of the composition dynamics of the column is used to define the manner in which this loss control will best be utilized.

Feedback Control

The analyzer, with the help of its microcomputer controller, can enter into feedback control of a unit operation. One area of great return is the furnace area of the ethylene plant. For example, the analysis of ethylene in the output of the furnace could be fed back to reset the set point of the coil outlet temperature of the furnace, so that reduced ethylene concentration could be compensated for by increasing the furnace temperature. In actual practice, the feedback is made a computer model of the furnace process, which adjusts several variables to maintain the ethylene output of the furnace.

Feedforward Control

When a change takes place in the feed to a column or a furnace in a plant, the control system of that unit responds slowly and too much to that change. Thus, a certain amount of oscillation of the unit takes place before the optimum response is obtained. To avoid the lost energy, production, and product quality brought about by this oscillation, feedforward control is used.

Feedforward control consists of measuring the inputs to a unit and predicting the time pattern of the output corresponding to the change in that input. The control system then follows the innate time lags of the unit to adjust the controls in the unit to absorb the change without oscillation.

A classical example is the feedforward control of a distillation column. For example, let us discuss an ethylene column, whose feed is primarily a mixture of ethylene and ethane. Ethylene is removed overhead, and ethane is brought out at the bottoms. In feedforward control, an analyzer is placed in the feed line, monitoring the ethane and ethylene in the feed. When the analyzer detects an increase in ethylene, the computer system gradually increases the overhead product flow to remove the increased ethylene; gradually increases the overhead reflux to correspond with the new ethylene loading; gradually changes the reboiler energy to correspond with the new column loading; and gradually decreases the bottoms product withdrawal to correspond to the reduction in ethane in the feed.

Adaptive Control

Adaptive control is a mode of control which is based upon an attempt to combine the adaptive capabilities of a human operator with the continuous on-line operation of a computer controlled system. It is the way the ideal operator would control the plant if he were fast enough and knew enough about the system. When such an operator controls a process, he provides not only the feedback and feedforward loops but an additional quality of "judgment," with which he can criticize his own success in maintaining control and correct his control measures to adapt to the changing situation.

What adaptive control adds to the control loop is a *critic;* the computer provides the means of adaption. The classical change to be made in a proportional, integral, derivative (PID) controller in a simple feedback loop is to change one or more of the following.

- The proportional band.
- The reset rate.
- The derivative rate.

A simple analogy of adaptive control in a jacketed batch reactor is to use a cascade temperature control. The batch reactor is operated by:

1. Feeding the reactants into the vessels.
2. Increasing the temperature until the reaction starts.
3. Decreasing the temperature of the jacket to remove the exothermic heat generated by the reaction.
4. Holding the temperature at a critical level to ensure completion of the reaction and formation of the selected product.

In this system, the *critic* is the temperature in the reactor; the slave control is the temperature in the jacket. During the course of the batch, it is necessary to manipulate the temperature in the jacket to obtain (at each stage) a desired temperature in the reactor. As repeated batches are made in the reactor, the walls of the reactor are coated by material, reducing the heat transfer efficiency through those walls. Hence, for each condition of each batch, a different temperature in the jacket is required. The cascaded system reacts to this changing condition, and tries to set the jacket temperature to the appropriate value to obtain the required temperature in the vessel, but the standard analog control system will over-react, causing oscillations in the control systems. If the temperature in the reactor is too low, as during the first phase of operation, the master temperature controller will increase the set point of the jacket heating system, probably to a too-high point, before reaction starts, because it takes time for the heat in the jacket to get through the coated walls of the reactor. By the time the reactor temperature sensor feels the increased heat, the reaction will have started, perhaps to explosive velocity, and the control system will have failed.

Computer control has built into it the logic of each step, as well as a time model of the reactor system (which automatically modifies itself, or adapts to, the changes in the heat transfer coefficient of the reactor wall). Whereas, after the first near explosion, the operator placed the conventional cascade controller on manual and changed the temperature set points himself, the new computer controlled batch system adapts to changing conditions just as the operator would have on his best day, and with optimum knowledge of the dynamics of the system he is trying to control.

Adaptive Control of Product

The above example illustrates the concept of adaptive control. A further refinement is to use an analyzer in the batch to watch the formation of products and to "adapt" the control system to force the formation of desired products and inhibit the formation of undesired products. Such adaptive techniques pay off in polymerization reactions where a molecular weight distribution of polymerized product is required, and where such can be obtained based on a chemical-kinetic model of the process, and adaptation of the dynamics of the control system.

Efficiency Monitoring

Efficiency monitoring can be based on any of several definitions of efficiency. Is efficiency measured in terms of energy usage, conversion or yield of a reactor, or quality of separation of a distillation column? In each case, once the "figure of merit" is decided upon, the computer can collect the data to measure that quantity, and report the efficiency of each unit operation of the process in the plant. Often, the analyzer is required to provide the composition data required to measure the efficiency (e.g., an ethylene analysis is required to calculate the energy usage per pound of ethylene produced by a furnace, or of a separating column, or of an entire plant).

Optimization Techniques

Optimization is a mathematical technique which incorporates a control scheme that forces a system to operate at its optimum possible state. In a chemical plant this means one or more of the following.

- Produce the maximum amount of product per pound of raw material.
- Produce the optimum quality product.
- Make the maximum use of equipment available.
- Use the minimum amount of energy.
- Provide the maximum profit per pound of product.

When using a computer for this technique, optimization can be either off-line or on-line, differentiated as indicated below.

Off-Line Optimization

- Data is continuously collected and stored in a local computer.
- Data is brought to the optimizer computer by data link, or by feeding in cards.
- The optimization calculation is made in a larger computer, periodically.
- The off-line computer lists changes in process set points to be made by the operator.

On-Line Optimization

- Data is continuously collected, stored in a local computer and fed by data link to the optimization computer.
- The optimization run is made periodically by the central computer.
- Output data is fed by data link to the local computer.
- The local computer automatically upgrades the set points of local control loops.

Design of Optimization System

Design of a computer optimization system must include the following.

1. A *criterion for optimization.*
2. An *objective equation* relating the variables of the plant to this criterion.
3. A *method* of driving the plant state variables to this criterion following the objective equation.

The analyzer computer system is the most popular way of obtaining the essential plant data to carry out local, area and plantwide optimization. A distributed hierarchy of local microcomputers (shown in Figure 6-21) which feed into area minicomputers feeding into a plantwide maxicomputer is the optimum way to optimize a plant.

The optimization routine can be shared over all the units in the plant, down-loaded from decisions made by the central computer to feed farther into loop set points of local computers. The local analyzers feed the status of local processes to their local computers,* which analyze the data and feed important criteria to the plantwide optimizer, which is now ready when optimizing time comes to make management decisions for the entire plant.

The optimizing computer system is often justified by the savings obtained by maintaining the process plant at *optimum* level, rather than having the plant run at some random level *below* optimum.

Techniques of Mathematical Analysis

The techniques of mathematical analysis can be used to drive a system to its optimum. It is the development and use of these techniques that has made computer control popular. It is also the possession of these techniques that has made certain

*Alternately, a central minicomputer can be used for all analyzers. It can be connected to local, area and central computers to provide the information needed for optimization.

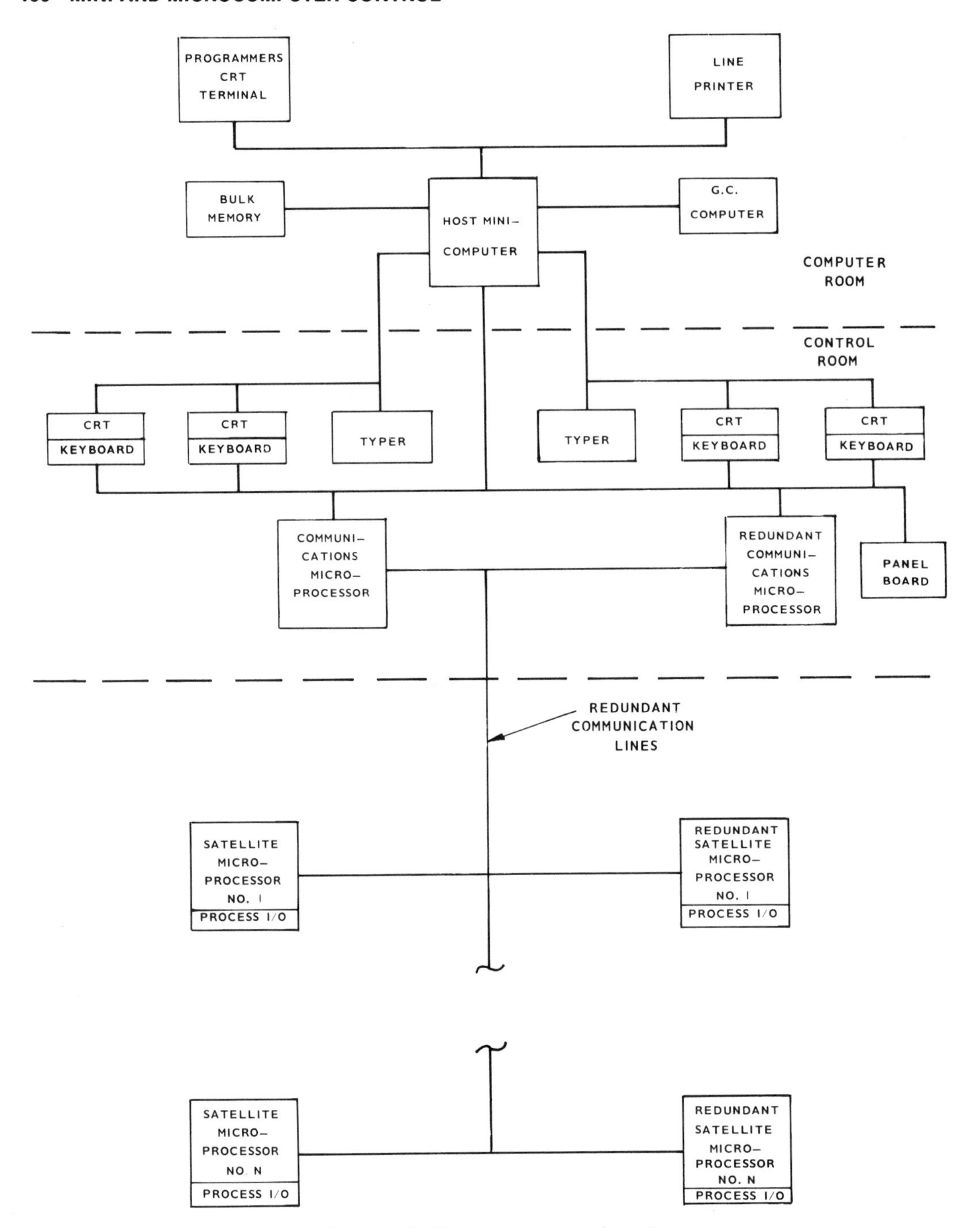

Figure 6-21. Typical computer hierarchy.

companies, users and computer companies so well fitted for the use of computer control in optimization. The step from feedback control to optimization control has been due to the use of these techniques.

Mathematical methods for optimization can be classified as follows.

Method of Partial Differentiation

Having established the profit equation, $P = F\ (F_1, F_2, X_3)$, constraint equations establish limits on F_1, F_2 and their sum.

Conversion equation: $C = \phi\ (F_1, F_2)$

Product equation: $X_3 = \phi\ '(C)$

We then take the partial derivatives of the profit equation, P, with respect to F_1 and F_2 and set each one equal to zero to find the relationships that produce maxima in P. The optimum maximum then is obtained by solving the maxima equations simultaneously.

The Method of Lagrange Multipliers

We could have solved the same problem without solving explicitly for the variables we needed to control by using Lagrange multipliers. These are arbitrary constants.

$$\phi = P + \text{sum } \lambda_Q Q$$

In the above, Q is of the same form as P, but each variable is multiplied by λ_Q. If we now take the partial of ϕ with respect to each variable, we have a set of equations between the λ's which can be solved with the constraint equations to provide the solution.

Inequality Constraints

Each variable must necessarily have an upper and lower limit; these are the inequality constraints of our system. These constraints usually require some trial and error operations to find a solution within the constraints for the mathematical equation for the optimum.

Dynamic Processes

The previous comments apply only to continuous processes. In optimizing batch processes, we have a dynamic system whose state is changing from beginning to end. The performance, or *profit,* equation of this type of process must be specified in terms of maximizing the *time average* performance. We know intuitively that a batch process that takes two hours is a lot more economical than one that takes three hours, if the results are the same. Hence, we must integrate the performance over the entire period of the batch and divide the integral by the total time.

Calculus of Variations

The calculus of variations provides an analytical technique for finding a function of $y(x)$ which has an extremum (will optimize our profit equation) between given boundary values. The problem is expressed in terms of an integral:

$$I = \int_{X_1}^{X_2} f(y,y',x)\,dx$$

where x is the variable representing our profit equation, P, y is the function to be determined and y' is its derivative, dy/dx. The solution of this integral is obtainable by formula, and will give us the function that will optimize our system.

Dynamic Programming

This technique is directly adaptable to a digital computer, and hence is the most popular of the available techniques. The problem is expressed as a series of decision processes based on a general principle of optimality enunciated by Bellman. It states, in short, that as you proceed with more and more trial and error, you arrive closer and closer to the optimum point. Dynamic programming has been applied to:

- Control of an exothermic chemical reactor.
- Optimum catalyst replacement program.
- Optimum multi-stage cross-current extraction.
- Optimum temperature gradients in a chemical reactor.

SUMMARY

In this chapter, we have shown how the continuous process analyzer fits into the overall control of a chemical process. We have seen how special instruments have been developed for the on-line measurement of composition, using electrochemical and spectrochemical methods. We have also discussed how the process chromatograph developed and how it took over many of the monitoring functions required of an analysis instrument. The mass spectrometer then appeared on the scene, and we have discovered that it may become the process analyzer of the future. The reliability of solid state circuits has made most analyzers, including the mass spectrometer, more reliable, and the incidence of the microcomputer has made on-line calculation exact and inexpensive.

We have seen how the analyzer fits into the control of a typical olefins plant, and how advanced control of such a plant now becomes feasible by the interaction of the information produced by on-line analyzers and the information processing capability of a digital system, whether maxi, mini or micro. Thus, we climb the ladder from material balance control to feedforward control, adaptive control and on-line optimization.

The marriage between the process analyzer and the microprocessor has produced a powerful tool for the improved control of chemical and petrochemical processes.

7. Computer Systems Specification Design

Merrill G. Thor

Senior Control Systems Engineer
Stone & Webster Engineering Corporation
New York, N. Y.

INTRODUCTION

In this chapter, we will first examine the function of a specification in a computer project and its importance in the project. Next, we will look at a typical specification for an ethylene plant computer system. Several notes interspersed with the specification present the background and reasoning behind it, which the reader may find useful in augmenting his own thought patterns during the writing of a specification for a process control computer system.

In order to understand the importance of a specification to a computer project, a basic understanding of the tasks involved in a project is essential. Figure 2-3 (p. 14) shows the task milestones. The specification has two major functions: it serves as an aid in selection of the vendor and as a way of determining exactly what the selected vendor must provide.

A specification is a very important element in the competitive bidding procedure. It allows the buyer to determine which vendor can provide a system that will do the job required for the least money. It forces each vendor to bid similar equipment and to spell out any areas where he cannot meet the buyer's requirements. This makes a comparison of prices more meaningful to the buyer, since he can compare the prices of systems performing the same functions; i.e., he can compare "apples to apples" instead of "apples to oranges."

A specification should be primarily concerned with the functions of a computer system. It should deal with what a system should do rather than how a system should do it. There are often several ways to perform a given function, and each bidder may propose a different method. As long as all vendors are qualified, the bidder who can meet the specification for the lowest price is the one who should get the order. Some may argue that a vendor who exceeds the specification and whose price is only slightly above the lowest bidder should be awarded the contract, but if the specification is written properly, any performance exceeding its requirements will be unnecessary.

The reader may ask: "How can I know if the level of performance required by the specification will meet or exceed the funds I have available for a computer system?" The best way to determine this is to discuss your requirements with several vendors before you write the specifications. They will be happy to tell you about the systems they have that will meet your needs. They will suggest additional features and functions you may not have thought about and will give you approximate or "budgetary" prices for the systems. This will allow you to decide at what size and level of performance the system should be set and to decide how big your budget for the project must be. These initial discussions will also be very helpful in determining the bidders list.

Equally important, this effort will allow specification of a system which is available from several vendors and which is already field-proven.

After the bidder has been selected, the specification should be revised to include any additional functions which were included in the vendor's proposal and to incorporate any exceptions to the original specification. The revised specification should be included in the purchase order. The purchase order can then govern over the vendor's proposal and the buyer's original specification. In this way, the specification is used to describe what a user or buyer expects to receive from the seller. As such, it allows both the buyer and the seller to come to a meeting of minds concerning what is included in the purchase. This protects both parties by eliminating surprises and disappointments during the execution of the project. A good "spec" tells the buyer exactly what he is getting and tells the seller exactly what he must provide. The typical table of contents is shown in Table 7-1.

TYPICAL SPECIFICATION FOR A PETROCHEMICAL PLANT PROCESS CONTROL DISTRIBUTED COMPUTER SYSTEM

1.0. Introduction

1.1. Scope

This specification covers the requirements for a process control computer system to be installed in a 300,000 MTA ethylene plant in Anytown, U.S.A. The plant will go on-line in the summer of 1982. The computer shall be capable of data acquisition, alarming, logging, process calculations, data display, advanced control and overall plant optimization. (See *Note A.*)

Table 7-1. Typical Specification Table of Contents.

1.0. INTRODUCTION
- 1.1. Scope
- 1.2. Bidder qualifications
- 1.3. Bidding instructions

2.0. FUNCTIONAL DESCRIPTION
- 2.1. Data acquisition and storage
- 2.2. Alarming
- 2.3. Logging
- 2.4. Data display
- 2.5. Advanced control of individual units
- 2.6. Overall plant optimization

3.0. HARDWARE CONFIGURATION
- 3.1. Overview
- 3.2. Host minicomputer
- 3.3. Bulk memory
- 3.4. Line printer
- 3.5. Card reader
- 3.6. Control room operator console
- 3.7. CRT display
- 3.8. Input keyboard
- 3.9. Typer
- 3.10. Gas chromatograph computer system
- 3.11. Communications microcomputer
- 3.12. Satellite microcomputer
- 3.13. Process input/output equipment

4.0. PROGRAMMING LANGUAGES
- 4.1. Fortran IV
- 4.2. Linear programming
- 4.3. Report generator
- 4.4. Continuous process control
- 4.5. Batch process control
- 4.6. Display builder

5.0. OPERATING SYSTEM SOFTWARE AND DIAGNOSTICS
- 5.1. Real-time executive
- 5.2. File manager
- 5.3. Input/output
- 5.4. Job control programs
- 5.5. Diagnostics

6.0. DOCUMENTATION

7.0. ACCEPTANCE TESTING
- 7.1. Factory acceptance test
- 7.2. Field acceptance test

8.0. MAINTENANCE, TRAINING AND SPARE PARTS
- 8.1. Installation
- 8.2. Maintenance
- 8.3. Spare parts
- 8.4. Consumables

9.0. PROJECT MANAGEMENT

10.0. TERMS AND CONDITIONS

1.2. Bidder Qualifications

1.2.1. All hardware and software shall be field-proven. Bidders shall provide a user's list as part of their proposal. The user's list shall contain a description of each computer system the vendor has supplied for a petrochemical application. The description shall cover the functions performed by system, the name of the user, the geographical location of the equipment and the model computer used. (See *Note B*.)

1.2.2. The vendor must be capable of providing maintenance field service on the system. The vendor must also have a stock of spare parts, in addition to those to be stocked by the buyer. The bidder shall describe his maintenance capability, including the number of employees, if any, with prior maintenance experience on the type of computer equipment proposed. The bidder shall also describe his existing spare parts stock facilities. *(See Note C.)*

1.3. Bidding Instructions

1.3.1. Bids shall be delivered by November 30, 1979. Late bids will not be considered.

1.3.2. Bids shall be addressed to Mr. John Doe, Smith and Jones Engineering Corporation, 8 Fourth Avenue, Anytown, U.S.A.

1.3.3. This specification is meant to show the functional requirements for the computer control systems. Bidders are encouraged to bid their standard field-proven products rather than custom-designed equipment. However, all exceptions to this specification must be clearly noted. *(See Note D.)*

Notes

A. The scope section is included at the beginning of the specification in order to give prospective bidders an idea of the size and type of system to be supplied. From reading this section, a prospective bidder should be able to determine whether this is the type of system for which he would like to make a proposal.

B. In petrochemical industry applications, a field-proven system is a must. High reliability and short start-up efforts are far more important than state-of-the-art performance. The software as well as the hardware should be field-proven. Reliability of computer hardware has reached the point where software errors cause as many system problems as hardware failures. A user's list is evidence that a vendor's hardware and software are truly field-proven.

C. The ability to supply adequate field service and spare parts is a prerequisite for a vendor. A vendor who cannot provide these services should not be considered unless the installation is in a location at which no vendor can provide service and spare parts.

D. During the preparation of a proposal, a vendor will often have to decide whether to meet a portion of the specification by designing a special piece of equipment or to take exception to that portion of the specification and bid standard equipment. Since field-proven equipment is so important, and since the specification is concerned primarily with the functions of the systems and secondarily with how the functions are accomplished, the vendor should bid his standard system whenever possible.

2.0. Functional Description

The system shall perform the following functions:

Data acquisition and storage
Alarming
Logging
Data display
Advanced control of individual units
Overall plant optimization.

2.1. Data Acquisition and Storage

2.1.1. The system shall scan analog and digital information and store the information in memory.

2.1.2. The control room operator shall have the capability to suspend the scan of any analog input and to substitute a fixed value for that variable. (See *Note E.*)

2.1.3. The scan for analog inputs shall be as follows. (See *Note F.*)

Flows	2 seconds
Levels	2 seconds
Temperatures	16 seconds
Pressures	2 seconds
Miscellaneous	4 seconds

2.1.4. Digital inputs shall be scanned at least once every second.

2.1.5. A historical data base shall be established for up to 200 process variables. The data base shall store each variable every 6 minutes. It shall retain the 10 6-minute readings for the previous hour, and the average hourly reading for the previous 24 hours. (See *Note G.*)

2.2. Alarming

2.2.1. Each analog input shall be checked to see if it is within high and low electrical signal limits (i.e., a check for reasonability). If it is outside the alarm limits, the control room operator shall be notified via a message on a CRT, and a record of the alarm shall be printed on the alarm typewriter. (See *Note H.*)

2.2.2. Provision shall be made for rate-of-change alarming for every analog input. If the rate-of-change is above the rate-of-change alarm limit, the control room operator shall be notified via a message on a CRT, and a record of the alarm shall be printed on the alarm typewriter.

2.2.3. Each analog input shall be checked to see if it is within its high and low alarm limits. If it is outside the alarm limits, the control room operator shall be notified via a message on a CRT, and a record of the alarm shall be printed on the alarm typewriter.

2.2.4. If a digital input indicates that an alarm condition exists, the control room operator shall be notified via a message on a CRT, and a record of the alarm shall be printed on the alarm typewriter.

2.3. Logging

2.3.1. The following logs shall be printed on the line printer:

Hourly log
Shift log
Daily log
Monthly log.

2.3.2. The operator shall have the ability to insert missing data needed to calculate the shift, daily or monthly logs. (See *Note I.*)

2.3.3. The user shall have the ability to create customized reports using a report generation software package. (See *Note J.*)

2.4. Data Display

2.4.1. The following types of displays shall be available on the system:

Unit graphic display
Unit status display
Current alarms display
Trend display
Single point display
Single alarm display
Control loop display.

2.4.2. The unit graphic display shall consist of a line diagram of a unit with important operating parameters displayed. These operating parameters shall be updated at the same rate as the data base is updated.

2.4.3. The unit status display shall be an alphanumeric listing of important operating parameters for a unit. These operating parameters should be updated at the same rate as the data base is updated.

2.4.4. The current alarms display shall show all alarm conditions which currently exist in the plant.

2.4.5. The trend display shall show the trends of up to 4 variables for the past hour or for the past 24 hours. Each of the trends shall be shown in a different color. The operator shall have the ability to select the range for each variable trended.

2.4.6. The single point display shall be shown at or near the bottom of any display. It shall show the current value or status of any single point selected by the operator.

The display shall be updated at the same rate at which the data base is updated.

2.4.7. The single alarm display shall appear at or near the bottom of all displays. It shall show the oldest unacknowledged alarm. When that alarm is acknowledged, the display shall be updated to show the oldest unacknowledged alarm, if any. The single alarm display is initiated by the occurrence of an alarm or the acknowledgment of an alarm by the control room operator. If there are no unacknowledged alarms, the display shall be blank.

2.4.8. The control loop display shall show the important parameters in a control loop. The operator shall have the ability to change tuning constants via this display. The display shall provide access to both feedback and feedforward control parameters, such as PID controller tuning constants and dynamic compensation constants (deadtime, lead, lag and scale factor). *(See Note K.)*

2.5. Advanced Control of Individual Units

2.5.1. Advanced control shall be provided for the following units:

Furnace
De-methanizer
De-ethanizer
De-propanizer
Ethylene column
C_2 acetylene hydrogenation unit
Cracked gas dehydrator.

(See Note L.)

2.5.2. Advanced control on the furnace shall perform the following:

Automatic start-up and shut-down of each furnace

Automatic steam de-coke

Constant severity and constant selectivity control.

2.5.3. Advanced control on the de-methanizer, de-ethanizer, de-propanizer and ethylene column shall perform the following: 1) Adjustment of reboiler duty, and reflux flow in order to make on-spec product without over-separating or over-refluxing; and 2) Feedforward control of the column based on feed rate and feed composition in order to stabilize operation of the column.

2.5.4. Advanced control of the C2 acetylene hydrogenation shall perform feedforward control of the reactors based on feed rate and feed composition in order to reduce energy consumption and increase ethylene gain.

2.5.5. Advanced control of the cracked gas dehydrators shall sequence the active and stand-by units. The computer shall predict when it is time to regenerate the active unit from a record of previous operation of the unit. The computer shall then switch the stand-by unit into active service, regenerate the exhausted unit and place it in stand-by status when regeneration is complete.

2.6. Overall Plant Optimization

2.6.1. A linear programming package shall be supplied to allow the performance of overall plant optimization in the host computer. *(See Note M.)*

2.6.2. Overall plant optimization shall determine the most economic operation of the plant, taking into account feedstock availability, throughput requirements, desired ethylene to propylene ratio, energy costs and interactions between units.

2.6.3. Vendor shall state the largest size matrix which the linear programming package can solve.

Notes

E. When an analog input goes out of range (e.g., when a transducer fails), the control room operator is alerted. If the input is used in a control loop, the loop is immediately taken off control. In some cases, the operator may wish to continue control of that loop because he is confident that he knows the correct value of the failed analog input.

F. The scan periods shown are for an ethylene plant. Scan periods required by other processes may be different.

G. The historical data base is used for

hourly, shift and daily logs. It can also be used to show the trend of any of the stored measurements for the previous hour or the previous 24 hours. This can be useful in the review of plant upsets.

H. For example, if a transducer which normally provides a 4 to 20 milliamp signal is sending a 3 milliamp signal, the situation will be detected by the resonability check.

I. If the computer is shut down, or an input signal has failed, all or part of the information in a log may be unattainable from the computer. To make the log a finished document, the missing data should be included, if known. Therefore, the control room operator, or, preferably, the shift foreman, should have the ability to insert missing data.

J. The need to develop new reports and to modify existing reports will arise many times during the life of the system. This necessitates that an easy procedure be available for specifying the information to be included in a report and the format of the report. A report generation program is a good solution to this problem. A report generation program allows the user to specify new logs and reports or changes in existing logs and reports via a high-level language.

K. The control engineer needs to change tuning constants as plant conditions change and particularly during the initial start-up of the plant. The control loop display will enable him to accomplish this quickly and without the need for computer programming skills.

L. The advanced control functions shown here are for a typical ethylene plant.

M. Overall plant optimization can only be accomplished after a great deal of operating data has been accumulated. It is usually a task which requires years of effort. Therefore, this specification requires that the vendor supply only a linear programming package and not a finished program which performs overall plant optimization. Optimization of a single area in the plant is not as formidable a job. The furnaces in an ethylene plant, for example, are a prime target for optimization. A mathematical model of furnace operation (referred to as a yield predictor) is required. Using linear programming, the optimum operating parameters for each furnace (such as hydrocarbon feed rates, dilution steam feed rate, residence time and coil outlet temperature) to provide the desired yield from the furnaces can be determined. Although the furnace model may contain non-linear equations, a linear program can determine the optimum solution using mathematical iterative techniques.

3.0. Hardware Configuration

3.1. Overview

3.1.1. The hardware configuration shall be as shown in Figure 2-5 (p. 16). Computer processing shall be distributed between a host minicomputer and six microprocessors. Each microprocessor shall be backed up by an identical redundant unit. (See *Note N.*)

3.1.2. All programming shall be accomplished in the host. The system shall allow programs to be down-loaded from the host to the satellites. (See *Note O.*)

3.1.3. The host shall be able to retrieve data from the satellites, as shown in Tables 7-2 and 7-3.

3.2. Host Minicomputer

3.2.1. The host minicomputer shall have an instruction word-length and a memory word-length of at least 16 bits.

3.2.2. An automatic bootstrap loader shall be provided.

3.2.3. Power failure detection and automatic restart shall be provided.

3.2.4. Main memory shall consist of a minimum of 64K 16-bit words. Memory shall be expandable to at least 128K words in increments no larger than 16K words. Memory cycle time shall be 800 nanoseconds or less. Sufficient main memory shall be provided to perform all the functions outlined in this specification.

Table 7-2. Typical Process Input/Output Requirements for an Olefins Plant.

SYSTEM	HIGH-LEVEL A.I.	LOW-LEVEL A.I.	CONTACT SENSE D.I.	CONTACT D.O.
Host minicomputer	400	250	50	50
De-methanizer microcomputer	20	10	20	20
De-ethanizer microcomputer	20	10	20	20
De-propanizer microcomputer	20	10	20	20
Ethylene column microcomputer	20	10	20	20
C_2 acetylene hydrogenation microcomputer	30	20	20	20
Cracked gas dehydrator microcomputer	20	5	100	100

Table 7-3. Typical Analog Data Acquisition Requirements for an Olefins Plant.

SYSTEM	FLOW	LEVEL	TEMPERATURE	PRESSURE	MISCELLANEOUS
Host minicomputer	100	0	250	250	50
De-methanizer microcomputer	5	5	10	5	5
De-ethanizer microcomputer	5	5	10	5	5
De-propanizer microcomputer	5	5	10	5	5
Ethylene column microcomputer	5	5	10	5	5
C_2 acetylene hydrogenation microcomputer	10	5	20	10	5
Cracked gas dehydrator microcomputer	5	5	5	5	5

3.2.5. Main memory shall be provided with protection features to prevent unauthorized changes in memory locations.

3.2.6. Main memory shall be either core or semiconductor. If semiconductor memory is supplied, a battery back-up power supply sufficient for 30 minutes of operation shall be provided.

3.2.7. Hardware multiply/divide and hardware floating point shall be included. Firmware multiply/divide and firmware floating point shall also be acceptable. A floating point divide instruction shall be executible in less than 15 microseconds.

3.2.8. A watchdog timer feature shall be included in the host computer. If the timer is not reset within 1.0 second, an audible alarm shall be sounded and all control programs executing in the host shall be suspended.

3.2.9. A real-time clock, with a resolution of 0.1 second shall be provided. The clock shall be accurate within 1 second per day. The frequency of the AC power must not be used as the time standard. (See *Note P.*)

3.2.10. A data link to the gas chromatograph computer shall be provided. This data link shall be compatible with the data link specified in Section 3.10.

3.2.11. The host computer shall contain at least two DMA (Direct Memory Access) channels. A separate DMA channel shall be used for the bulk memory. The CRT displays shall be connected to the host computer via a DMA channel.

3.2.12. The host shall control the furnace area of the plant via its process input/output equipment. (See Table 7-2.)

3.3. Bulk Memory

3.3.1. Bulk memory shall be fixed head drum, fixed head disk or moving head disk. More than one type of bulk memory may be provided (e.g., both fixed head disk and moving head disk). The bulk memory access time and transfer rate shall be sufficient to support the functions outlined in this specification.

3.3.2. The bulk memory size shall be 125% of the bulk memory required to fully implement the functions described in this specification (i.e., there shall be 25% spare bulk memory at the completion of the project).

3.3.3. Moving head disks shall include at least one removable disk pack.

3.3.4. A write-protect feature shall be provided for all tracks in bulk memory.

3.4. Line Printer

3.4.1. The line printer character set shall consist of at least 64 characters.

3.4.2. There shall be at least 132 characters per line and the horizontal spacing shall be 10 characters per inch. Vertical spacing shall be 6 inches per inch. Characters shall be formed from a 9 by 7 matrix.

3.4.3. The printing rate shall be at least 250 lines per minute.

3.4.4. The line printer shall have a facility for testing the printing mechanism without the use of the central computer. This facility shall include the ability to print test patterns. (See *Note Q.*)

3.4.5. A buffer register, sufficient to hold at least one line (132 characters), shall be included in the line printer electronics.

3.4.6. The print motor shall start upon receipt of the first character in a message and shall turn off no more than 10 seconds after the receipt of the last character in a message.

3.4.7. The impact force shall be sufficient to make an original and at least three copies simultaneously.

3.5. Card Reader

3.5.1. The card reader shall read 300 80-column cards per minute.

3.5.2. The capacity of the hopper/stacker shall be at least 500 cards.

3.6. Control Room Operator Console

The seller shall provide two identical control room operator consoles. Each console shall contain two CRT monitors, an input keyboard and a typewriter. Sketches of the consoles shall be shown in the bidder's technical proposal.

3.7. CRT Display

3.7.1. The CRT display shall employ dual input/output ports to allow communication with either the host computer or the communications microprocessor. (See *Note R.*)

3.7.2. The CRT display shall consist of seven colors: red, blue, cyan, white, green, yellow and magenta.

3.7.3. The diagonal measurement of the CRT screen shall be at least 19 inches.

3.7.4. The display shall be comprised of at least 48 lines. There shall be at least 80 characters per line.

3.7.5. Each character location shall have independent control of color, blink, inverted or normal video.

3.7.6. The character set shall consist of at least 64 alphanumeric and punctuation characters, and 64 special graphic characters.

3.8. Input Keyboard

3.8.1. The input keyboard shall employ two input/output ports to allow communications with either the host computer or the communications microprocessor. (See *Note R.*)

3.8.2. The input keyboards shall be designed for use by a petrochemical plant control room operator and shall be the vendor's standard.

3.9. Typer

3.9.1. The typers shall be KSR (Keyboard Send Receive) devices. Although the typers will be used for printing only, a keyboard shall be included to facilitate maintenance.

3.9.2. The typers shall print at a rate of 30 characters per second or greater. Spacing shall be 10 characters per inch (horizontal spacing) and 6 lines per inch (vertical spacing). A line shall be at least 120 characters long.

3.9.3. The paper shall be greater than 12 inches in width.

3.10. Gas Chromatograph Computer System

3.10.1. The gas chromatograph computer system shall handle up to 25 gas chromatograph analyzers. The gas chromatograph computer shall provide the timing for the analyzers and shall interpret the data from the analyzers. This analysis data shall be sent to the host minicomputer over a data channel. (See *Notes S, T.*)

3.10.2. The analyzer interface unit shall be supplied for each analyzer and shall include all functions needed to operate, calibrate and service the chromatograph under the control of the computer, or under the control of a serviceman, when service is needed. Among the functions included in this unit shall be:

1. Detector amplifier with auto-ranging circuitry.
2. Analog to digital converter and communications unit compatible with computer interface.
3. Instrument control console, containing manual instrument valve select switches with indicators; calibration gas select switch with indicator; detector bridge output jack for test recorder; detector zero balance control and current meter; power switch with indicator; manual/computer master control switch; and regulated bridge power supply with on/off switch.

3.10.3. The computer interface unit shall contain all components needed to scan the analyzer interface units, and to convert and retain the signals obtained for use as required by the computer main frame. Among the features included in this unit shall be:

1. Multiplexer-switching high-level signals from each detector at the rate of 10 samples per second.
2. Real-time clock providing 120 interrupts per second, synchronized with power frequency.
3. Fixed gain buffer amplifier.
4. Visual status checks.
5. Manual input/output to the computer.
6. Audible and visual alarm connections.
7. Manual maintenance checks.

3.10.4. Vendor shall supply, load and test all software required to operate the gas chromatographic system unattended.

- The software shall establish the calibration and analytical procedures for each chromatograph.

- The automatic stream selector program shall in turn select each stream for analysis; actuate solenoid valves in the sample handling system; actuate indicating lights that identify which stream is being analyzed; and advance automatically to the next stream after sample injection. A manual advance interrupt shall be provided, to allow the operator to advance the selector program and to select any stream out of sequence when desired. A manual override switch shall be provided to eliminate streams.
- Provision shall be made for a manual command that will allow switching the output signal of any chromatograph to a calibration recorder.
- As each peak appears on each chromatograph during normal operation, the computer shall read, scale and digitize the signal, recognize the peak and integrate the area, using appropriate base line and calibration correction. Peak information shall be stored in common memory chained into lists associated with individual chromatographs.
- When an analysis is terminated, the computer will process its list of peak blocks to generate an analysis report. Peak areas shall be corrected for base line drift; matched with component names specified; and corrected for relative response. Compositions shall be calculated and an analysis report shall be printed.
- The system shall execute automatic calibration by standard gases for the analysis calculation described above. The system shall also permit an operator to calibrate it by entering an appropriate signal thru the typer, or by operating a switch on the analyzer interface unit.
- Analysis methods shall be loaded by the vendor, but shall be available for review and modification by qualified analysis personnel. The analysis methods shall be sets of directives defining the time sequence during a run; e.g., injection of sample, opening and closing of each component gate, activation of column switching valves, calibration factors and method of integration of each peak. A device shall be provided to lock this information into memory so that it cannot be modified by other than authorized personnel.
- The program shall include the provision for high and/or low process alarms selectable by the operator for any component of any stream. The program shall also include required system malfunctioning alarms, such as alarms indicating that the instrument is too far out of calibration; that there has been a shift in the elution time of the peaks or an inordinate base line drift; that analyzers are malfunctioning; that a peak did not appear; and that there is a lack of information transfer to the computer. Other alarms are power fail and restart, watch dog timer unit and limit checks for each component.

3.10.5. The data channel shall conform to a previously published standard and shall be compatible with the host minicomputer.

3.11. Communications Microcomputer

3.11.1. There shall be two identical communications microcomputers. One shall be active and the other shall serve as an on-line redundant spare. In the event of a failure in the active communications microcomputer, the spare microcomputer shall take over the functions of the failed unit.

3.11.2. The communications microcomputer shall control the flow of data between the satellite microcomputers and the host minicomputer and the CRT's. The communications microcomputer shall also control the flow of data between satellite microcomputers.

3.11.3. The satellite microcomputers shall be connected to the communications microcomputer via two identical communications lines. One line shall be active and the other shall be a redundant spare. In the event of a failure in the active communications line, the communications microcomputer shall switch to the redundant spare communications line.

3.12. Satellite Microcomputer

3.12.1. There shall be twelve satellite microcomputers. Six shall be active and six shall be redundant spares. Each redundant spare shall be identical to its counterpart active satellite microcomputer. The six satellite microcomputers are as follows.

De-methanizer
De-ethanizer
De-propanizer
Ethylene column
C_2 acetylene/hydrogenation unit
Cracked gas dehydrator.

The functions of the six satellite microcomputers are specified in Section 2.5.

3.12.2. Each of the satellite microcomputers, including the redundant spares, shall have its own dedicated process input/output equipment. Process input/output requirements are given in Section 3.13. (See *Note U.*)

3.12.3. The microcomputers shall have a word-length of at least 8 bits and a memory cycle time of 1 microsecond or less. Memory shall be of the RAM (Random Access Memory) type. Memory shall be expandable to at least 64K bytes.

3.13. Process Input/Output Equipment

3.13.1. The analog input section of the process input/output equipment shall meet the specifications below.

1. The scan rate for high-level analog signals shall be at least 200 points per second.
2. The scan rate for low-level signals shall be at least 100 points per second.
3. The input circuit shall be two-wire differential; the input impedance shall be at least 10 megohms.
4. The analog to digital converter shall use at least 12 bits plus sign.
5. Accuracy shall be ±0.15% of full scale.
6. Common mode rejection shall be at least 100 db at frequencies between 0 and 60 Hertz.
7. Thermocouple open-circuit detection shall be provided. (See *Note V.*)

3.13.2. The digital input section of the process input/output equipment shall meet the specifications below.

1. Three types of digital inputs shall be available: contact sense, digital events and pulse counters.
2. Digital inputs shall be scanned once each second.
3. Digital event inputs shall provide a priority interrupt to the computer when an input changes state. Digital event inputs shall latch until reset by the computer.
4. Each pulse counter input shall be counted in a dedicated register which stores at least 1024 counts.

3.13.3. The digital output section of the process input/output equipment shall meet the specifications below.

1. Four types of digital outputs shall be available: contact, voltage level, pulse contact and pulse voltage level.
2. Contact outputs shall be totally isolated from ground.

3.13.4. The termination facilities in the process input/output equipment shall meet the specifications below.

1. Each system shall contain termination cabinets which can accommodate the inputs shown in Table 7-2 and 15% spares in addition.
2. Termination facilities for thermocouples shall include RTD measurement of cabinet temperature to allow the computer to perform reference junction compensation.
3. The cables between the termination

facilities and the input/output multiplexer shall be prefabricated with connectors at both ends to ensure easy installation.

3.13.5. The process input/output equipment shall provide the numbers of inputs and outputs shown in Table 7-2 and at least 15% additional of each category in each of the seven systems.

Notes

N. The hardware configuration provides complete redundancy for all microprocessors. The host is not backed up because it is performing supervisory control and the gas chromatograph information is required only for advanced control functions. Thus, if the host fails, automatic control will continue.

O. The capability to down-load a program from the host to a satellite eliminates the need for peripherals (such as a CRT, line printer or card reader) to be connected directly to each satellite computer.

P. The frequency of the power in developing countries is often not accurate enough to be used as a time standard.

Q. Most maintenance performed on a line printer pertains to the mechanical printing mechanism. It is very desirable to be able to check and adjust the printing portion of the line printer without requiring the computer to be operational. This allows the major portion of the line printer maintenance to be performed without interrupting operation of the host computer when it is on-line. Also, start-up is made easier, because most of the line printer check-out can be performed before the computer is in operation.

R. Two ports for input/output are required for each CRT display and keyboard. In this way, the operator can communicate with the satellite computers via the communications microprocessors when the host is not functioning. This allows the operator to retrieve data from the satellites and to change the control set points in the satellites. Therefore, automatic control can continue when the host fails.

S. This hardware configuration uses a single minicomputer to accommodate all the gas chromatographs. (See Figure 7-1.) For an ethylene plant application where 20 to 25 gas chromatographs are used, this should be the most economical solution. Other possible configurations employing microcomputers could be used as well. A microcomputer can handle approximately four gas chromatographs. Two alternate configurations are shown in Figure 7-2. Alternate *A* entails the replacement of the single gas chromatograph minicomputer shown in Figure 7-1, with several microcomputers, and for applications where a small number of gas chromatographs are used, alternate *A* may prove to be the most economical solution. In Alternate *B*, the gas chromatograph microcomputers are connected to the common data bus. Analysis results are transferred to the satellites over the data bus, and the host can obtain gas chromatograph analysis data via the communications microcomputer. A third alternative is to perform the functions of the gas chromatograph computer in each of the individual satellite microcomputers. This approach eliminates the need for separate computers dedicated to interpreting gas chromatograph analysis data. However, for some control applications, the computational burden of performing control of a process unit and interpreting data is too great for a single microprocessor.

T. Analyzers other than gas chromatographs might also be employed. The mass spectrometer, for example, is much faster than a gas chromatograph, and a single mass spectrometer can replace up to eight gas chromatographs.

U. The process input/output equipment can be located either in the control room or in the field. If the equipment is located in the field, a considerable portion of the field wiring costs will be eliminated. (This subject is discussed further in Chapter 8.)

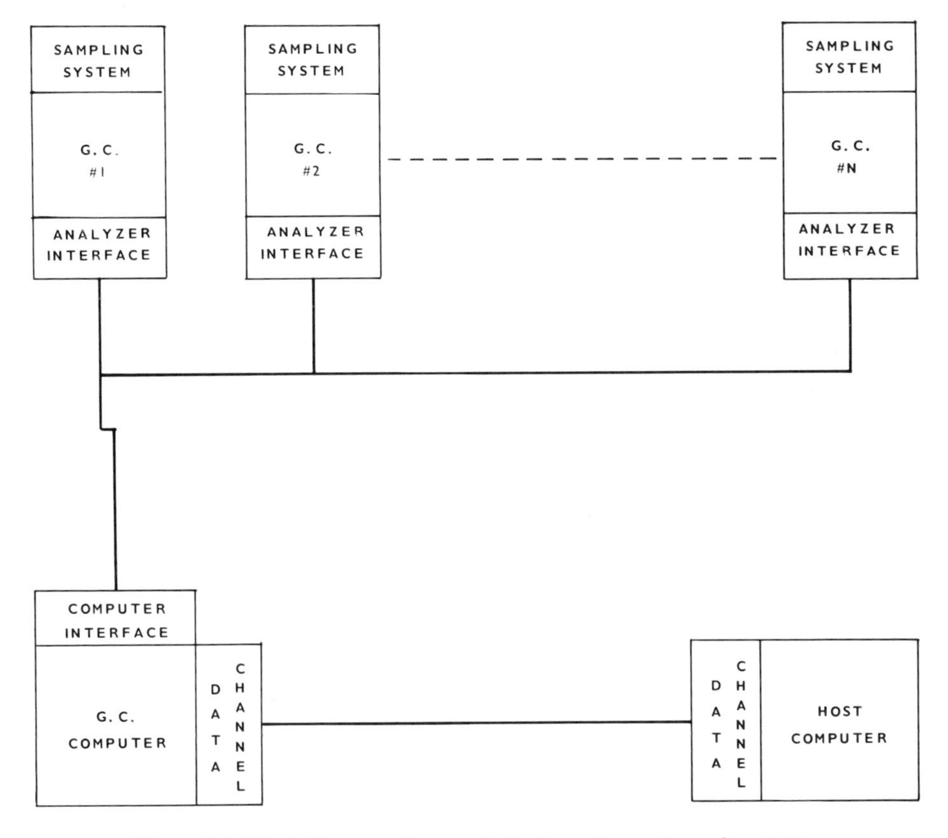

Figure 7-1. Block diagram of computerized gas chromatograph system.

V. Thermocouple open-circuit detection is very important in systems which use a flying capacitor analog input system. If a thermocouple input suddenly open-circuits, the capacitor stays charged at the last closed-circuit reading. Without open-circuit detection, the temperature will appear normal for many hours, until the capacitor gradually discharges.

4.0. Programming Languages

The following programming languages shall be included in the system:

Fortran IV
Linear programming
Report generator
Continuous process control
Batch process control
Display builder.

4.1. Fortran IV

The Fortran IV compiler shall conform to ISA (Instrument Society of America) Standard S61.1 regarding features for process input/output.

4.2. Linear Programming

A linear programming package, using the simplex algorithm, shall be provided. The program shall be capable of solving a matrix of 100 by 100.

4.3. Report Generator

A report generation program shall be provided. This program shall allow the user to develop special reports to be printed on the line printer.

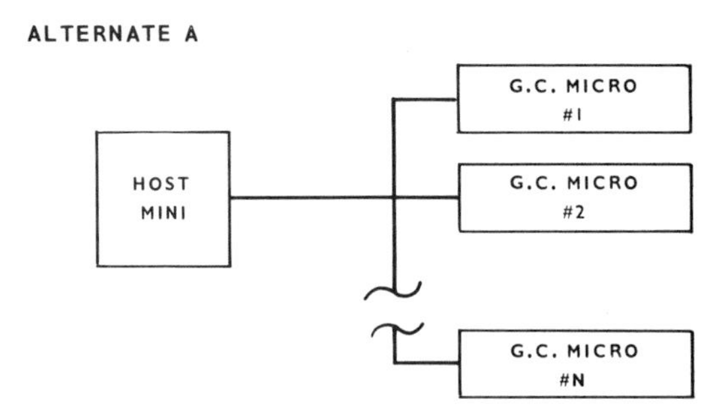

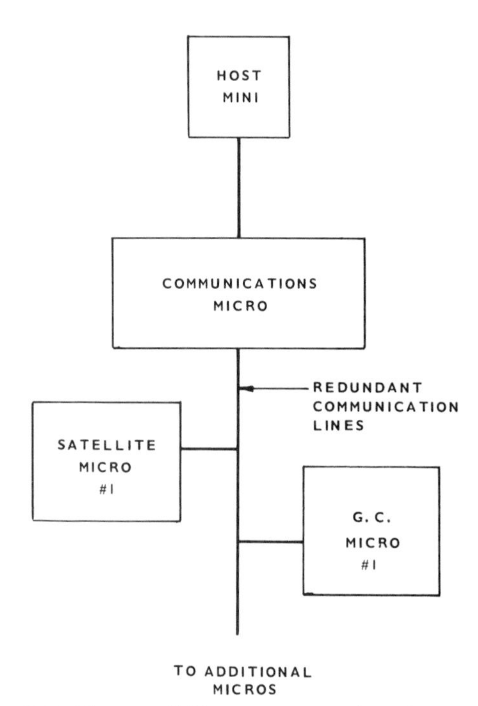

Figure 7-2. Alternate G.C. computer configurations.

4.4. Continuous Process Control

A process control language, suitable for continuous processes, shall be provided. The language shall be capable of formulating feedforward and feedback control loops using a "fill-in-the-blanks" procedure. (See *Note W.*)

4.5. Batch Process Control

The vendor shall supply a process control language for batch or sequential control. This language shall be suitable for the control of furnace start-up and shut-down and furnace de-coking, as well as the cracked gas dehydrator sequencing.

4.6. Display Builder

The display builder program shall allow a display to be defined by the user by building the image from the CRT keyboard. The background portion or static portion of the display shall be taken directly from the image input from the keyboard. The foreground or updated portion of the display shall be input symbolically.

Notes

W. A "fill-in-the-blanks" language is a language by which the user can create a control function by specifying certain parameters which describe the function. For example, to create a PID control algorithm, the user would specify parameters such as the loop name, the loop number, the multiplexer addresses for the controlled variable, the controller set point feedback, the output address, the proportional gain, the integral action, the derivative action and the frequency at which the algorithm is to be executed. When the initial data base is being set up for a system, these parameters can be assigned by filling in a computer coding sheet. The term "fill-in-the-blanks" comes from this procedure. Once data base generation has been completed, these parameters can be assigned, using a CRT console.

5.0. Operating System Software and Diagnostics

Operating system software shall include the following:

Real-time executive

File manager
Input/output
Job control programs.

The operating system shall be a disk operating system utilizing multi-programming.

5.1. Real-Time Executive

The real-time executive shall allocate computer system resources, including main memory, bulk memory transfers, CPU time and input/output facilities.

5.2. File Manager

The file manager shall be capable of providing expandable random access files as well as fixed-length contiguous files.

5.3. Input/output

The input/output program shall control the sending and receiving of data between main memory and the input/output devices.

5.4. Job Control Programs

The job control programs shall process communications between the operator and the computer system. It shall read and interpret control statements, initiate job requests and issue error messages.

5.5. Diagnostics

5.5.1. The vendor shall provide on-line and off-line diagnostic programs.

5.5.2. The on-line diagnostic programs shall consist of a peripheral exerciser program and a malfunction check program. The peripheral exerciser shall print a test message on any of the printing devices upon operator demand. The malfunction check program shall run continuously and shall check operation of the CPU, memory and input/output equipment.

5.5.3. The off-line diagnostic programs shall consist of a library of programs used to verify proper operation of a device or to aid in the trouble-shooting of a device. Off-line diagnostics shall be provided for every device in the system.

6.0. Documentation

The seller shall provide the complete sets of documentation. Complete listings, descriptions and flowcharts for all advanced control programs shall be included in the documentation provided. The documentation shall also include all hardware installation drawings, theory of operation manuals, maintenance manuals, wiring diagrams and any other documentation the seller normally provides. (See *Note X.*)

The seller shall be diligent to ensure that the revision level of the documentation furnished matches that of the equipment furnished.

The seller shall provide instruction manuals for the use of all diagnostic programs.

The seller shall provide a complete list of process inputs and outputs.

Notes

X. Documentation is too often neglected in computer systems projects. Neglecting documentation is a serious mistake because a system is only as good as its documentation. After all, a system is an organization of many simple elements. The key to the organization of a system is the documentation, and without sufficient documentation, a system cannot be properly understood, installed, maintained or modified. Documentation must be delivered at the same time as the equipment, if not before (otherwise, system installation could be severely delayed). The fixed cost of putting the documentation together is much higher than the variable cost of additional sets of the documentation. Therefore, the specification should call for enough sets to ensure that it will not be necessary to place an order for additional sets of documentation later on.

7.0. Acceptance Testing

(See *Note Y.*)

7.1. Factory Acceptance Test

7.1.1. The factory acceptance test shall be held at the seller's system assembly location. The seller must provide notice of the test to the buyer at least 20 days before the start of the test.

7.1.2. The test must be witnessed by the buyer or his representative and performed by the seller. The system shall not be accepted until the entire factory acceptance procedure is completed successfully.

7.1.3. Details of the test shall be agreed upon after selection of the vendor, but before the finalization of the purchase order. The test shall require approximately ten working days, not taking into account the time required to fix any problems encountered.

7.1.4. During the test, the vendor shall use the actual diagnostic card decks which will be shipped with the system.

7.1.5. The factory acceptance procedure shall include inspection of all documentation, spare parts and test equipment.

7.1.6. Hardware testing shall conform to ISA Standard RP55.1.

7.2. Field Acceptance Test

7.2.1. The field acceptance test shall be held on-site after the system has been installed. The test must be witnessed by the buyer or his representative and performed by the seller. The system shall not be accepted until the entire field acceptance test is completed successfully.

7.2.2. Details of the test shall be agreed upon after selection of the vendor, but before the finalization of the purchase order. The test shall require approximately five working days, not taking into account the time required for the correction of any problems encountered.

Notes

Y. Thorough acceptance tests, both in the manufacturer's plant prior to shipment and on-site, will help ensure a successful project. Usually, a portion of the payment is withheld until each acceptance test is completed. The factory acceptance test is by far the more important of the two. It is much easier to fix a problem in the factory than in the field, since in the factory the full resources of the vendor are at hand. A small problem in the system, not corrected when it is in the factory, can become a major problem when the system reaches the field. Therefore, the factory acceptance test should be as rigorous as possible, within practical limits. Usually, a thorough factory acceptance test will require five to ten working days. An investment in effort on the factory acceptance test will pay dividends in smooth start-up and operation of the system.

At one time, "heat-tent" runs were popular for the factory acceptance tests. This involved constructing a tent around the system, controlling the temperature in the tent (usually to 105° F) and running diagnostic tests (for 24 hours). This procedure was used in order to detect marginal components in the system. The use of integrated circuits in systems has reduced the popularity of heat-tent tests. Many people feel that running a system at high temperatures reduces the life of the components. Since most integrated circuits are thoroughly tested by the manufacturer, the number of marginal components revealed by heat-tent tests is not worth the effort.

Six month availability runs have also decreased in popularity. They were used to show that a system could meet or exceed a given availability (e.g., 99%) over a six month period. Some availability runs also required that a given MTBF (mean time between failures) be met by the system. The vendor was required to repair the system free of charge until the six month run was completed successfully. As hardware reliability increased, demonstrations of availability became less and less important. However, some buyers continued to require availability runs because they had the effect of a free six month maintenance service contract. Today most vendors price a six month availability run as if it were a six month service contract. Buyers requiring service of their systems are advised to include a maintenance contract in the specification, rather than an availability run.

This gives the buyer the service he requires and eliminates the bookkeeping required in performing an availability run.

The factory and field acceptance tests should include the running of all hardware diagnostics and running of the total system program. Often, the system program will reveal hardware problems which are not revealed by the hardware diagnostics. These problems are usually caused by the interaction of the elements in the system. Therefore, the hardware diagnostics, by themselves, are not sufficient to show that there are no hardware problems.

In addition to the hardware, the factory acceptance test should also check the software, system documentation, diagnostic programs and spare parts. Documentation should be checked for correctness and completeness (Does the revision number of the equipment agree with the revision number of the documentation? Is the correct number of copies being provided?). Similarly, the spare parts should be examined for correctness and completeness. Too often, vendors ship a printed circuit board which is revision *E* for a system which requires a revision *D* board, and the discrepancy isn't discovered until the system is down and the proper spare is not available.

When the diagnostics are run, the actual card decks, or tapes, to be shipped with the system should be used. This will prove that a working copy of each diagnostic is being provided.

One method of double-checking a factory acceptance test is to take a copy of the system specification and go through it, paragraph by paragraph, making sure that each requirement has been met.

Since a portion of the vendor's payment is normally withheld until the factory acceptance test is completed successfully, the vendor will be very anxious for the system to be accepted. The buyer wants his project to proceed according to schedule and is usually anxious to accept the system, too. Small problems are sometimes listed on a punch list and the system is accepted under the condition that the vendor will eventually remedy them. The system should never be accepted if any design problems or other major problems exist. These problems should be corrected in the factory, regardless of the delays caused.

8.0. Maintenance, Training and Spare Parts

8.1. Installation

The seller shall provide trained engineers and programmers to supervise the installation and start-up of the computer system. The buyer shall pay the seller's prevailing per diem rates at the time of installation.

8.2. Maintenance

The seller shall provide separate prices for a one year on-call maintenance contract. The seller shall provide prices for both 24 hour per day coverage and for business hours only coverage. The seller shall state the distance of the nearest maintenance service office from the job-site. The prices shall include stocking of spare parts, tools and test equipment at the job-site and the parts and labor for replacement of defective parts. (See *Note Z.*)

8.3. Training

The seller shall state the number of weeks of training required for full hardware and software training for one person. The seller shall provide a separate price for full hardware and software training for six people. (See *Note AA.*)

8.4. Spare Parts

The seller shall provide a separate price for all recommended spare parts, test equipment and tools. The seller shall state the spare parts depot nearest to the job-site, as well as which of the recommended spare parts are currently stocked at this depot. (See *Note BB*.)

8.5. Consumables

The vendor shall provide a one year supply of line printer paper and ribbons, typewriter paper and ribbons and punch cards.

Notes

Z. Maintenance requirements depend very much on the individual user. As a general practice, spares are taken at 10% in the continental U.S. and other heavily industrialized areas of the world, and at 20% or more for remote areas with no access to spares. Before a specification for a system is written, a user should determine what his maintenance philosophy will be. Some users have a good deal of experience in computer maintenance and can maintain the equipment through their own organizations. They may already have equipment in operation from one of the vendors. Other users have no experience at all. In order to determine a maintenance philosophy, the following questions must be answered:

1. How much of the hardware maintenance will be provided by the vendor?
2. If the vendor will provide only part of the maintenance, will he be required to be available 24 hours a day, seven days a week, or will availability during business hours, Monday through Friday, be sufficient?
3. What portion of the required spare parts will be purchased and what portion must the vendor stock nearby?
4. How many people must be trained in hardware and software maintenance?

Once these questions have been answered, the maintenance philosophy has been established. Then the portion of the specification dealing with maintenance, training and spare parts can be written.

AA. A computer system does not run entirely unattended. It requires routine or preventative hardware maintenance and corrective hardware maintenance. A system which meets the continuing needs of the plant also requires software maintenance. Preventative hardware maintenance involves such tasks as cleaning the filters in the cabinet blowers; oiling electromechanical parts; and periodic adjustments which keep the system running properly and prevent problems from occurring. Corrective hardware maintenance involves fixing the system when a failure has occurred. Maintenance programming, or software maintenance, involves the adding or modifying of the functions of the system program. Assuming the user decides to perform all of the maintenance himself, he must train people to become hardware and software specialists. It is important to note that the hardware specialist has some training in software as well as in hardware. Likewise, the software specialist must have an appreciation of hardware as well as software. The solution to most system problems requires a joint effort between the hardware and software specialists, and it is often difficult to initially determine whether a problem in the system is due to a hardware or a software problem. This is especially true of intermittent problems. Also, some hardware faults are not detectable with the diagnostic programs, and the system program must be used as an aid in hardware trouble-shooting. Because a joint effort is required, each specialist must have an understanding of the other's discipline. Therefore, it is recommended that hardware and software specialists be adequately trained in both areas.

BB. The specification should require the vendor to provide a recommended list of spare parts and respective prices. The purchase of spare parts should be included in the purchase of the total system. This has three advantages. First, at this stage, the vendor will keep his price lower for spare parts because he is competitively bidding the job. (This will not be the case if the vendor is pricing the spare parts after the system has been purchased.) Second, many spare parts will be purchased or manufactured by the vendor at the same time as the parts that will be used in the system. (This will reduce the chance of a spare part being of the wrong revision number.) Third, this procedure will ensure that that the spare parts will be available for shipment with the system. It is very difficult to start up and commission a system without spare parts, and it is important that the spare parts be shipped no later than the system is shipped.

9.0 Project Management

The seller shall assign a full-time project manager for the duration of the project.

The project manager shall be the technical and contractual liaison between the seller and the buyer. He shall also be responsible for the seller's project schedule.

The project manager shall provide the buyer with monthly progress reports in writing.

10.0. Terms and Conditions

(See Note CC.) Progress payments shall be made as follows.

- 30% of contract price within ten days after placement of purchase order.
- 30% of contract price six months after placement of purchase order.
- 20% of contract price after factory acceptance.
- 20% of contract price after field acceptance.

Vendor shall state his standard warranty.

Notes

CC. The commercial terms and conditions are greatly dependent on the corporate policies of each user. Normally, progress payments are used. This involves a few payments during the life of the project and allows the vendor to invest in hardware at the beginning of the project without having to borrow money. It also allows the user to withhold payments at the end of the project until all problems are resolved and the system is accepted.

CONCLUSION

The key to a good specification is planning. Before a specification is written, the entire project must be well thought out. The specification then transfers this planning into words. It should specify the functions the system must perform as well as vendor support services (such as maintenance, documentation, training, acceptance testing and spare parts). It should give the bidder all the information he needs to price his equipment and services and to write a technical proposal. This information includes installation dates, payment terms, import-export license requirements and other commercial considerations.

It is hoped that this typical specification and the annotated comments and explanations are useful to the reader in the planning of a computer project and the writing of the system specification. It is also hoped that this chapter gives the reader a feel for the size of the effort involved in a process control computer project. The effort is a large one, especially if it is an organization's first attempt at such a project. Perhaps this is why many users require assistance from engineering contractors and systems houses.

REFERENCES

1. *Hardware Testing of Digital Process Computers,* ISA-RP55.1, Pittsburgh, Pennsylvania, Instrument Society of America, 1971.
2. *Industrial Computer System Fortran Procedures for Executive Functions and Process Input/Output,* ISA-SG1-1, Pittsburgh, Pennsylvania, Instrument Society of America, 1972.

8.
Plantwide Multiplexing Systems and Computer Networks

Merrill G. Thor
Senior Control Systems Engineer
Stone & Webster Engineering Corporation
New York, N. Y.

INTRODUCTION

This chapter will discuss the use of digital data communications in both remote signal multiplexing and computer networks. First, different methods of digital data communications will be examined. Next, we will discuss different aspects of remote multiplexing and computer networks, including distributed control. Finally, we will look at the impact of distributed control on system reliability.

Computer control began to become popular in the petrochemical industry at about the same time that split architecture analog control systems began to appear. The configuration of equipment that has evolved is shown in Figure 8-1. A single computer is used to perform supervisory control and process monitoring. The computer monitors the process by reading analog and digital signals from the field via the process input/output equipment. It changes an analog controller set point (i.e., performs supervisory control) by sending either incremental signals (raises or lowers pulses) or an analog signal from the process to the controller. Field analog signals, such as flows, levels and chromatograph analyses, which are used by the computer and the analog panel instrumentation, are connected to both in parallel. The control room console consists of CRT displays, input keyboards and typewriters. Not shown in the figure are the various peripherals required by the computer, such as a card reader, bulk memory, line printer, etc.

This traditional configuration is still used in many installations today and it has had a profound influence on the technical and human aspects of computer control. For example, a considerable amount of effort and money has been put into the development of process control software for a single computer configuration and into the development and marketing of analog control instrumentation.

Plant operations personnel have been conditioned to think that a large panel

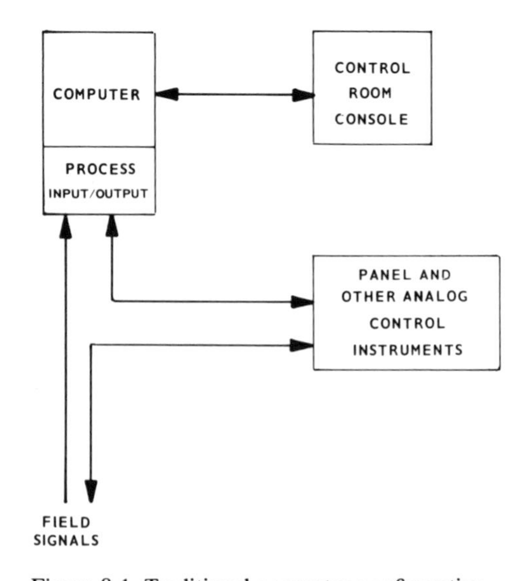

Figure 8-1. Traditional computer configuration.

board in the control room is a necessity. Also, both equipment manufacturers and operating companies have divided the responsibilities for computer control between two departments—one responsible for the computer system and the other responsible for the analog instrumentation. Because of inertia, a move to state-of-the-art equipment configurations has been slow to take place, but presently it seems to be picking up speed.

The traditional configuration is sometimes referred to as host multiplexing, to distinguish it from remote multiplexing. Before delving into remote multiplexing and computer networks, a discussion of digital data communications is in order.

DIGITAL DATA COMMUNICATIONS

Examples of communications via the transmission of digital data are widespread and have permeated into all aspects of society. When a check is cashed at a bank, a teller types the account number on a telephone touch-tone pad and a computer replies with the balance in the account. Large computer-driven scoreboards are visible at the most recent sports stadiums. Transactions on the large stock exchanges are transmitted throughout the country as they happen and can be seen in stock brokers' offices and on some television stations. Yet, visible uses of digital data communications are only a very small portion of all existing applications. Digital data communications is growing rapidly and has a deep influence on the lives of the people in the civilized world. This section will deal with the fundamentals of digital data communications, with emphasis on areas applicable to remote multiplexing and computer-to-computer communications networks.

Any facility which allows signals to be sent from one end to the other can be considered to be a communications channel. A leased telephone line, a co-axial cable and a microwave radio link are all examples of communications channels. Many different types of signals can be sent over such channels. For example, voice, video and digital data are commonly sent via communications channels. A communications channel provides a usable frequency range or band-width. Voice signals require a band-width of 1000 to 3000 Hz, depending on the desired fidelity. Commercial television signals require band-widths of up to 6 MHz. The band-width required by digital data transmission is a function of the rate at which the data is sent.

Digital data are normally transmitted on a channel as modulated signals. At the receiving end, the signals are demodulated back into digital logic signals. The transformation of digital logic to modulated signals and the transformation from modulated signals to digital logic is performed by a modulator-demodulator, commonly referred to as a modem. This is represented in Figure 8-2. Computers, teletypes and CRT terminals are examples of digital devices that use digital communications. It should be noted that communications over distances of only a few feet do not require modems; digital logic signals can be used for such short distances.

Communications channels can be divided into three types: simplex, half-duplex and full-duplex. Data are only sent in one direction on a simplex channel. Such a channel might be used by a receive-only (RO) teletype. A half-duplex channel allows data to travel in both directions, but in only one direction at any instant. Data can travel in

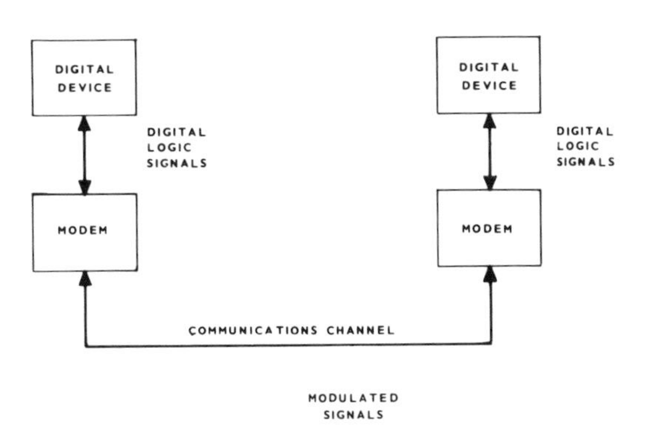

Figure 8-2. Digital communications.

both directions simultaneously on a full-duplex channel.

Modems can be classified by the type of modulation employed and also by the communications protocol adhered to. The protocol dictates the sequence in which data is to be transmitted, the structure of the message, the method of synchronization of the transmitter and receiver, and the error-checking procedure. Some popular protocols are IBM's Binary Synchronous Communications (BISYNC), DEC's Digital Communicatons Message Protocol (DDCMP) and IBM's Synchronous Data Line Control (SDLC). The use of communications protocols has alleviated much of the problem of data transfer between digital devices.

Data communications may be asynchronous or synchronous. Asynchronous data usually occur in short bursts. Synchronous data usually occur in long streams. Asynchronous data messages are usually preceded by a start signal, which serves to synchronize the transmitting and receiving modems. Synchronous data usually contain a special bit pattern which is used for synchronization. The pattern is defined by the communications protocol.

Since communications channels are subject to intermittent noise and signal failures, it is important to determine if the transmitted message is received without error. This is especially important in process control applications, where undetected errors could cause erroneous control action. One obvious solution would be for the receiving station to re-transmit the message received back to the transmitting device. This is an inefficient way of utilizing the communications channel since the message must, in effect, be transmitted twice. However, transmission rates of up to 1 million bits per second can be achieved today. In some applications, the balance between security of data and efficient utilization of the communications channel may tip towards data security. In applications where efficiency is important, there are several methods of error-checking that allow the receiving device to determine with a high degree of confidence that a message has been received correctly. The simplest but least secure method is a parity check. This involves adding an extra bit, called a parity bit, to each block of data in the message. This bit is set or reset so that the total number of binary ones in the block is always odd. (This method is referred to as "odd parity." Similarly, some systems use "even parity," where the number of binary ones in the block is always even.) Parity checks do not provide sufficient security for process control, and for that, other methods are used.

Three popular methods of error-checking are Vertical Redundancy Checking (VRC), Longitudinal Redundancy Checking (LRC) and Cyclic Redundancy Checking (CRC). Usually a combination of these checks are employed, depending on the message protocol. In general, a message is comprised of several blocks of binary data. The data may represent alphanumeric characters or numerical values. VRC involves adding a parity bit to each block of data. LRC adds an additional block to the message. Each bit in the check block is obtained by performing a logical exclusive or on each corresponding bit in the other blocks in the message. The receiving device also calculates what the contents of the check block should be from the data received. If the check block received is the same as the one calculated, then it is presumed that the message has been received correctly. Figure 8-3 shows an example of a combination of VRC and LRC. A more complicated

	BIT 0	BIT 1	BIT 2	BIT 3	VRC BIT
BLOCK 0	0	0	1	0	1
BLOCK 1	0	1	0	0	1
BLOCK 2	0	1	1	0	0
BLOCK 3	1	0	1	1	1
BLOCK 4	1	1	0	0	0
LRC CHECK BLOCK	0	1	1	1	1

Figure 8-3. Example of vertical redundancy checking and longitudinal redundancy checking.

method for calculating the check block is utilized in CRC. A polynomial is determined from the stream of data. This polynomial is divided by a constant polynomial. The remainder from the division is used as the check block. Bose-Chauduri is another type of cyclic polynomial code often used.

REMOTE MULTIPLEXING

The installed cost (i.e., the cost including materials and labor) of wiring and cabling has been increasing, while the installed cost of electronic digital systems has been decreasing. Remote multiplexing has evolved because it can drastically reduce the amount of field wiring necessary in a process plant. In fact, remote multiplexing systems are often called "wire replacers." Most often, remote multiplexing is employed in a manner similar to that shown in Figure 8-4. Several remote multiplexers are located throughout the plant. Analog and digital signals are wired to the nearest remote multiplexer (also referred to as a remote "Mux"). Each analog input to the remote multiplexer is assigned an address, as is each word of digital inputs. In the case of an analog input, the remote multiplexer converts the signal to a digital word or block, usually 16 bits long, by means of an analog-to-digital converter. The control unit sends signals to the remote multiplexer, which then requests the unit to send a particular block of data or a group of blocks. The remote multiplexer then sends the data requested to the control unit.

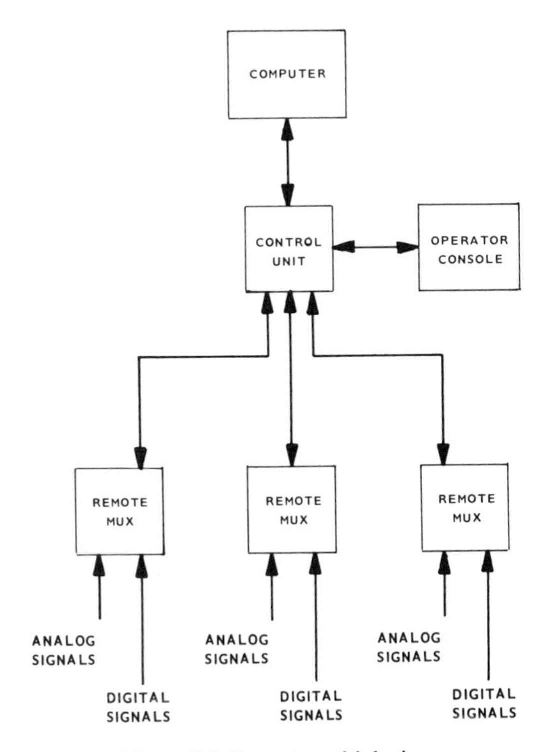

Figure 8-4. Remote multiplexing.

The control unit normally scans the data sequentially. As data are received, they are sent to the computer. Data can also be displayed on the operator console, at the request of the control room operator. If the operator or the computer requests a single word of data, the control unit interrupts its sequential scan and retrieves the required word of data from the proper remote multiplexer.

Variations to the above scheme are sometimes made. For instance, remote multiplexing can be employed in plants where there is no computer. The major benefit of remote multiplexing—reduction of field wiring costs—is still realized without a computer. In some cases, only analog signals are remote multiplexed. In these situations, a less sophisticated remote multiplexer, which sends the data to the control unit as analog signals, can be used. The analog to digital conversion is then performed in the control unit.

The remote multiplexing discussed above is concerned only with signals originating in the field and flowing to the control room. The scheme shown in Figure 8-4 does not show signals going in the other direction, from the control room to the field. This is because remote multiplexing is currently used for process monitoring but not for process control. It is used for non-critical signals; i.e., signals not used in a control loop. However, the reliability of remote multiplexing is high enough today so that it may be employed for critical signals as well. This can be accomplished by the equipment configuration shown in Figure

8-5, which we will refer to as "Total Remote Multiplexing."

Total Remote Multiplexing

This scheme accommodates both analog and digital signals flowing either to the control room from the remote multiplexer or *vice versa*. In the case of a flow control loop, for example, a flow transmitter provides an analog input to the remote multiplexer. The remote multiplexer converts the analog signal, usually 4 to 20 milliamps, to digital data by means of an analog-to-digital converter (ADC) and transmits it to the control unit. The control unit converts the digital data back to a 4 to 20 milliamp analog signal by use of a digital-to-analog converter (DAC) and a sample and hold circuit. The analog signal is connected to an analog instrument, such as a flow controller. The control output of the controller, usually a 4 to 20 milliamp signal, is converted to digital data in the control unit by means of an ADC. The control unit sends the data to the remote multiplexer. A 4 to 20 milliamp analog output is formed at the remote multiplexer, using a DAC and sample and hold circuit.

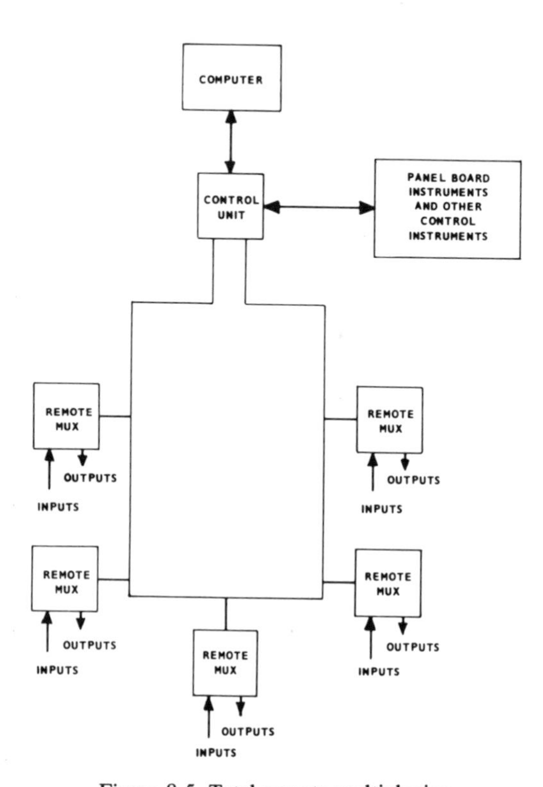

Figure 8-5. Total remote multiplexing.

Remote multiplexing equipment available today transmits data at a rate sufficient to update each analog value once a second. In petrochemical applications, analog values which are updated every second are, for practical purposes, the same as continuous signals. In the case of the flow control loop example, the flow controller is a standard analog controller. The flow transmitter sends a 4 to 20 milliamp signal and the controller receives a 4 to 20 milliamp signal. The accuracy of the ADC and DAC are high enough that conversion errors are negligible.

Pneumatic Multiplexing

Even though electronic instrumentation technology is increasing at a rapid rate, many petrochemical plants which employ penumatic instrumentation are still being built today. This is because pneumatic instruments are thought to be safer in hazardous locations than electronic instruments. Also, many plants in the past have successfully used pneumatic instruments and management is hesitant to make any major changes in a proven design.

There are many pneumatic-instrumented plants that use computers. There are two methods of interfacing pneumatic instruments to a computer system. The first is to use P/I (Pneumatic-to-Current) and I/P (Current-to-Pneumatic) converters. This approach provides improved performance because the output of the converters are continuous signals.

The second method is to replace the P/I converters with a pneumatic multiplexer. This can be accomplished with either a host or remote multiplexing configuration. A typical pneumatic multiplexer can handle six pressure inputs per second. Each pneumatic input signal, usually 3 to 15 psi, is converted to an analog DC voltage.

The analog signal is sent to a multiplexer, which is connected to a computer. The computer controls the stepping of the pneumatic multiplexer from one input to the next. Another approach is the use of a separate device to accept the analog signal from the pneumatic multiplexer and to control the stepping action. This reduces the need for special software in the computer.

The cost of a pneumatic multiplexer is considerably less than the cost of individual P/I converters.

COMPUTER NETWORKS

At first, computers were used for process monitoring only. Then closed-loop control became popular and overall plant optimization followed. The computer in the traditional configuration (Figure 8-1) became larger and larger as the number of functions performed increased. In a sense, this paralleled the growth of large main-frames used by the data processing industry. In data processing, there has recently been a trend away from the use of a large, single, central processing unit and toward the use of several small processors. These processors are usually located in close proximity to one another and are connected together in a network. The data processing industry refers to this approach as distributed processing. Most often, the processors are not dedicated to any single function, but are continuously assigned jobs to perform by a processor which oversees the network. The popularity of computer networks in data processing, along with the increased use of remote multiplexing, has given impetus to the wider use of distributed control in the process control industry. Distributed control is somewhat different from distributed processing, since it also makes use of remote multiplexing. The processors communicate with field-located remote multiplexers. Each processor is usually dedicated to performing the same job in an on-line environment. This section will deal with the basic principles of computer networks, as applied to distributed control.

The Point-to-point Network

The simplest network configuration is the point-to-point, shown in Figure 8-6a. A host processor (designated by the larger box) is connected by one line to a satellite processor (designated by the smaller box). The line between the processors is referred to in network topology as a link. A link is any communications channel, such as a coaxial cable, leased telephone line or fiber optic cable. Each device in a network is referred to as a node. Hence, the two processors are nodes.

The Multi-drop Network

The multi-drop configuration shown in Figure 8-6b is also commonly called a Data Bus, Data Highway or Multi-point configuration. The host processor controls the flow of data on the highway. Data can flow between any two nodes. Therefore, any satellite can communicate with the host or

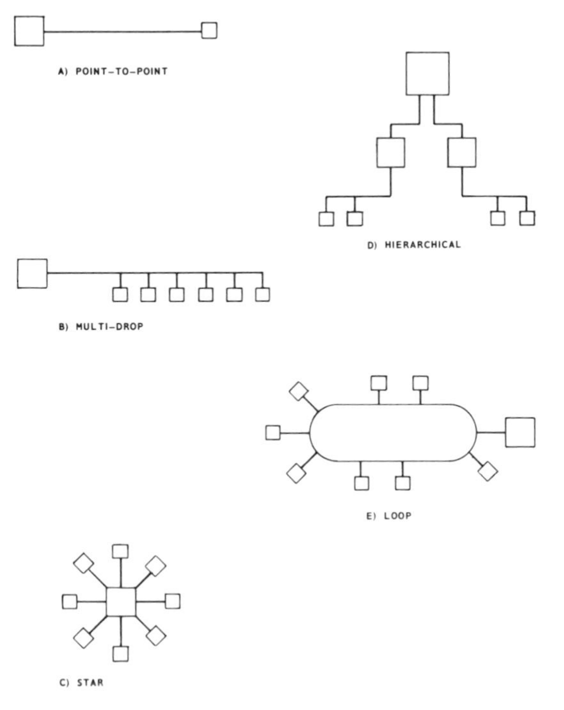

Figure 8-6. Computer networks.

any other satellite. Since there is a common communications channel, only two processors can communicate at any given time.

The Star Network

Figure 8-6c illustrates a Star configuration. In this type of network, the host processor is the center of the system and each processor communicates with the host. Communication between satellite processors is done by way of the host. This is also referred to as a Radial or Centralized configuration. This configuration lends itself to easy control by the host, and more than one satellite can talk to the host at the same time. However, the host is burdened with supervising data flow between satellites. Where large distances are involved, the extra cabling required by this scheme becomes a factor.

The Hierarchical

A three-tier Hierarchical system is shown in Figure 8-6d. An example of such a configuration is a computer control system used for two plants. The top tier is a large data processing computer. Two supervisory control minicomputers, one in each plant, occupy the middle tier. The bottom tier is filled with several microprocessors, which are used for direct digital control.

A second example of the Hierarchical configuration is a system employing two minicomputers as front-ends or data concentrators. A host minicomputer is on the top tier. The front-end minicomputers are in the middle tier and several microprocessors are connected to each front-end in a multi-drop configuration. This scheme offers increased reliability. If the top tier fails, the remaining processors will continue in operation.

The Loop Network

The Loop configuration depicted in Figure 8-6e is popular in remote multiplexing. If any single link breaks, all nodes can still communicate. This is also referred to as a Ring configuration.

DISTRIBUTED CONTROL

Approaches

There are two basic methods of distributed control: the Loop approach and the Unit Operations approach.

In the Loop approach, single satellite microprocessors perform a fixed number of functions. For example, a microprocessor might perform eight selectable functions. Examples of a function are P.I.D. control, square root extraction, lead/lag, dead time and curve following; these are basically the same functions performed by the standard elements of an analog control system. Conceivably, a single microprocessor could possibly perform a single function in eight different loops. However, normally, a microprocessor is dedicated to a single loop. In this way, a single failure will cause the loss of only one loop.

The Unit Operations approach involves the design of a separate control system for each unit in a petrochemical plant. A microprocessor is assigned to each unit. For example, a microprocessor might be assigned to the control of a distillation column or a furnace.

Advantages of Distributed Control

Some of the benefits of distributed control are obtainable by either approach. Other benefits can only be gained using the Unit Operation approach.

One of the principal advantages of distributed control over the traditional configuration is the reduction in field wiring material and labor. Since the microprocessor's input/output equipment is located in the field, signal wires must only be run short distances. This provides a significant savings in installation costs. Since the wire runs are much shorter, there are also fewer problems regarding induced interference on low-level DC signals.

The elimination of the need for a large control panel is another advantage. Using distributed control, the operator can run the plant from a CRT console. By eliminating the control panel, a large amount of instrumentation will no longer be needed, and, as a result, the control room can be considerably smaller. However, there is a great deal of controversy regarding this feature of distributed control. Some experts feel that a control panel is essential and that the operators cannot properly run a plant without one. Pro-panel forces argue that during a plant upset, more operators can operate simultaneously by use of a panel board. Those experts in favor of the CRT method feel that an operator can access more information and perform more control functions in a shorter period of time by sitting at a CRT console than he can by standing at a control panel, especially when the operator is concerned with more than one area of the plant. It is also felt by this faction that enough CRT consoles can be included in the system so that a sufficient number of operators can operate the plant at any given time, even during a plant upset.

The argument goes on. It appears that distributed control may someday be widely accepted and that a plant can be run without a panel board. The change may take place gradually, with panel boards growing smaller in size as operation from a CRT console gains acceptance. It certainly will not happen overnight. As one observer recently remarked: "If the petrochemical industry had been in charge of putting a man on the moon, they would have built the world's longest panel board and filled it with strip-chart recorders." Of course, a panel board can still be used in conjunction with a distributed control system.

The reliability and maintainability of a system using distributed control is improved. The cost of microprocessor hardware is so low that all processors in a Unit Operations configuration can be backed up by an on-line spare (See Figure 2-5, p. 16). This means that if one of the microprocessors fails, a spare will automatically assume its functions. Thus, the failure will have no effect on the operation of the plant. Since microprocessors are built on a single printed circuit board, they are easily repaired. It is usually only a matter of replacing one board.

The Unit Operations distributed control system is inherently modular. Computer control can be added to the units in a plant one at a time, without disrupting control of those units already on computer control. This is particularly attractive to an existing plant operating without computer control. The unit which is the most likely candidate for computer control can be established after a process study. The unit chosen may turn out to be the plant bottleneck, a large energy user, or one which is complicated to operate. This unit can usually be put on computer control with a modest capital outlay. If the effect is successful and the payout period is short enough, computer control can be added to other units in the plant. This piecemeal strategy is popular with users who have little experience with computer control.

Such projects can be started with a small investment. As technical experience on the first unit is acquired, the merits of computer control can be demonstrated to management and further investment in computer control can be justified.

REFERENCES

1. Digital Equipment Corp., Introduction to Minicomputer Networks, 1974.
2. Jenkins, Donald W., Centralized vs. Computer Control: What are the Trade-offs?, Instrument Society of America Conference, Niagara Falls, N.Y. October, 1977.

9. Compressor Surge Control

Aaron R. Kramer
State University of New York
Maritime College
Fort Schuyler, N.Y.

INTRODUCTION

Gas compression trains in the chemical industry consist of large turbine- or motor-driven dynamic centrifugal compressors ranging in size from 2500 to 50,000 horsepower. These compressors represent investments of millions of dollars per installation. Their operation is critical to the production efficiency of the process of which they are a part.

Many excellent articles have been written about the control of gas compressors and their problems.[1,2] These articles are generally concerned with various control strategies as a result of the engineering required for the number of compressors and the physical configuration of the compressor installation, for the process requirements and for prevention of compressor surge. The literature does not, as a rule, include discussions on computer application to compressor control, with the exception of discussions on simulation. It is the purpose of this chapter to present information on the causes of surge, to outline possible solutions and to show how mini-micro computer applications may assist in the implementation of these solutions.

WHAT IS SURGE?

Compressor surge may be defined as that point in the operating characteristics of the machine where a strongly pulsating flow occurs. The result is serious hammering and vibration. If this condition is not corrected, it could destroy the unit. This point exists when, at a given compressor speed, as flow through the compressor is reduced and the discharge pressure is increased, the pressure ratio exceeds the capability of the machine and a momentary breakdown of flow occurs. This flow breakdown is actually a periodic flow reversal, with reversing pressure differences across the compressor. These differences create forces which act on the impeller wheels, setting up vibration. An example of a surge condition analysis is given in the References at the end of this chapter.

Figure 9-1 shows a compressor characteristic curve. Consider the unit operating at a specific design point given by a flow, speed and pressure ratio (point *A* on the curve). The characteristic curves show that if flow is reduced by a closing off of some discharge valve(s), the pressure will in-

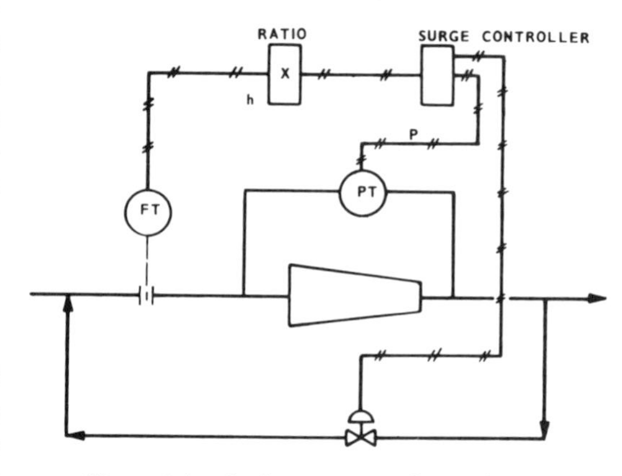

Figure 9-1a. Anti-surge control recycle gas.

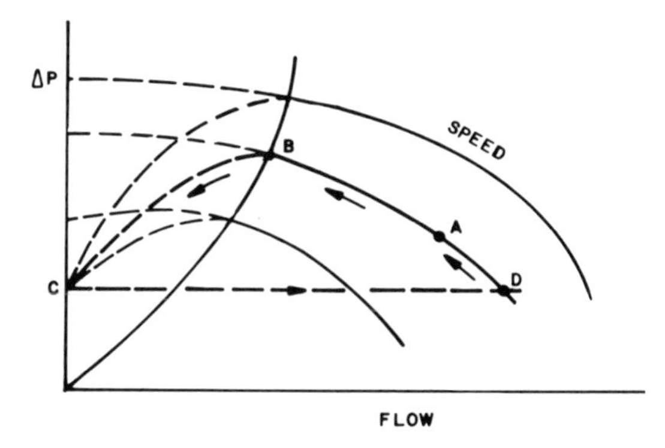

Figure 9-1. Compressor characteristic curves.

crease until the surge point is reached (point *B*). At this point, the flow pattern is destroyed and the pressure falls to zero (point *C*). When this occurs, due to loss of throughput, the compressor will attempt to deliver a new flow at point *D*. If the resistance of the system has not changed, the process will again move through the initial path, *D* to *A* to *B* to *C* to *D*. This indicates that there are two flow-rates available that can satisfy the pressure speed condition. The compressor, however, cannot choose an operating point; consequently, it cycles between two points. This is the surge condition.

SURGE DETECTION

The onset of the surge condition, when it occurs on an operating compressor, can be painfully obvious. The problem, however, is not the detection of surge when in progress, but the detection of incipient or potential surge.

The detection of a potential surge condition offers many obvious advantages to the control of the compressors and the plant and to the operating personnel. Among them are:

- Prevention of compressor damage.
- Prevention of product losses resulting from down time.
- Safety of personnel.

There are several techniques for the detection of incipient surge conditions. The most general methods are given below.

- Establishment of a surge control line based on differential pressure across the compressor (ΔP), differential of the inlet flow orifice meter (Δh) and a constant (K). The surge line is then defined by $\Delta P = K \Delta h$. If temperature variations are a factor, then $\Delta P = K\ \Delta h\ f(T)$, where $f(T)$ = temperature function correction.
- Establishment of a surge control line (assuming constant speed operation), using power input to the prime mover and the differential of an inlet flow orifice meter. The relationship $\Delta KW = K \Delta h$ remains linear for large temperature variations.
- A relatively new method of incipient surge detection is the measurement of a noise or fluidic pressure pattern.[3] This method utilizes pressure pulsations at the "incipient surge point" to detect this condition and to initiate corrective action.
- Another method of incipient surge detection, currently under investigation by the author, is the use of an adjustable narrow band frequency analyzer to detect characteristic frequency signatures indicative of surge. The characteristic frequency and its power spectrum are obtained via testing the machine. Once these characteristics are known, they are stored and used as comparisons for day to day operation. If a frequency characteristic is detected that equals the incipient surge data stored, corrective action is taken.

There are many other methods discussed in the literature concerning detection of surge conditions,[4,5,6] all of which have the same inherent disadvantage; namely, that of predicting an incipient surge point without a direct measurement.

Direct measurement techniques, such as those described above, provide some advantages over methods which approximate the surge condition by calculation.

Some of the advantages are as follows.

- Operation closer to the surge point is made possible.
- Dedicated hardware is furnished for this purpose.
- Overall loop gain compensation (stability) is more easily achieved.

COMPRESSOR SURGE IN SINGLE COMPRESSOR OPERATION

Causes of Compressor Surge

The circumstances under which a compressor can go into a surge condition may be broadly classified into three categories:

1. A process related category.
2. A control system related category.
3. Operator related categories.

The process related causes of compressor surge depend on the process-compressor configuration and on operational requirements. Generally, the process requires either a variable quantity of gas at a constant pressure, or a constant quantity of gas with limited pressure variation. Any change in the resistance of the process as seen by the compressor (such as a valve position change) constitutes a change in the load demand. This change causes a transient condition to occur in the pressure and flow parameter in the compressor. If the frequency of the transient is at or near the resonant process-compressor frequency or is near the surge point, the compressor may temporarily go into a surge condition.

Control system related causes of compressors going into a surge condition depend on all of the process related points discussed above and on the characteristics of the control system. The process-compressor interaction is fixed by the requirements of the process and its operating condition. This interaction is modified as dictated by the application of various control systems.

Installations of these control systems and subsequent tuning of the controllers fix the overall process-compressor control loop gain. The compression of the gas, system resistance and compressor speed are all non-linear functions, and as such, they cause the overall system gain to change. Fixed gain controllers cannot compensate for changes in loop gain, and should the loop gain increase, the tendency toward instability will also increase. If this occurs at or near the surge point, a surge condition may result.

Operator related causes of compressor surge condition are mainly operator inexperience; lack of knowledge of process-compressor interaction; and carelessness. A foolproof method to combat these shortcomings has yet to be devised. A thorough training program using computer based simulations would probably attenuate some of these problems.

Possible Solutions to Compressor Surge Problems

Solutions for the prevention of surge when operating close to the surge line have been presented in many publications.[1,2,3,4,5,6] This section outlines some of these solutions and offers commentary where appropriate.

Process related causes of compressor surge, as discussed previously, can never be totally eliminated. By design or definition, the process requirements are independent of the compressor. The compressor must perform up to its capacity as dictated by these requirements. It is assumed, for purposes of this discussion, that the *process* is the independent variable and the compressor itself can be considered to be the dependent variable. There are three general types of process demand-load classifications:

1. Frictional resistance variation in the process.

2. Constant pressure for the process.
3. A combination of both.

The three methods commonly used to regulate, or control, this demand are as follows.

1. Variable speed control.
2. Variable inlet vane angle control.
3. Variable inlet pressure control.

If the process demand is such that the compressor can operate between the surge capacity and the capacity at rated discharge pressure (stability limit), then surge may not become a problem.

If the capacity is reduced such that the surge condition is approached or, in fact, is reached, then a corrective action must be taken. Each of the three methods of demand regulation have their own advantages and disadvantages, as indicated. Therefore, key requirements for compressor control are based on minimum and maximum capacity limits.

Variable Speed Control

Given a manufacturer's compressor flow, speed and head "map," speed reduction or regulation provides a greater flexibility for capacity rangeability. However, a decision to use variable speed control is a function of compressor and process characteristics. If large fluctuations in load prevail, and the impedance output is small relative to the compressor, speed regulation is difficult and instability may occur. A compressor map with a flat heat capacity curve suggests that a small change in speed would give rise to a large change in capacity. If the process could accept a constant mass flow instead of a constant pressure requirement, then speed control offers a higher degree of stability.[2]

Variable Inlet Vane Angle Control

Variable inlet vanes operate by changing the gas flow quantity passing through the compressor on the inlet side. The characteristics of inlet guide variations at constant speed are much the same as that of variable speed. The surge line in this situation is the inverse of a variable speed surge curve. Thus, at constant speed, the dynamic range (stability range) is slightly larger than that of variable speed control.

Variable Inlet Pressure Control

Throttling inlet pressure changes the capacity in much the same way as variable speed control, with the exception that as the capacity is reduced, the outlet pressure remains constant. One consequence of this is that more is required for the same capacity.

Control Systems

The surge control systems most commonly used are combined with demand controls. These controls maintain the compressor capacity above the surge point by either venting to atmosphere any excess gas or by recycling the excess gas back to the compressor inlet. The methods to measure the surge point and surge line have been discussed earlier. Some methods of control have been suggested by M. H. White.[7] (See Figures 9-1, 9-2 and 9-3.)

In each of the three possible control configurations suggested, the primary consideration is the establishment of the surge line and its slope. ($\Delta P = Ch$). Where ΔP is the pressure rise across the compressor, h is the head developed by a flow meter and C is the slope of the surge line. When the calculation is complete, the system provides an output to the recycle valve which maintains compressor capacity and thus prevents surge. A. Rutshtein and N. Staroselsky[8] offer some criticism of conventional surge control. Based on tests they conducted using a computer simulation, they concluded that the conventional methods of control have incipient interactive characteristics which may, in fact, cause the compressor to go into a surge condition

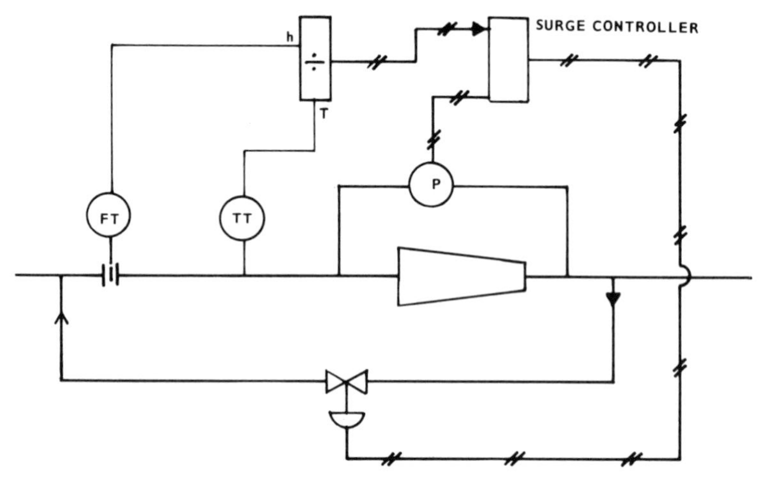

Figure 9-2. Anti-surge control with temperature compensation recycle gas. Reprinted with permission from *Chemical Engineering*, p. 60 (Dec. 25, 1972).

even though the actual surge point has not yet been reached. They attribute this phenomenon to the fact that the dynamics of the controllers (both capacity and surge), control valves, compressor prime mover (speed), and load disturbances cause first and second order effects to push the compressor into a surge condition. These dynamics cause the flow capacity controller to try to maintain capacity during a transient condition by driving the compressor operating point closer to the surge point, while the surge control attempts to keep the operating point on the surge control

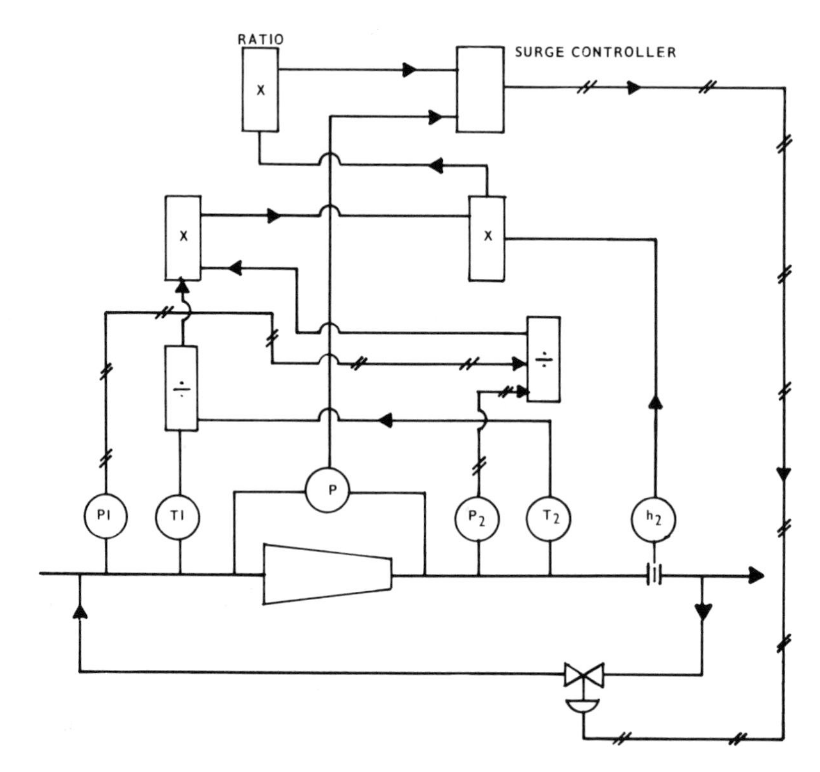

Figure 9-3. Surge control with flow device in compressor discharge.

line established by test. The result is a typical case of interaction between independent controllers. They further state that a control system with an independent anti-surge control loop cannot guarantee reliable protection even if the operating point is well away from the surge control line. Their solution, based on the simulation test, was to construct an integrated control system which de-coupled the interactive effects of the capacity control and the anti-surge control to maintain the operating point at or near the surge control line during steady state and dynamic conditions.

An illustration of a non-interacting control system offered by Rutshtein and Staroselsky is shown in Figure 9-4. According to their design, the protective module (6) and control module (5) control the recycle valve to maintain required mass flow rate. The protective module and performance module (7) control compressor speed according to the surge control line. In the system, according to the authors, the modules support each other instead of counteracting their functions.

Other systems for surge control are outlined in the literature. There is no so-called "universal best method" of centrifugal compressor control. The keys to good control are to define the objectives of the process; design an appropriate control system; simulate the system; and revise as necessary.

Operator Related Solutions

This author, involved in industrial control and education and training, has found that the majority of accidents in industrial plants are due to operator error. This conclusion is borne out by a great many insurance studies and by federal investigation of serious accidents. To prevent these operator

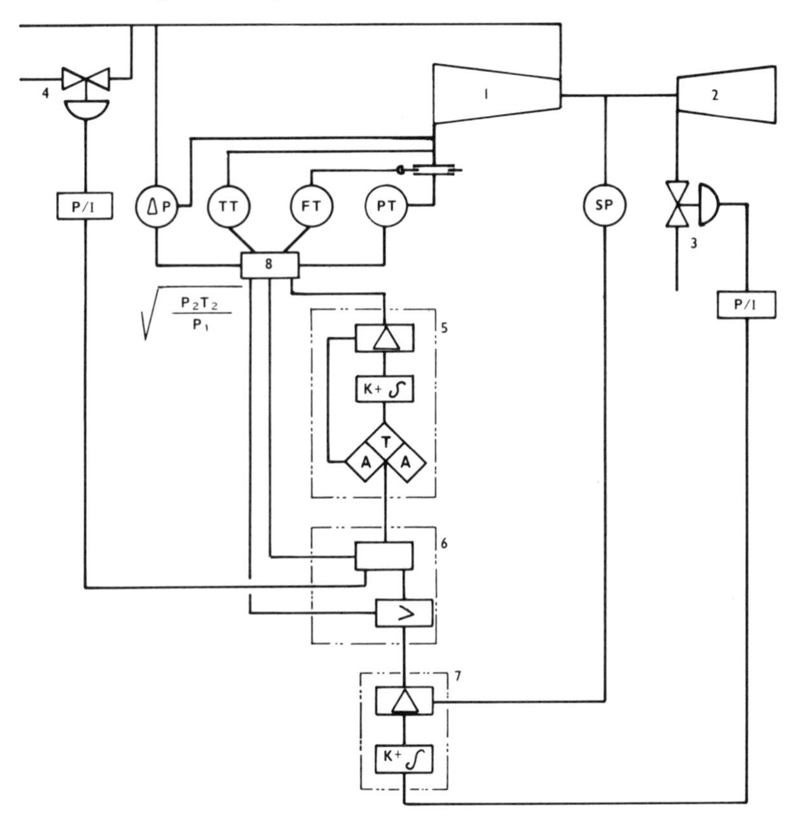

Figure 9-4. Non-interactive compressor discharge. Reprinted with permission from *Instrument Society of America Transactions,* Vol. 16 (No.2) 7.

errors, a serious effort should be made to train operators in compressor control. To this end, a training plan is offered (Figure 9-5). It is by no means comprehensive, but it does provide the necessary basic steps to qualify an operator for compressor supervision. The training program works best with a simulation the operator can manipulate. However, simple logic circuits with "yes" and "no" indications could work as

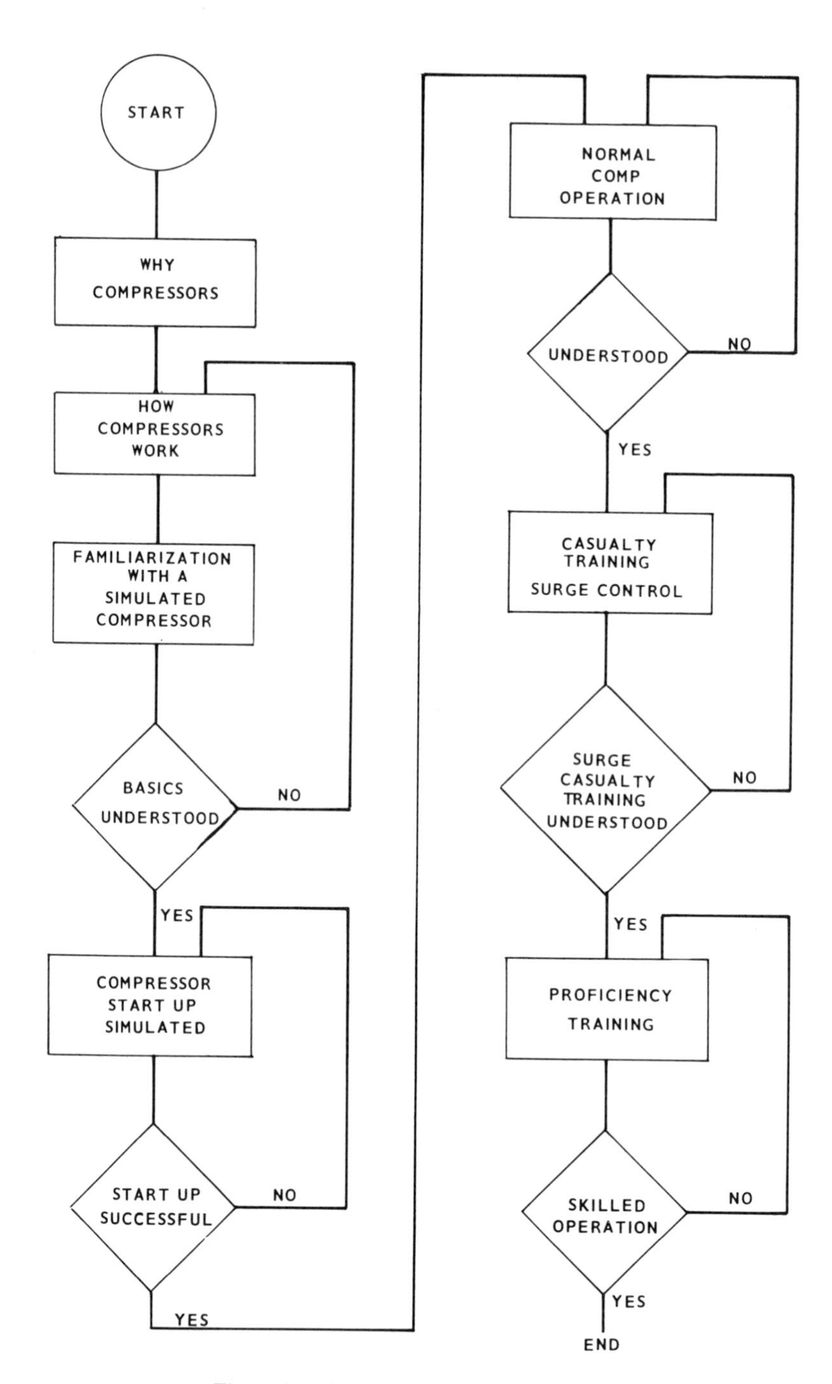

Figure 9-5. Compressor training program.

effectively, should simulation not be feasible.

The cost of compressor down time in any process far exceeds the cost of the training and equipment necessary for it. The industry can ill afford the luxury of allowing compressor operators to learn on-the-job. Therefore, the simulation approach (off-line) is a much safer and better approach to operator training.

Mini- and Microcomputer Applications

A compressor, as the name suggests, is a device which increases the pressure of a gas while delivering a quantity of this gas. The prime consideration in compressor operation is the prevention of surge conditions which—if not limited or acted upon with great speed—could destroy the machine.

The various control systems outlined earlier can all be implemented with a "mini" or "micro" configuration. The levels of involvement of these computers are much the same as those for a fractionating column (see Chapter 4), and are repeated here for convenience.

- Supervisory or set point analog control.
- Direct digital control.
- Communications interface with a host minicomputer.
- Optimization and other advanced control techniques.

The details of computer implementation and the explanation of each of the applications are covered in Chapter 4. The greatest potential, however, for the use of computer technology in compressor application lies in the flexibility and adaptability to advanced control techniques of such systems and optimization.

The operation of a compressor in a stable process condition offers no particular difficulty for its control. As the compressor capacity is reduced and the operating point approaches the surge line, control of any nature requires the use of an adaptive control mode to compensate for changing gain characteristics of the process and the compressor. This cannot be accomplished with conventional analog equipment. Digital equipment such as mini- and microcomputers can utilize developed algorithms which can perform these adaptive functions.

The mini- or microcomputer system can also provide the communications tool for preprocessing data related to detection of incipient surge. The minicomputer, for example, can perform a frequency spectrum analysis of the noise generated as a result of the compressor approaching the surge point. It can communicate with the microcomputer, which then can take whatever corrective action may be necessary.

Should the compressor become a governing factor in any energy conservation program, mini-micro computer technology will become an integral part of an optimization program. Optimization of the compressor with respect to efficiency, energy use, capacity or other parameters then becomes possible.

Mini- and microcomputers with advanced mathematical and control techniques have reached a degree of sophistication where the application of a control philosophy to prevent surge must be carried beyond empirical engineering. The advent of these techniques, along with the computer technology, offers the potential of providing closer control of the compression process along with greater reliability and safety.

COMPRESSOR SURGE IN MULTIPLE COMPRESSOR OPERATION

Causes

The conditions under which compressors go into surge conditions (discussed earlier) obviously also apply to multiple compressor installations. Process, control and operator related conditions are the factors which must be attended to in order to prevent surge conditions, whether the compressor is a single unit or is part of a number of compressors supplying process requirements.

The system difference between a compressor operating independently and a multi-unit installation is the interaction between the machines via their associated piping and process requirements. There are two basic compressor installation configurations: series connections and parallel connections. Figures 9-6 and 9-7 show these two configurations.

The causes of surge in multi-installation of compressors are compounded by the dynamics of the interconnecting piping system.

Series

In the case of a series connected compressor installation (Figure 9-6), two possible causes of surge exist.

1. If the compressors are operated close to the surge point, as dictated by the process requirements, and are close coupled (piping between them is short), a disturbance transient on one machine could set up pressure transients on the second, sufficient to push the machine into surge. This situation is caused by the lack of capacitance between the machines, in the form of piping, which cannot attenuate pressure variations. Conceivably, the series compressor installation may set up a second order underdamped effect that will cause surge problems and possible instability.
2. If the dynamics of the controls, particularly the natural frequency of the system, are closely related to the frequency response of the compressors and interacting piping, the creation of resonant circuit is a possibility which can cause oscillating difficulties. This, again, is due to the capacitance character of a compressed gas in a volume, and its energy storage capabilities. This energy, when released at the natural frequency of the system, will obviously cause considerable problems of oscillation.

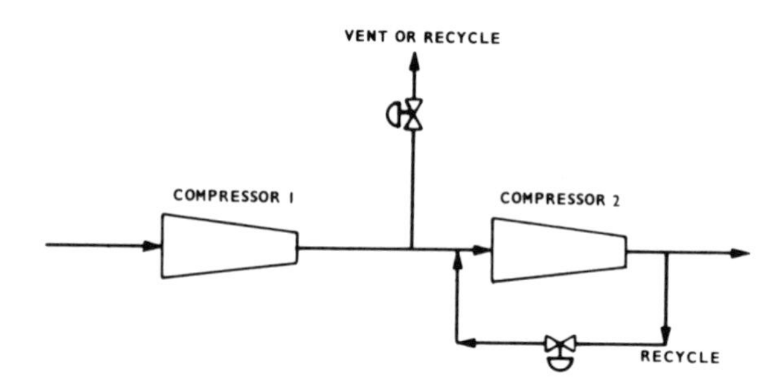

Figure 9-6. Series installation.

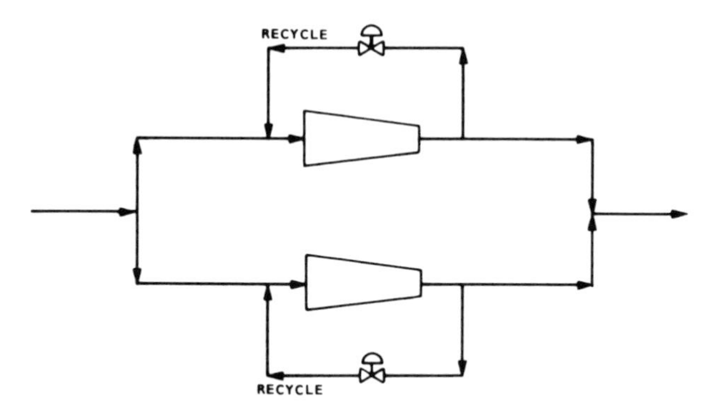

Figure 9-7. Parallel installation.

Parallel Compressors

Problems associated with parallel compressors have been studied at length.[9,10] Process related conditions which could cause surge conditions are primarily associated with start-up and load imbalance.

Start-up conditions in a parallel situation are, at best, difficult and dangerous. Because a compressor cannot begin to deliver a quantity of flow until its discharge pressure is equal to (or slightly in excess of) the pressure in the heater, the machine is prone to a surge condition before it gets started. The obvious solution is to recycle the gas of the machine being started.

The second problem of load imbalance occurs when the more efficient of two theoretically similar machines takes the entire load. The less efficient unit then operates with its recycle valve open in order to prevent surge.

A third problem is that of surge prevention through the use of recycle valves. The recycle flow from one compressor directly affects the flow and pressure in the entire system. The result could be a severe pressure transient, and again a possible surge condition may be created.

Possible Solutions

Process Related Solutions

The possible solutions for prevention of surge have been discussed previously with respect to single compressor installations. The process related solutions to multi-installation compressor surge problems still must follow the fundamental requirements of maintaining demand as dictated by the process; that is, overcoming frictional resistance, maintaining constant pressure or a combination of both.

The demand on each machine can, in theory, be regulated by variable speed, variable inlet vanes or variable inlet pressure.

One fundamental method of possibly reducing the effects of transients on multi-installations is based on knowledge of the dynamics of the capacitive effects of interconnecting piping. Engineering analysis via simulation techniques has been suggested and, in fact, has been performed many times.[4,10] Simulation in the analysis of these process interactions with multi-installations would be an invaluable tool in determining the stability of operation.

Control Related Solutions

Surge control systems are generally combined with capacity controls to operate the multiple installation compressor systems, either in parallel or in series, above the surge control point. M. H. White[7] and A. E. Niesenfeld[10] show various methods for control of both series and parallel arrangements (Figures 9-8, 9-9 and 9-10). G. Korn-

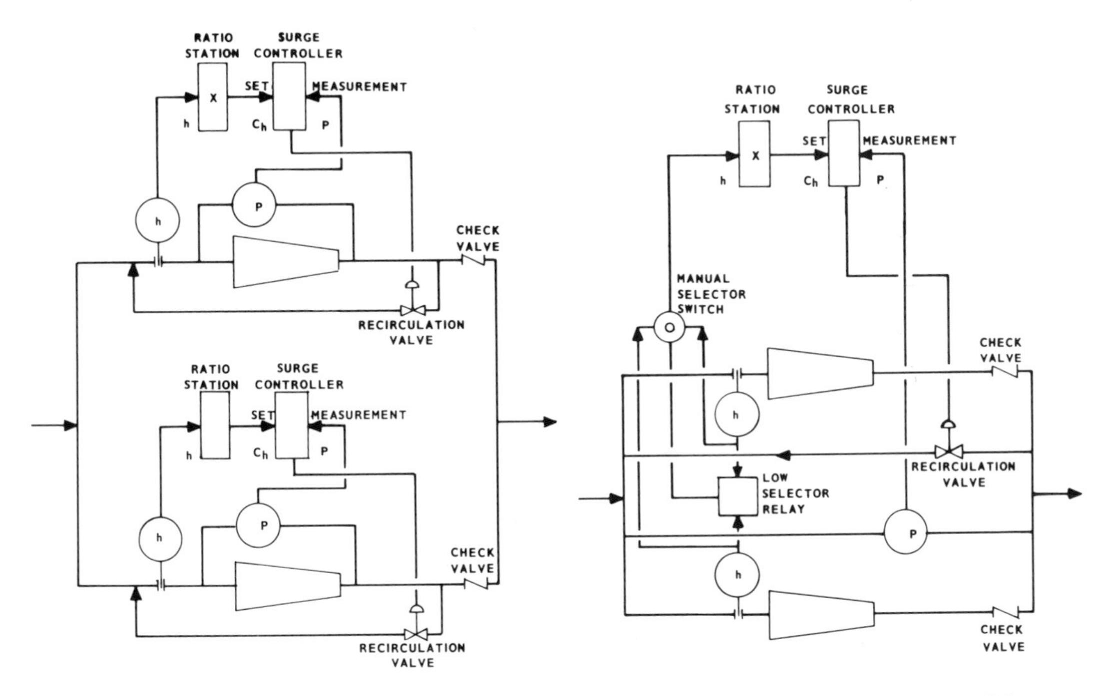

Figure 9-8. Alternate control arrangements for centrifugal compressors operating in parallel. Reprinted with permission from *Chemical Engineering* (Dec. 1972).

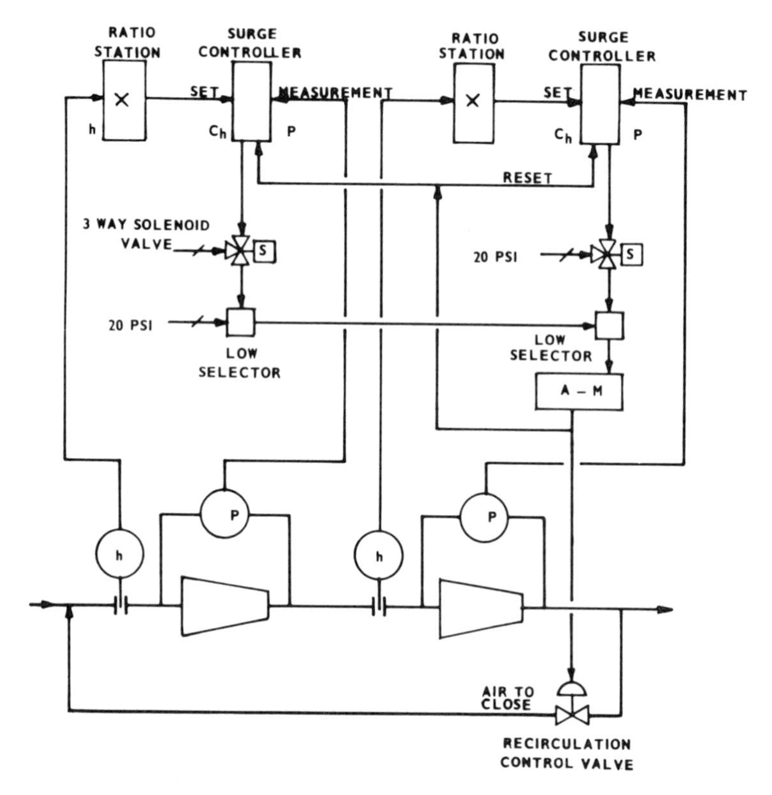

Figure 9-9. System for controlling compressors in series. Reprinted with permission from *Chemical Engineering* (Dec 25, 1972).

reich[11] shows a control system for a series-parallel system based on base loading all machines on demand, excepting one (and allowing that one to float). The surge controls are essentially independent of capacity control (See Figure 9-11).

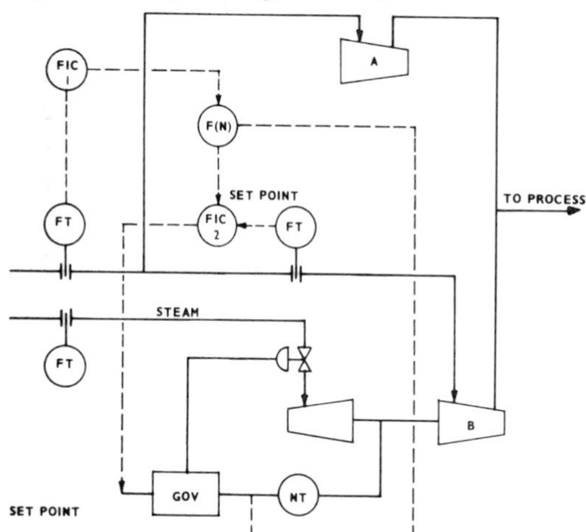

Figure 9-10. Load balancing control system. Reprinted with permission from *Hydrocarbon Processing* (Feb., 1978).

In all cases, the systems shown appear feasible and in keeping with the intended purpose of maintaining capacity and minimizing potential surge. The one ingredient missing from these systems is the consideration of piping configuration and its capacitance. Niesenfeld states: "There are no universal techniques for designing compressor control systems, especially for parallel compressor systems. Cook book solutions cannot be provided." Niesenfeld also strongly recommends that control systems for compressors, particularly for parallel compressors, be designed by dynamic simulations.

Operator Related Solutions

Solutions for operator related causes of compressor surge in multi-unit installations are not substantially different than those for a single unit operation. The exceptions are that with multiple units, the operator

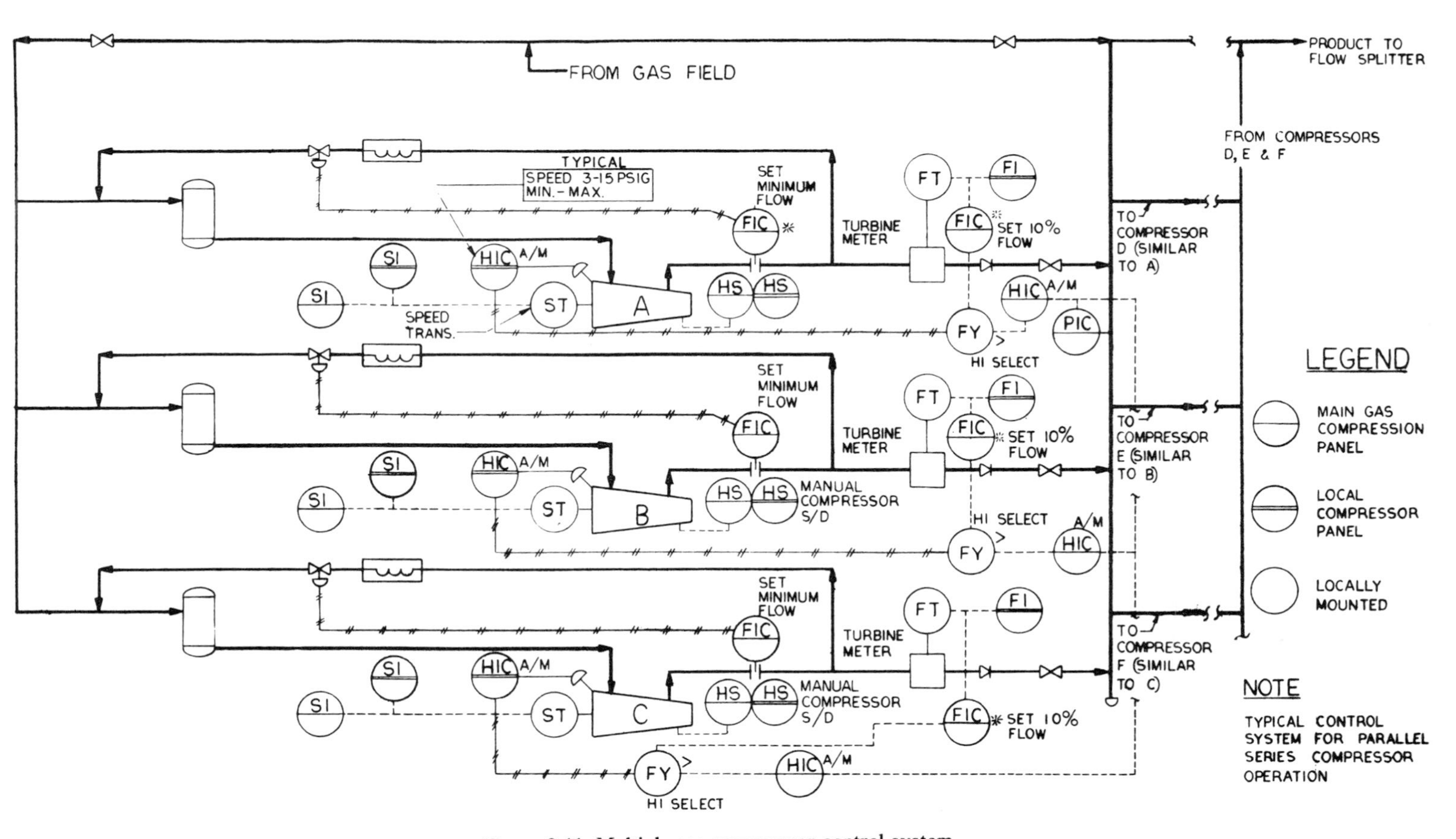

Figure 9-11. Multiple gas compressor control system.

must be well trained in recognizing interaction patterns which may occur as a result of the capacitive coupling effects of piping. This training and recognition procedure could be implemented in the proficiency phase or casualty phase of the training program outlined in Figure 9-5. Again, compressor down time and the resulting lost revenue far exceed the cost of training. The industry must consider operator education and training if it is to compete successfully in world markets.

MINI- OR MICROCOMPUTER APPLICATIONS

The general applications of mini- and microcomputers have been discussed earlier in this chapter. The specifics of computer applications to multi-unit compressor installation include a wide variety of special control schemes, optimization functions and special data preprocessing manipulations. This section outlines some of these applications.

Special Control Schemes

All control schemes shown earlier can be implemented with mini-microcomputer hardware and software. The prime advantage of digital implementation is the flexibility and adaptability of the software for variable gain control of capacity and recycle. The computer software also has the capability of decision-making for selection of a proper control strategy based on process demand, safety or approaching surge conditions.

The variable gain control is necessary to compensate for the pressure-flow rate compressor map, which is a non-linear function. General analog controllers operate with fixed gains throughout their operating range. At the high flow rates, the gain of the compressor map is high; that is, 2 $\Delta P / \Delta Q$ has a large negative slope. At the lower flow, the slope is almost flat; that is, $\partial \Delta P / \Delta Q$ is approximately 0. Since the recycle (bypass) control is activated by the ΔP across the compressor or a flow meter in the compressor discharge, the loop gain varies as the compressor map characteristic. The gain of the loop, therefore, should be characterized according to process demand. Variable gain control is difficult to achieve with conventional analog control. Safety, control strategy selection and detection of approaching surge conditions are functions of the control engineer's imagination, process considerations and plant management requirements.

Optimization Functions

Many installations involving compressors contain unit operations both upstream and downstream. The plant throughput is a function of all unit operations, their operating efficiencies and their characteristics. In a multi-unit operation which is an integral part of the total plant, it may be desirable (if not essential) to in fact integrate the control of these machines as a function of an overall plant optimization strategy.

Optimal control of the compressor is limited to its function within the strategy of the plant. In multi-unit installations with compressors of different capacities and characteristics, it is possible to distribute the load on each compressor such that demand is satisfied in total, with the least possible energy consumption by the compressor. The method is similar to the large utilities' energy distribution networks, where individual power generation units contribute a portion of demand power such that the total energy consumed in the generation of this power is minimum.

This method of load distribution does not imply that each compressor is operating at its own lowest energy consumption level or at its greatest efficiency, but rather that all compressors are operating so that the total energy consumption is minimal.

Data Processing Manipulations

Today, management of chemical plants require quantities of information in order to make intelligent decisions about modes and degrees of operation. Communication of shift data, maintenance data and performance of the plant are all expedited with the advent of computer technology. (These aspects are discussed in detail in Chapter 4.)

Of particular interest, in the application of a micro-minicomputer application in a compressor installation, is the ability of the computer to perform mathematical computations of a specialized nature, relating to signal analysis and conditioning.

Earlier in this chapter, the concept of a narrow band adjustable filter for frequency analysis of a compressor signal was introduced. A microcomputer, programmed with the well known Fast-Fourier Transform algorithms (FFT) could perform a spectral analysis of the frequency character of the compressor. Communication with a host computer that has access to a frequency signature of the compressor at an incipient surge condition allows comparison of the two signals and determines the likelihood of a surge condition occurance. Should the computer determine that surge is about to occur, corrective action is then initiated.

CONCLUSIONS

We have discussed the problems of surge in single and multiple compressor installations, focusing primarily on surge detection, causes of surge, solutions to problems and the applications of mini- and microcomputers. The surge characteristic is an unfortunate physical phenomena and is a severe limitation to the operation of a compressor installation. Since the principles of physics prevent engineers from eliminating surge, attention must be paid to its detection, prevention and control.

Detection of surge is generally accomplished by establishing a surge control line with a factor of safety somewhat to the right of the actual surge line. The control system then strives to control within the limits of this fictitious surge line.

Intrusive physical surge detection is currently being investigated. Two methods of such detection are the measurement of a fluidic pressure pattern as the compressor approaches surge, and a frequency analyzer which exhibits amplitude spectrums related to incipient surge.

Surge in single and multiple compressor installations stems from variations in process demand and the interaction between compressors due to capacitive coupling effects of connecting piping.

Several possible solutions outlined in this chapter indicate that prevention of surge requires the development of a surge control line, along which the compressor must operate. This method implies the existence of a safety margin, which (it is hoped) will mitigate movement into a surge condition during transient process upsets. Other possible solutions include the use of variable gain controllers, which compensate for changes in system gain as the compressor map changes with load.

At this point, it does not seem feasible to optimize a single compressor, because of its interaction with the entire process train. It *is* feasible to optimize the entire train with the compressor integrated into the program. Load sharing and energy management complete the optimal process with respect to compressors.

Mini- and microcomputer hardware and software offer alternatives to conventional analog control and provide the flexibility and capability to perform data preprocessing, variable gain control and communications with a host machine for many auxiliary purposes.

It is well known in the industry that computer technology has the potential for providing almost limitless methods for application to compressor control, both in single and multi-installations. These applications are limited only by the creativity and imagination of the control engineer. The ever-increasing complexity of processes and process control strategies and the increasing cost of energy almost assure the future application of this technology.

REFERENCES

1. Arant, J.B. and Crawford, W.A., "Compressor Control Problems Case Histories," *Procedings of the 15th Annual ISA Chemical and Petroleum Instrumentation Symposium, February, 1974,* Instrument Society of America, Pittsburgh, Pennsylvania.
2. Gaston, J.R., "Centrifugal Compressor Control," *Ibid.*
3. Process Controls, Inc., New Product, Houston, Texas.
4. Sweet, Stuart W., "A Look At Two Surge Control Systems." *Ibid.*
5. Hopkins, Kenneth L., "Centrifugal Compressor Control Problems." *Ibid.*
6. Shinskey, F.G., *Process Control Systems,* McGraw-Hill, New York, 1967.
7. White, M.H., "Surge Control for Centrifugal Compressors," *Chemical Engineering,* December 25, 1972.
8. Rutshstein, A. and Staroselsky, M., "Some Considerations on Improving the Control Strategy for Dynamic Compressors," *Instrument Society of America Transaction* **16,** No. 2, 1976.
9. Jerekjuin, E.A., "How to Control Centrifugal Compressors," *Hydrocarbon Processing,* July 1963.
10. Niesenfeld, A.E., and Cho, C.H., "Parallel Compressor Control—What Should Be Considered," *Hydrochemical Processing,* February 1978.
11. Kornreich, G., Stone & Webster Engineering Corporation, Private Communication.

10
Interfacing With The Systems House

Ralph E. Bothne

President,
EMC Controls, Inc.
Cockeysville, Maryland

INTRODUCTION

This chapter presents a definitive description of the functions of a systems supplier company versus the supplier of digital equipment (i.e., Digital Equipment Corporation (D.E.C.), IBM, Modular Computer (MODCOMP), Data General (D.G.), etc.). The systems houses of today had their start in the early 1960's, when computer equipment became available for industrial applications. IBM introduced the IBM-1800 systems, G.E. introduced the GE-4045 system and a small company in Massachusetts introduced a PDP-4 system. The computer age had become a reality to the American Industrial Community. This highly competitive industrial community was faced with the problem of types of applications and equipment selection. But the most serious problem faced was where to find the level of technical expertise necessary to implement a successful system. This talent was not available in the early 1960's, requiring the industrial companies to train their own people to become computer systems engineers. The large companies had essentially created the systems house concept by establishing central systems engineering groups that were responsible for all corporate computer installations. This philosophy existed until the early 1970's, when the talent began to break away and form small systems houses that could address computer installations for the smaller corporations unable to economically justify a systems house within their own corporate structures. The systems houses formed had certain characteristics that were dependent on the industrial background experience of the groups starting the companies. Thus, Systems House *A* could have in-depth experience in installing a computer system in a refinery and Systems House *B* could have in-depth experience in installing a computer system in a steel mill. Both could be successful in their related fields and equally disastrous in their unrelated fields. Since this systems house characteristic still exists, an understanding of the experience background of each systems house supplier considered is essential prior to selection of any one supplier. The functional requirements of the desired system should be evaluated relative to the experience level of the systems house. These functional requirements at a minimum should include the following.

1. A basic description of the process.
    - Serial continuous process.
    - Parallel continuous process.
    - Serial batch process.
    - Parallel batch process.
2. System functional requirements.
    - Data acquisition requirements.
    - Operator interface requirements.
    - Reliability requirements.

- Availability requirements.
- Control requirements (direct digital control, supervisory control, control interlocks and batch (time sequence of events) requirements).

The above basic functional requirements are necessary to determine the type of systems house to be selected. The technical level of the user should determine the degree of support required from the systems house. Technical support of systems houses can be defined as the level of system responsibility that specific houses will accept. The following is a method of differentiating this area of responsibility into system types.

Standard system. The user will use standard applications packages, designed and implemented by the systems house, usually running with a high-level operating system (e.g., D.E.C.'s RSX-11M). The user will take the responsibility of applying the standard system to his process.

Semi-applied system. The customer will generate the control and data acquisition configurations to be implemented by the systems house. The responsibility for the proper control philosophy and actual operation of the plant still belongs to the user.

Applied system. The systems house will generate the control configurations based on an in-depth knowledge of the process and control requirements. The actual physical installation will still be the responsibility of the user.

Turn-key system. The systems house will accept the responsibility for an applied system, along with the installation of the system. The supplier (in some cases) will be responsible for the selection and installation of process interface instrumentation.

The successful project requires that there is a complete understanding between the user and the systems house as to the level of system responsibility the user will accept and the level of system responsibility the systems house will and can accept. The best method to establish this understanding is for the user to define the system requirements and level of responsibility required in a detailed functional specification.

This chapter will discuss in detail the basic requirements necessary to define the functional requirements of the system; how to evaluate the systems house prior to negotiating a contract; and how to evaluate the systems house's performance during the life of the contract.

SYSTEM AVAILABILITY CONSIDERATIONS

Prior to the generation of the functional specifications, the user should select the type of system with regard to the requirements of the process which dictates reliability and availability. The classical definitions of today's available systems are given below.

- Single computer system.
- Resource sharing system.
- Redundancy system.
- Distributed process control system.

Resource Sharing System

The avoidance of resource sharing should be a prime consideration in the selection of any control system. This requires a basic understanding of both control system and process resource sharing concepts. The resource sharing control system is defined as a system in which each processor shares the load and is not redundant. A failure of any process controller could overload and cause a failure of the entire system (see Figure 10–1). The reliability of this type of system is the product of the reliabilities of the individual controllers.

$$R_{TS} = R_{U1} \cdot \mathrm{R}_{U2} \cdot R_{U3}$$

where

mR_{TS}= reliability of the total system
R_{U1}= reliability of the first controller
R_{U2}= reliability of the second controller
R_{U3}= reliability of the third controller

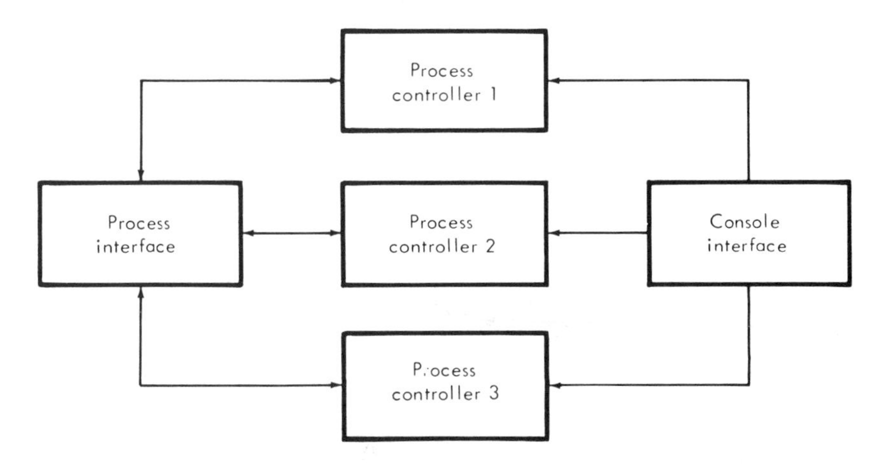

Figure 10–1. Resource sharing control system in which failure of any controller results in failure of system.

Using a reliability factor of 0.95,*

$$R_{TS} = 0.95 \times 0.95 \times 0.95$$
$$= 0.857.$$

This low reliability figure is intolerable for control of most industrial processes.

This concept of resource sharing not only applies to controllers, but also, in many cases, to the process itself. For example, in a process system with three unit processes in series (Figure 10–2), the overall reliability of the process is again the product of the individual reliabilities of the process units. If there were a separate controller for each process unit, in a distributed fashion, with a reliability factor of 0.95, the total reliability of the controllers

*If you think this number is too low, try getting a written guarantee from a computer vendor for a higher one.

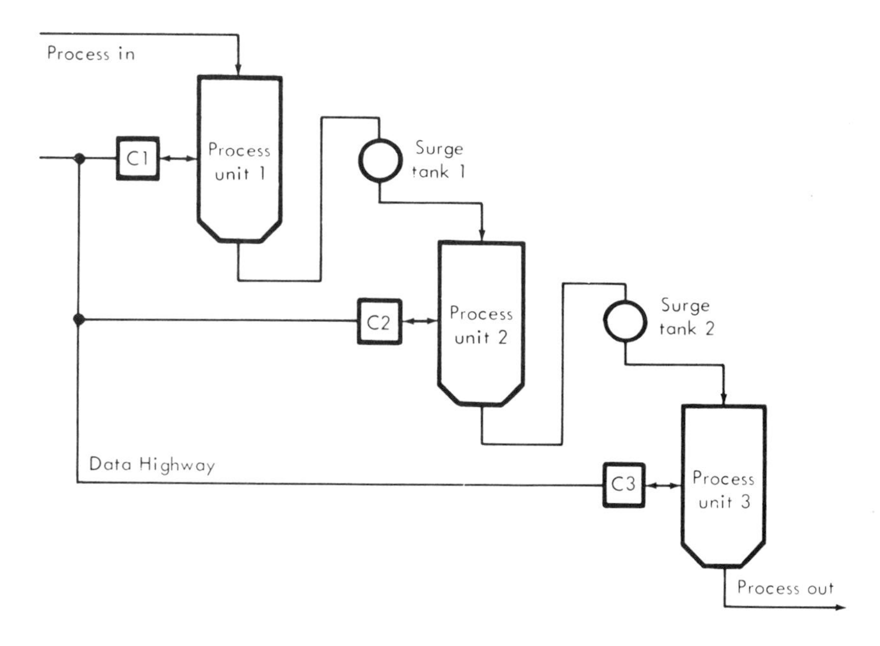

Figure 10–2. Serial processing units with distributed controllers result in multiplicative reliability factors and reliability similar to conventional analog controller practice.

would only be 0.857. With only one controller, the reliability factor would be 0.95. Thus, with a serial process, it appears that using a single computer system could result in a significantly higher degree of reliability.

Redundancy System

If two controllers in a redundant configuration were used to control the serial process, the reliability would be significantly increased (Figure 10–3). The total system reliability would be

$$R_{TS} = 1 - (1 - R_U)^2$$

where

R_U = the reliability of the controllers. In the case of two redumdant controllers,

$$R_{TS} = 1 - (1 - 0.95)^2 = 0.9975.$$

The redundant system reliability is much greater than that of the distributed system. It thus appears that, in a critical process, a redundant control system results in greater process reliability.

Distributed Process Control System

The current emphasis in industrial control is toward distributed control systems, in which one specific controller is responsible for only a well defined portion of the process. The advantage of this control concept is that a failure of a single controller will affect only the area with which that controller is related. Moreover, the advent of microprocessors has permitted design of distributed controllers with a high degree of modularity. This can greatly improve the availability of the system.

System Availability—The Important Factor

Availability is a critical factor that must be considered in the selection of the process control system. The availability (expressed in precent of real time) is the period of time that the system will be controlling the process. The availability of the controller is related to the reliability—which is directly related to the mean time between failures (MTBF) and the maintainability, expressed

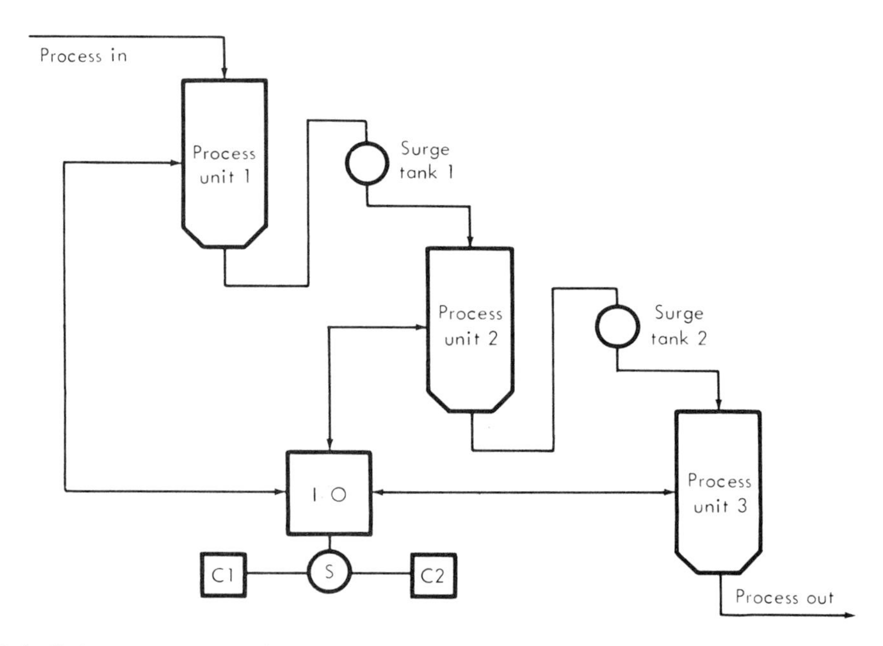

Figure 10–3. Series-type process with redundant central controllers has improved control reliability over conventional approach.

as the mean time to repair (MTTR) the controller.

This relationship is expressed as follows.

$$A_V = \frac{\text{MTBF}}{\text{MTBF} + \text{MTTR}} \times 100\ \%.$$

If the MTBF of a controller was 1000 hours and the MTTR was 1000 hours, the availability would simply be:

$$A_V = \frac{1000}{1000 + 1000} \times 100\ \%$$
$$= \frac{100}{2} = \qquad 50\ \%.$$

Assume the reliabilities of both the distributed and redundant controllers are the same, with calculated system MTBF's:

MTBF_D = 3000 hours

MTBF_R = 15,000 hours (for failure of both controllers at once).

But suppose the MTTR is 1 hour for the distributed controllers and 24 hours for the larger redundant controller. Then the total system availability would be

$$A_{VD} = \frac{3000}{3001} \times 100\ \%$$
$$= 99.90\ \%$$
$$A_{VR} = \frac{15{,}000}{15{,}024} \times 100\ \%$$
$$= 99.84\ \%.$$

The difference might seem insignificant. Expressed differently, however, the distributed system would be down 7.9 hours/year, while the redundant system would be down 14.0 hours—nearly double.

The MTTR ratios assumed above are considered typical, and the purpose of the whole exercise is to point out that the major factor in availability is not the MTBF, but rather the MTTR. With a basic understanding of MTBF, MTTR and availability, the control engineer should analyze his process to determine his best choice of control system type.

If the process is of a nature where a large cost is associated with a failure, regardless of down time, and a highly repetitive failure rate is intolerable, then serious consideration should be given to the redundant control concept. On the other hand, if down time (maximum availability) is the critical item, rather than the failure rate, then consideration should be given to the distributed control concept. If both the failure rate and the down time are critical items, the engineer should select a control system that will permit him to design a redundant distributed system.

The flexibility, modularity and good hardware communications modules offered today with the new microcomputer based systems permit the user to realistically consider his process control requirements. Serious analysis should be made of the microcomputer based control systems offered by control systems houses to determine:

- *The MTTR of each controller.* (Does the controller module permit plug-in replacement?)
- *Controller configurability into a redundant system.* (This feature should be a standard offering and not require any special integration design on the part of the user.)
- *Controller expandability to permit sizing to the process requirements.* (If the controller is limited to a small number of control loops or sequential logic steps, then multiple controllers may have to be used in a resource sharing configuration which will significantly lower the MTBF.)

Systems that provide all these features are available today from several reliable systems houses. The control engineer can now pay close attention to his process requirements, select the proper control system and generate the systems functional specifications. One of the most flexible and modular types of systems offered by several systems houses is a distributed control system. The functional description of a distributed control system, along with the type of operator interface displays available today, is presented in the next section.

TYPICAL DISTRIBUTED CONTROL SYSTEM

Figure 10–4 shows the general system architecture of one of several distributed systems available.

The true distributed control system should offer the user, as a minimum, the features described below.

Powerful, Versatile Process Control Modules (PCM)

Process control modules (PCM's) are microprocessor based controllers which may be located in close proximity to the process stage. These controllers should not be substitutes for conventional analog controllers. Each PCM should become a complete control entity unto itself after being down loaded once with the operational program from the host system. The down load of programs is under control of the communications control module (CCM). The operational programs should give the PCM the capability to perform continuous control, programmable logic control (time sequence of events) and data acquisition for all types of inputs (see Figure 10–5). Each PCM should be capable of communicating with other PCM's, but each PCM must not be affected by a failure within another PCM.

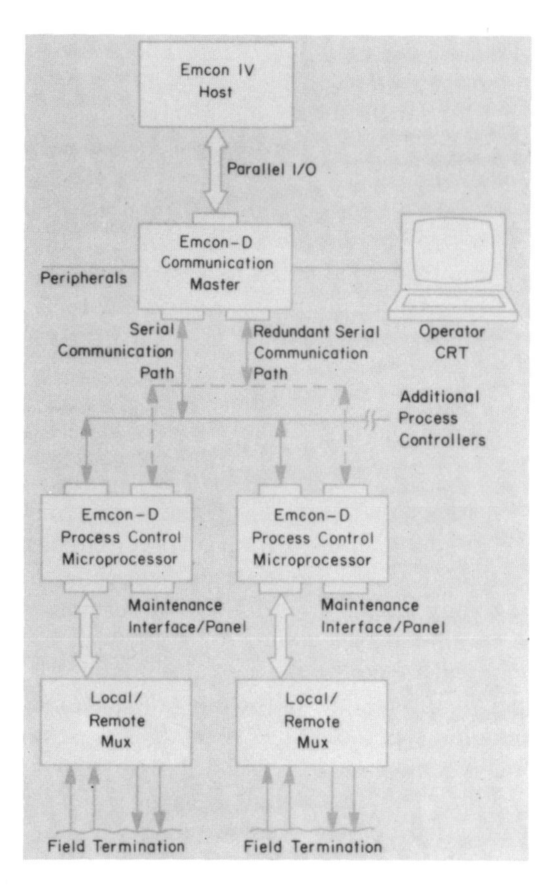

Figure 10-4. Emcon-D system architecture.

Distributed Display

This feature is one of the most important aspects of distributed control systems. The system should be capable of providing as many (or as few) operating positions as needed. Every operator position should have complete access to all system data, providing automatic back-up capability and improved security. (Beware of systems where one operator controls one group of loops and does not have access to others; losing control of half the process is really losing control of the entire process.) The interactive shared displays allow an operator/engineer to view every facet of a process in a format that provides maximum information and improved efficiency. The varied types of full color displays put the operator in control of the process in a way never before possible. All operator displays consist of real-time data which are continuously updated.

While improving operator efficiency, the shared display greatly reduces panel and instrumentation costs, along with related maintenance costs. The color CRT displays completely eliminate the need for controller face-plates, annunciator panels and pen recorders. Operator efficiency is maximized by the tremendous amount of information at the operator's fingertips, available through simple entry using dedicated keys on the CRT keyboard. Since all of the information is available at one central operator's console, the operator no longer

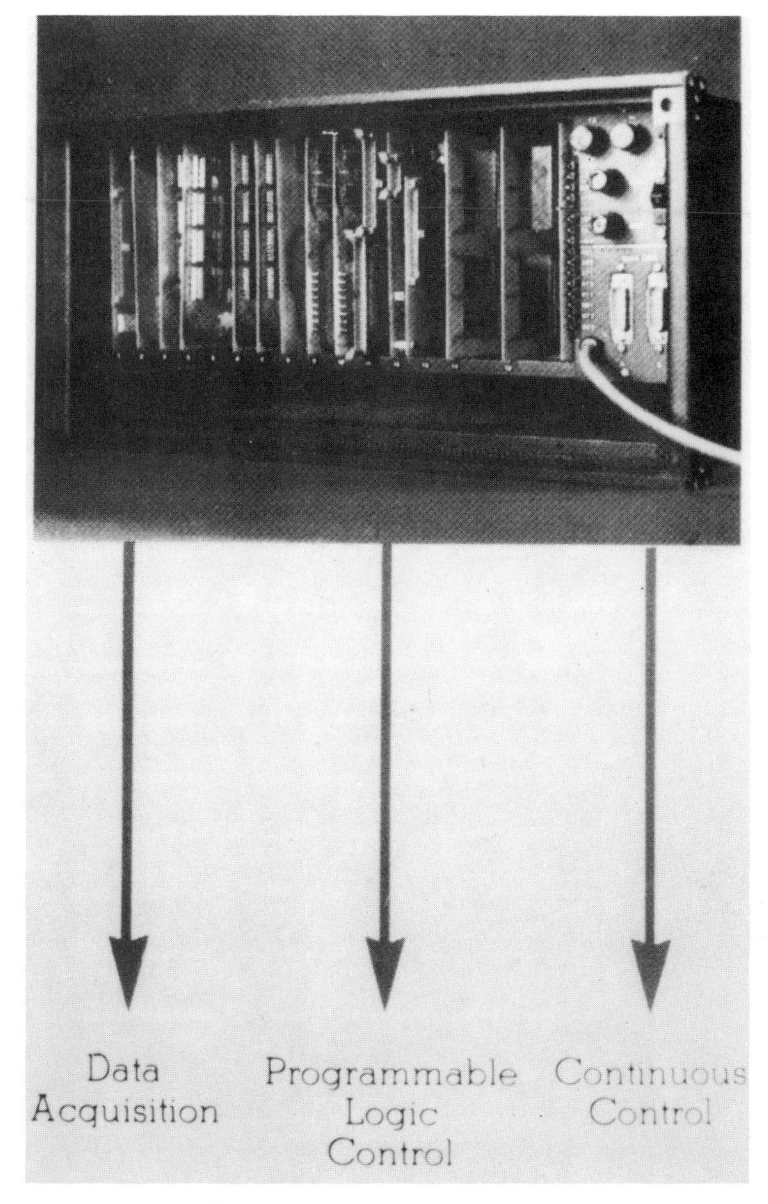

Figure 10-5. Process control module.

needs to run all around the control room, adjusting multiple types of controllers to control the process. Thus, the operator does not forget adjustments, makes fewer mistakes and keeps throughput and quality at a maximum, where they belong. Here are just three of the varied types of display formats that the selected distributed system should offer (Figures 10–6, 10–7 and 10–8).

While the operator interface is the most visible feature of the distributed system, the following features have given these systems' architecture universal acceptance.

Display of percentage of deviation from set point for 36 groups of 8 loops each (288 loops).
Percentage of deviation from set point, green for above set point, yellow for below. Zero deviation is displayed as a blank.
Prompting for operator to select one of the 8-controller group displays, if desired.
GRP
Deviation has exceeded low alarm limit (flashes)
Deviation has exceeded high alarm limit (Flashes)
Present time
8-FEB

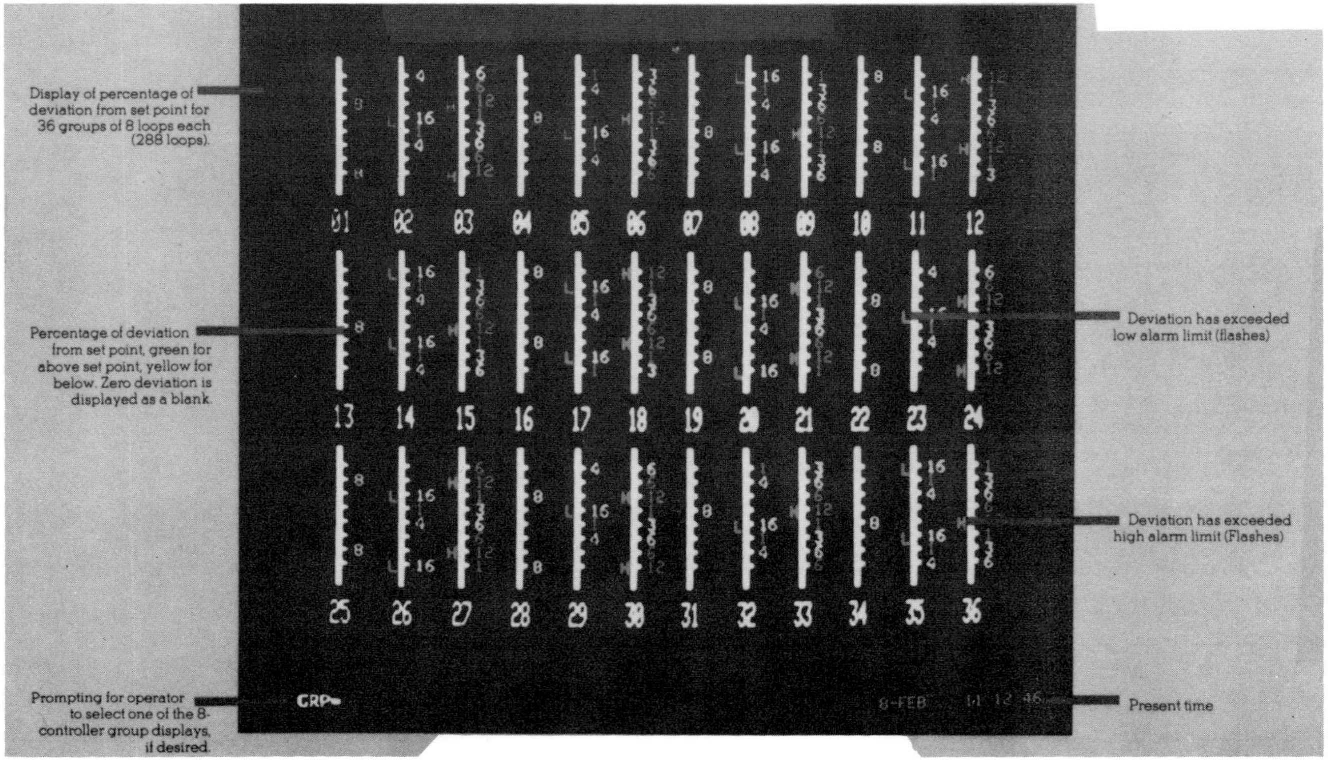

Figure 10–6. Total process summary.

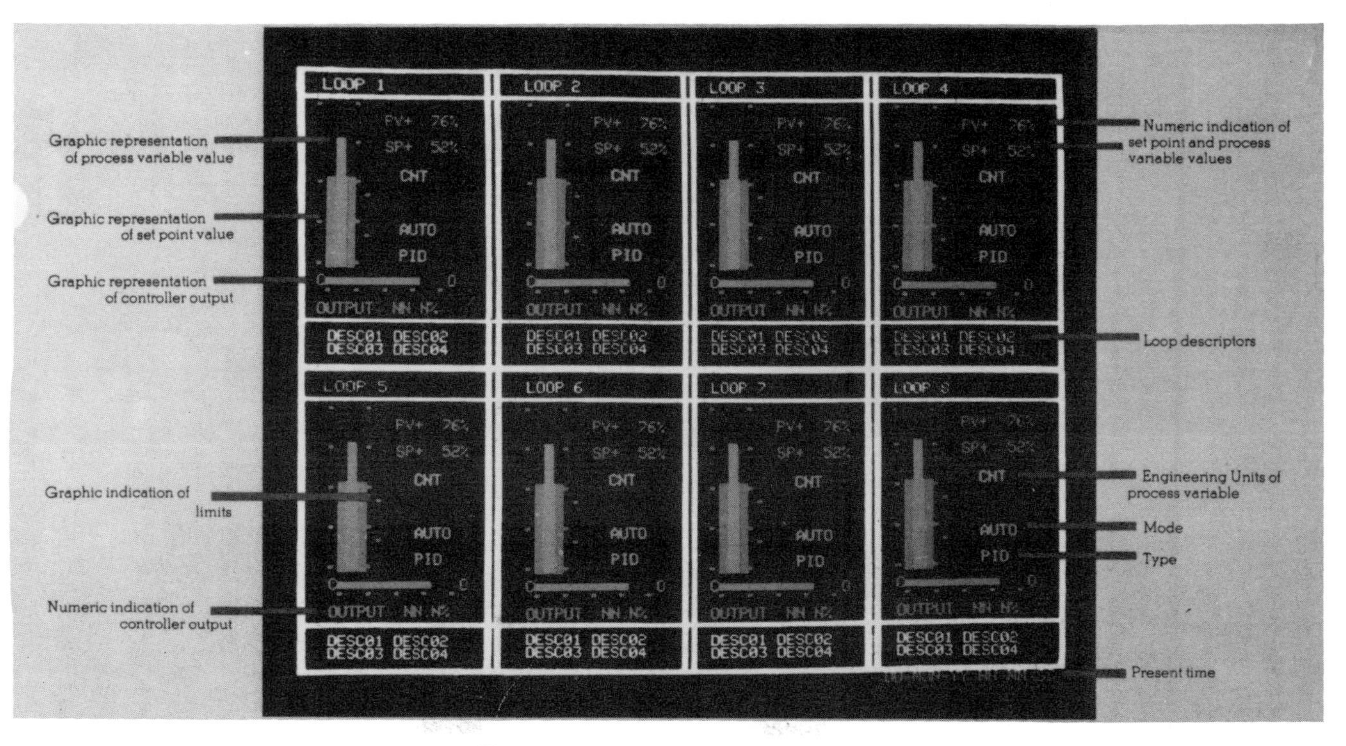

Figure 10–7. Eight-loop group summary.

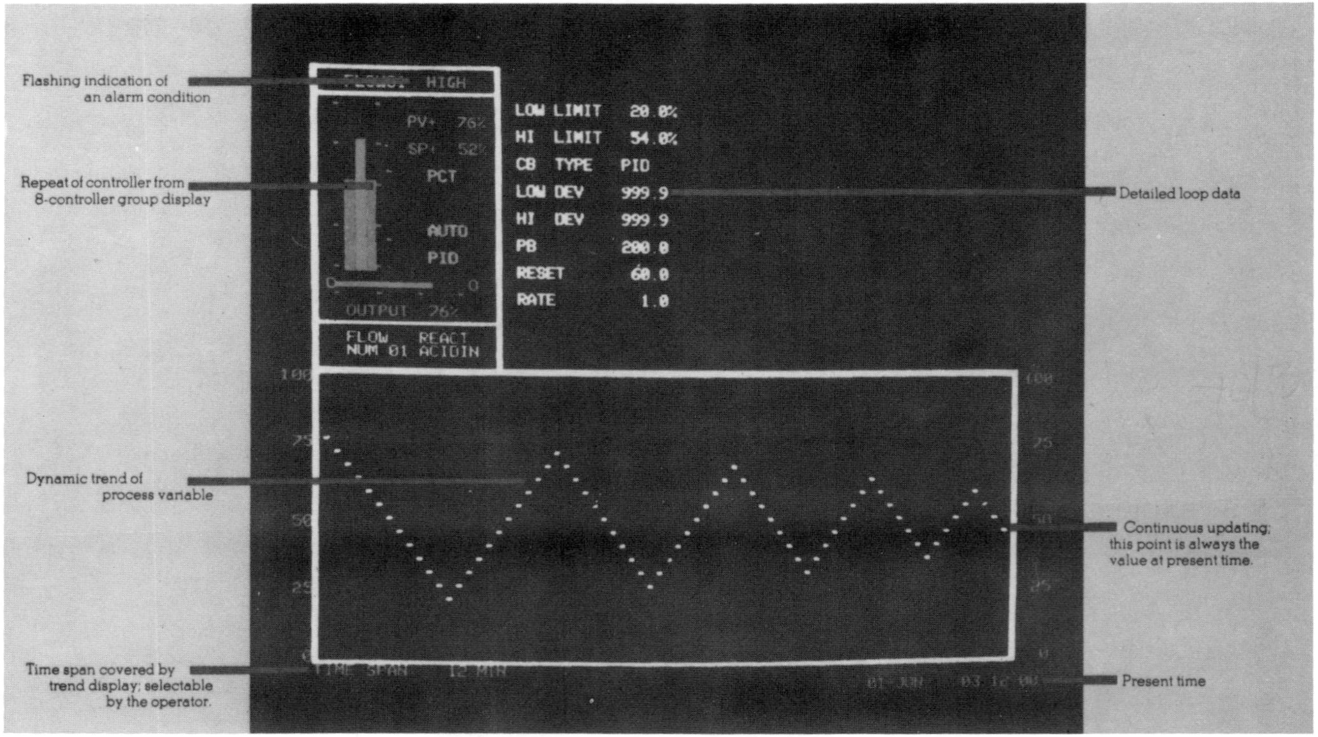

Figure 10–8. Single loop display.

- Distributed liability through the use of multiple independent process controllers (process control modules).
- Process control modules which perform data acquisition.
- Process control modules which perform batch and continuous control.
- Multi-level control utilizing the host computer.
- Host-independent intercommunication between process control modules.
- Analog back-up for critical loops.
- Peak operator efficiency through shared display.
- Total process visibility in full color—summary display, P.I.D.'s, profiles.
- Easy definition of data base and control strategies through user-oriented languages.

The basic attributes of a distributed control system is that a single failure cannot affect more than a limited, well defined area of the process. Consider the PCM's, each of which should be distributed to perform in a specific area. All data acquisition and data reduction functions—including scaling, compensation, linearization and limit-checking—are performed within the PCM. The PCM requires only communication to the operator at a shared display via redundant communications control modules.

The Host Computer

The distributed system should incorporate a powerful new generation computer (e.g., D.E.C.'s PDP 11/34 or higher, depending on system requirements). The host permits definition of the data base and control strategies in a conversational, user-oriented language, and also performs optimization routines in the background for downloading to the PCM's. Failure of the host affects the system only by temporarily removing these capabilities; the basic control system continues to function, unaffected.

Redundant Communications Control Modules (CCM's)

Communication between the PCM's and between the PCM's and the operator are accomplished through a communications control module. Thus, the CCM becomes a key function, and a good distributed control system should employ redundant CCM's. Reliability predictions indicate the mean time between failures of both modules—simultaneously—is in the order of years. The communications security between the CCM and the PCM should be attained by the use of a cyclic redundancy check for each data transmission. In addition, each transmission from the CCM to the PCM should be retransmitted back to the CCM and checked, bit for bit. This high degree of security is absolutely necessary for process control.

Low Mean Time to Repair (MTTR)

The system should include a built-in watch dog and other diagnostic testing features necessary to provide a rapid indication to the operator of a failed PCM. The changing of a failed PCM should only require removal of four wing-nuts and unplugging the process I/O cables and the communications link cables. This operation should have the PCM out of the rack and ready to be replaced by a spare in minutes. When the spare is energized, it should request a download from the host and, with a command from the operator, take control back from the standby analog controllers.

Ease of Definition or Change of Data Base and Control Strategies

At the system terminal, the engineer should be able to change a limit, define a new control strategy or download a new batch recipe to a PCM in seconds. The system should include a conversational, user-oriented language requiring no knowledge of software, hexadecimal codes or confusing mnemonics. These features are important in a distributed control system. Integration of these features in one complete system, capable of performing supervisory control, optimization routines and scheduling, provides the capability for payout limited only by the process engineer's imagination. Remember: employing microprocessors, utilizing a "data highway" or having a shared display do not automatically make a system a true distributed process control system. The system you are considering must be checked closely. If it lacks some of these features, consider the impact on your process due to lack of reliability, availability and capability.

Distributed micro based control systems as defined in this chapter are available as standard systems from several systems houses that are presently competing with the traditional instrumentation vendors (i.e., Foxboro, Taylor, Honeywell, etc.) for a share of the process control market.

GENERATION OF SYSTEM SPECIFICATION

The purchase of a control system from a systems house greatly differs from the purchase of equipment from the computer supplier (e.g. DEC, IBM, MODCOMP, etc.) in that the system to be supplied must be specified in detail. The equipment suppliers furnish equipment specifications for the user to evaluate prior to purchase. The systems house, in most cases, will purchase the equipment and add value to it—either additional hardware or applications software—and will price this added value as specified by the user. The interpretation of the specifications by the systems house is critical to the relationship between the parties; in addition, it will determine the general success of the project. The system specification is the basic foundation for good communication and agreement as to what is to be delivered for a given price. (A general basic specification for a shared display, distributed process control system as shown in figure 11-4 is given in the appendix.)

Generation of the Proposal

The systems house will review the specification with the user to determine the actual intent. (If the specifications are detailed enough, the systems house can eliminate this review.) The experienced systems house will question the user to determine whether he understands the specifications (in many cases, the user will actually copy a general specification without fully understanding the implications of various items specified). The systems house should determine the technical competence level of the user to determine the level of support required (e.g., training, applications engineering and project management) during the life of the project. The proposal by the systems house will reflect the specifications, whereas a very general specification (one without enough detail as to the actual system requirements) will require a very detailed proposal by the systems house.

The proposal is actually an implementation document. The user must require the systems house to indicate specifically where the proposal takes exception to the specification. If a special requirement is not detailed in the specification and is ignored by the systems house in the proposal, it could become a point of contention at a later date. In general, the proposal should contain the following, as a minimum.

1. System description.
2. Equipment list.
    - Maintenance and service support.
    - Recommended calibration equipment.
    - Recommended spare parts.
3. Basis for proposal.
4. Project execution.
5. System documentation.
6. Client training.
7. Pricing.
8. Terms and conditions.

System Description

The system description is a general overview of the proposed system. While it is general, it should detail all the systems submodules and describe the general functions of each. As an example, the PCM functional requirement may only be data acquisition, but the total capabilities of the PCM should be presented in this section. Then the user can determine which features could be available to him (with acceptable additional cost) at a later date. The following is a typical overview for a distributed micro based computer, as seen in Figure 10–9. The basic components of the system are listed below.

- Host (PDP-11/60 w/128K words) w/ LA-36 system terminal.
- Dual removable disk drives (7M words each).
- Redundant communications control modules (CCM's).
- Five LA-36 terminals (two driven by CCM's).
- Distributed process control modules (PCM's) and process I/O.
- Display control modules (DCM's) for back-up operator's interface.
- Redundant communications links from CCM's to PCM's.
- Four operator's color control consoles.

The basic design philosophy of the proposed system architecture is to provide the following.

- True distributed control (continuous and batch) and data acquisition at the first level of the system hierarchy (PCM's).
- Centralized operator's interface driven by the host, with back-up provided by DCM's.

The important advantages of this architecture are as follows.

- Redundant PCM's can be provided for increased system reliability.
- Fewer components are required at the first level than are required by other systems available because of the versatility and power of the PCM.
- The PCM is configured to meet the process, not *vice versa*.
- Single failure of host, CCM or DCM will not cause loss of any operator's CRT.
- Alarming is not dependent upon a specific CRT microprocessor but instead is performed at the PCM level and routed to the appropriate CRT's via the microprocessor/host network.
- Digital alarms are also monitored at the PCM level. Alarming is not dependent on the host. Digital points can also be defined such that when the point goes into alarm, a special applications program will be executed.

The host supports memory management (up to 128K of main memory); conversational down-line loading of microprocessors; alarms; displays; and many additional features that are beyond the scope of the specification. (The host is based on D.E.C.'s PDP-11/60 computer with 128,000 words of parity memory.)

The dual disks are used for system and user program storage, PCM data base images, historical data collection and process graphics. The system can be configured so the background disk is periodically updated by the primary disk on a regular basis, for system security. The dual disk configuration offers better system reliability with much more versatility than does one fixed head disk, for the reasons given below.

- All control functions are resident in PCM's and are not disk-dependent.
- All critical host programs can be fixed in the CPU memory to ensure continued execution independent of the disks (e.g., operator's interface).
- Reloading the operating system of the host is accomplished in a matter of seconds (instead of minutes).

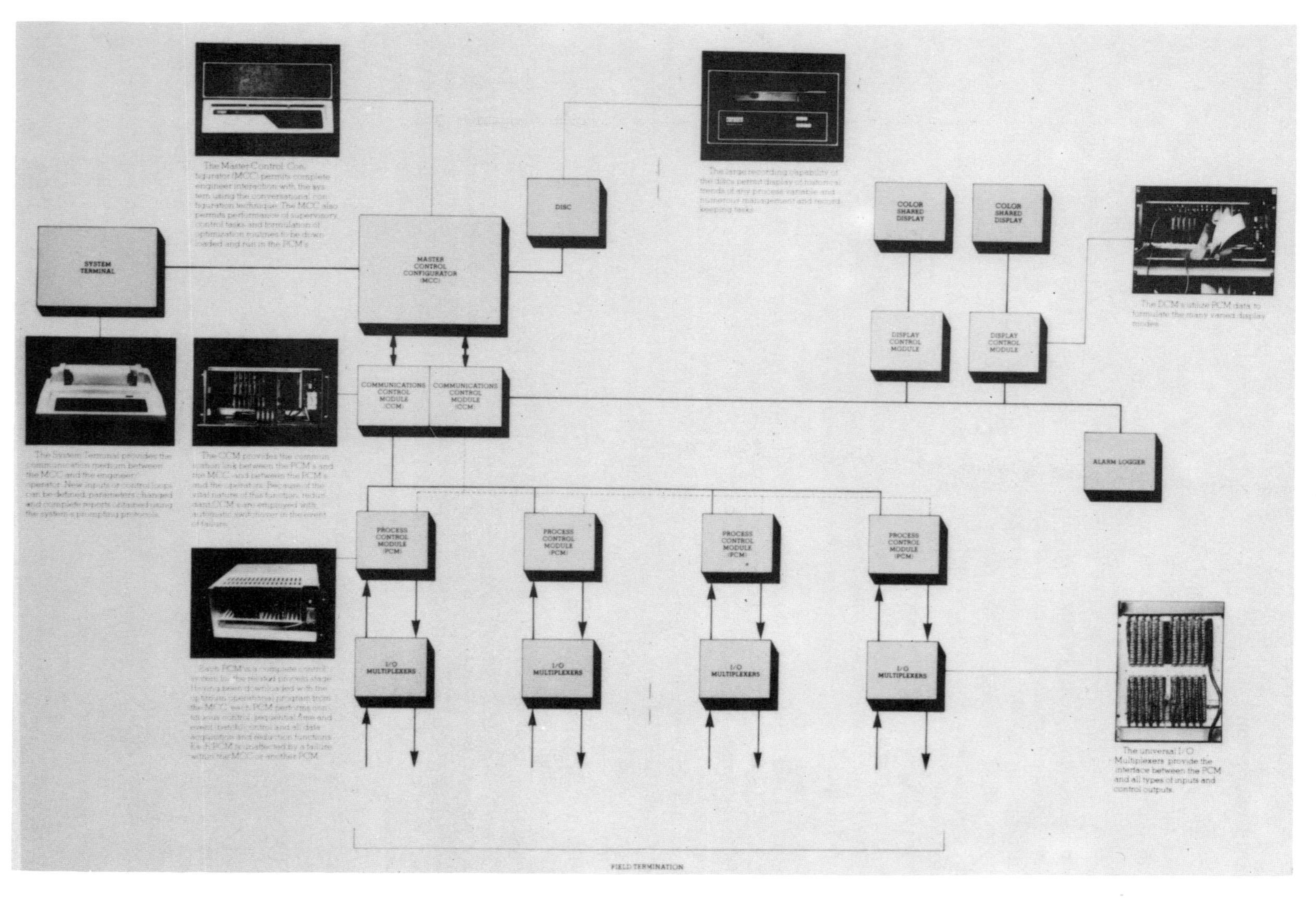

Figure 10–9. Distributed control system.

- Back-up copies of disks can be made.
- Secondary disk drive can be used as a back-up to primary drive by swapping the disk pack.

The proposed system concept is made possible by the advent of low-cost, highly reliable microprocessors utilized on a task basis to accomplish system objectives formerly relegated to large CPU systems. Within the distributed system concept, the system user is allowed freedom to structure his system from standard components in a manner best suited to his application. In essence, he may choose many or few system components, depending upon the goals he wishes to achieve. These goals may increase system reliability, improve efficiency of operation, reduce installed cost or facilitate ease of system expansion, or a combination of the above may occur, as well as other more subtle benefits of system distribution.

Equipment List

This section of the proposal should give a very detailed list of the equipment that is to be supplied. The following is the list of equipment to be supplied for the distributed system seen in Figure 10–9. This list should be a checklist for the user's acceptance engineer.

- D.E.C. hardware
  1 PDP 11/60 w/128KB MOS memory host
  1 RK611-EA controller and 2 RKO6 disk drives
  1 QU629-AT RSK-11M
  1 MS11-KF 128KB MOS memory board
  2 DR11-C parallel interface
  2 DL11-C serial line interface
  1 DZ11-C peripheral expansion multiplexer
  1 LA36 systems terminal
  5 LA36 terminal
  1 LP11-VA line printer
  1 DD11-C backplane
  1 FP11-E floating point processor
- Cabinets
  9 NEMA I 2-bay 76 × 33 in. cabinets
- Other hardware
  4 ISC color CRT's
  4 Switches
- DEC-9700 microprocessor
  2 communications control modules (CCM), each containing:
  1 D9700-4 microprocessor board
  5 D9712-12K static memory board
  1 D9722 priority interrupt control board
  2 D9740 serial communications board
  1 D9741 dual channel serial I/O board
  1 D9751 local parallel I/O board
  1 D9760-0-1 power supply
  1 9765-1 battery board w/batteries
  1 D9790 chassis
  1 D9795-1 front panel
  1 W76 cable, CCM to LA36's
  2 W78 cable, CCM to host
- 8 process control modules (PCM), each containing:
  1 D9722 priority interrupt control board
  1 D9700-4 microprocessor board
  3 D9712 18KW memory static memory board
  1 D9731 A/D converter assembly
  2 D9740 serial communications board
  1 D9750 remote parallel I/O board
  1 D9760–1–1 power supply
  1 D9765-1 battery back-up board w/ batteries
  1 D9790 chassis
  1 D9795-1 front panel
  2 D9797 data link loop through adapter
  8 W77 cable, PCM to MUX's
  8 W75 cable, PCM to MIP
- 2 display control modules (DCM), each containing:
  1 D9700-4 microprocessor board
  5 D9712 12K static memory board
  1 D9722 priority interrupt control board
  1 D9741 dual channel serial I/O board
  2 D9751 local parallel I/O board
  1 D9760-0-1 power supply
  1 D9765-1 battery back-up board w/ batteries
  1 D9790 chassis

1 D9795-1 front panel
2 W78 cable, DCM to CCM

- D9000 process I/O total equipment for each area, including:
  1 D9200 analog multiplexer (full)
  19 D9201 standard 3-pole dry reed relay boards
  1 D9200-1 I.S. sensor current board w/ test voltage
  1 D9221 local MUX address board
  19 06249-1 connectors
  4 D9100 digital multiplexer
  4 D9103 digital MUX address boards
  1 D9104 local driver board
  1 D9105 local receiver board
  18 D9110 solid state contact input boards
  18 D9145 electronically-latching Form A relay output boards
  14 D9160 D/A converter-manual back-up
  50 06249-1 converters
  8 D9500 A/C power distribution unit (remote)
  1 D9600 A/C power distribution unit (local)
  8 D9602 multiplexer interconnect panel
  8 D9616 power supply
  3 W7 cable, D/MUX to D/MUX

Many of the systems houses will be reluctant to give out a per unit price on the itemized equipment list, but the user should insist on these prices and establish, in writing, that these prices are valid for later additions or deletions. Be prepared for resistance, but a reputable house will comply.

Maintenance and Service Support

In addition to the normal hardware and software design, project management and systems test services provided during the term of the contract, the systems house should offer the following optional services.

Wiring and interface consultation. As soon as input and output requirements are defined by the user, preferably early in the contract, the interface considerations should be reviewed to ensure compatibility and system integrity. For this purpose, the supplier should provide as an option the services of equipment engineers to review these items: *primary power* (power source and characteristics, fault rating, fusing, distribution network, power wiring and back-up or UPS considerations); *sensor wiring* (customer supplied wiring specifications and/or wiring practices, grounding and shielding considerations for single and multi-core cables and wire-routing for noise immunity); *primary sensors* (platinum and copper RTD's and thermocouple selection, interconnection, grounding, shielding and wire sizing; contact closures analyzed to determine proper excitation voltage and current for reliable cleaning action); and *instrumentation and final control elements* (transducers, transmitters, controllers and recorders analyzed for I/O signal compatibility in normal and failure mode conditions, power supply isolation, ground loops and common mode potential). These interface considerations, if ignored by the user or not offered by the systems house, could contribute to the failure of the project.

Installation supervision. Provide consultation and direction, as required, for the proper system installation, including cabinet placements, wiring terminations and correct grounding and shielding procedures.

Start-up assistance. Provide single or multi-shift hardware and software personnel to assist in operation, debugging, maintenance and training during plant start-up. (The user should analyze his ability to perform these start-up tasks before accepting a proposal from a systems house that is not willing to make them available if needed.)

The Basis for Proposal

The basis for the proposal is the section wherein the systems house organizes all the events leading to the proposal being submitted. The events should include all the communications between the user and the systems house; i.e., meeting notes as well as any exceptions to the user's specifications. This section of the proposal should be reviewed very closely, and if any misunderstandings appear they should be cleaned up prior to the contractual commitment.

Project Execution

This section should identify the systems house's proposed task schedule. The schedule must identify the user's tasks and scheduling for the completion of these tasks (see Figure 10–10). The user must identify his commitments and be satisfied that he has the technical staff and facilities to comply with the proposed schedule. The user should not take these commitments lightly, for the successful and timely completion of the project is dependent upon them. A little more basic is the fact that if the systems house is running behind, he could rationalize the user's slippage as a means to gain time. The normal responsibility of the user is in the area of applications engineering. These applications engineering tasks generally include the following.

- *Installation wiring lists.* This is the physical layout of the equipment: location, type of device and engineering conversion (i.e., process range 4 to 20 milliamps + 100 gallons/time to 1000 gallons/time with a linear interpolation).
- *Data base requirements.* The total number of points and types in the system.
- *Develop control strategies.* These must be consistent with the process' operational requirements.

Again, the user must analyze and understand the project commitment he must make. If he is not capable of meeting this commitment, he must make arrangements for the systems house to accept this responsibility or to solicit an outside consultant.

Client Training

The training course offered by the systems house should include comprehensive documentation and descriptive literature, with detailed system control techniques and control analysis. The training sessions must be structured to be practical and productive, not abstract presentations of control theory mathematics.

Pricing, Terms and Conditions

This section of the proposal is usually evaluated relative to other proposals, but there are several subtle aspects of this section that should be discussed.

Progress payments. The cost of money will require the user to add to the cost of the project, dependent on the time of payment and the current cost of money. If the systems house insists on a payment of 50% of the contract price at the issuance of the purchase order (P.O.) and 5% per month for the next eight months, this could have a large impact on cost versus no progress payments.

Liquidated damages. Most systems houses will refuse to accept these terms. If they do, they will usually just add the maximum potential cost onto the price of the contract. Thus, a penalty for being late is converted to a reward for being on time. It is suggested that the user evaluate systems houses, find one that has a reputation for being on time and forget the liquidated damage clause.

Warranty. A return to factory warranty will usually add approximately 1 to 2% of P.O. price to the total project price. This type of warranty requires the users to pay for all field service and expenses. If this warranty is offered, the user should make sure the systems house quotes his standard rate subject to an agreed escalation clause.

After the review and acceptance of the proposal and agreement, the systems house will make project team commitments and start the implementation of the system. This will require constant referral to the

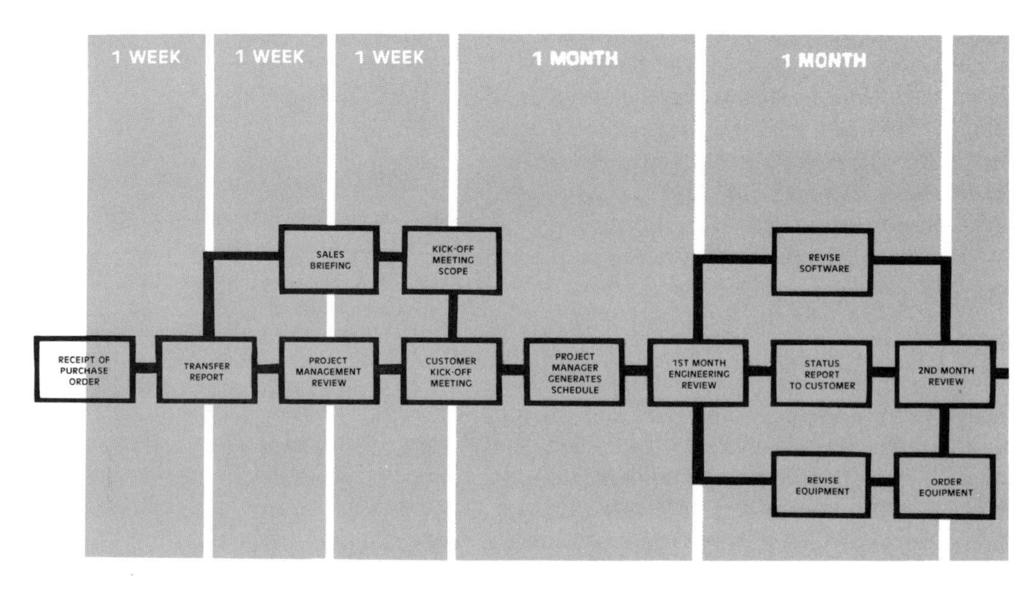

Figure 10–10. Project execution schedule.

specifications and the proposal. The user should interface with the systems house as to the method of communications during the implementation phase of the project, and should assign personnel with in-depth familiarity with these two documents for the duration of the project.

Three Party Project

The control system may be an integral part of the general instrumentation of a grass roots expansion or new facilities. The user may have hired an engineering contractor to design, build and instrument the new plant. The system may be specified by the engineering contractor, who will generate the specification, review the systems house's proposal and make recommendations or purchase the system; essentially, he becomes the user interfacing with the systems house. The systems house usually welcomes this arrangement with the engineering contractor because of the engineer's availability for generation of system specification and project implementation and his understanding of the actual process and its control requirements.

The type of systems generally supplied to knowledgeable engineering contractors are standard systems, with the major portion of the applications software generated by the engineering contractor. Many engineering contracting firms have built up substantial departments that will design, program and implement specific process applications programs. The engineering contractor should be able to demonstrate proven, significant increases in product profitability with his applications programs and control system designs. The user should, however, evaluate the engineering contractor to determine that this level of experience does exist and should not hesitate to check on references. (When checking references, the referred user, as well as the systems house generating the referred system, should be checked.)

PROJECT IMPLEMENTATION

The implementation of the project by the systems house should be achieved through a close, harmonious relationship with the user with respect to continuous specification review and management. The user should assign one official interface to the systems house project manager. This official interface (user system project manager)

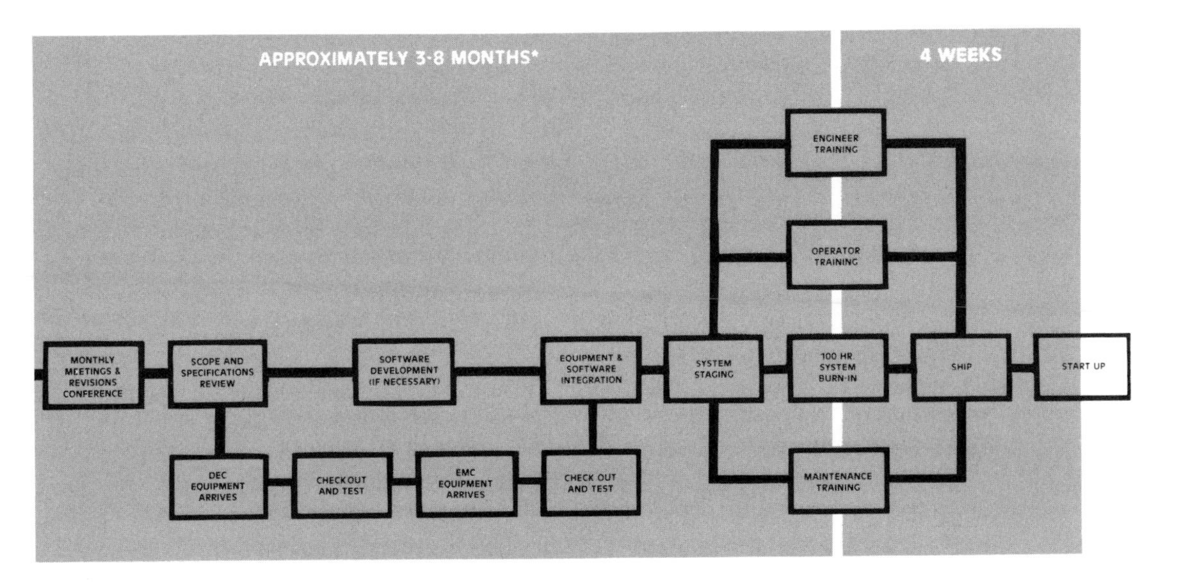

should be the only person to authorize any changes to the specifications relative to delivery or price. The user project manager should establish with the systems house a similar general schedule that reflects the schedule meetings, the user task commitments and major milestones.

Transfer Report

Within one week after receipt of the purchase order, the sales department should transfer to the project management department a document containing all external and internal information relative to the selling of the project. This document can (and usually is) quite extensive, since many selling efforts can cover a year's period on a large project. The following is a list of the typical information contained in this internal document.

- User's specifications.
- System proposal.
  System price.
  System cost (hardware cost and service cost signed off engineering estimates).
  Calculated profit margin.
- All pre-specification meeting notes.
- All post-specification meeting notes.
- General comments as to the technical competence of the user.
  Evaluation as to training requirements.
  Estimated special applications training requirements.
- Delivery requirements.

Experienced systems houses have found this to be a very important document. It requires the sales department to justify to the engineering department all commitments made to obtain the purchase order. This document is reviewed by the engineering groups at an internal staff kick-off meeting (commonly referred to as the Salesman's Roast).

This document is very seldom made available to the user, since it usually contains confidential information, but the user should determine how the project is transmitted to engineering and project management. It is important for the user to be assured that his project is profitable for the systems house, because if it is not, the project manager will be under pressure to either gain additional funds via changes in scope or to reduce the cost by not meeting all the requirements of the specification and proposal.

User's Kick-off Meeting

The systems house project manager should call the kick-off meeting to present not only the schedule, but also a complete description of the proposed system and any special technical considerations which need to be resolved. The kick-off meeting enables the user to meet the equipment and systems engineers who will be assigned to the project. The tone of this meeting will usually reflect whether the systems house project manager is in trouble over profitability. The user project manager should attempt to understand the financial position of the systems house project manager.

The user should make every effort to make sure the project begins with clear expectations and a distinct direction that is in keeping with the requirements of the process to be controlled. This meeting will permit the user to refine the requirements of the system before the project gets under way.

The conclusion of the kick-off meeting should result in approved plans and a detailed, inter-related schedule for the completion of the process control system. The user project manager should insist on the following.

- A detailed schedule showing all related tasks and indicating critical path tasks which are the indicators to whether the system is on scheduled time. The user project manager should review the schedule until he has identified every task start and completion date that could indicate that the system is in trouble.
- A review of the resumes of all project personnel, to make sure the system is not staffed with junior people.

Monthly Project Meetings

Depending upon the size and complexity of the system, monthly project review meetings may be required. If they are, the user project manager should insist on reviewing all tasks on the detailed schedule. The user should insist on witnessing the level of progress on critical tasks and, if not satisfied, request a review with the systems house management. If nothing else, this will establish the seriousness of the user's project. The project meeting should result in the systems house project manager commenting on the progress of the user's project team members. Again, it is critical that there be a free exchange of information at the monthly meetings. The user's people should not be intimidated by the system project team members. It is important that the user personnel begin to feel comfortable with the system and begin to request system time, if required, to generate special user applications programs. The user project manager should request training programs, even if they have not been budgeted, if he feels his project team is not progressing fast enough.

System Test Planning

The systems house should produce a test plan package containing procedures that will enable validation, test and check-out of the proposed system. This package should ensure that the system fully performs the functions outlined in the specification as implemented during the design phase.

To isolate any possible equipment and installation problems, the test package should be divided into separately conducted tests, each to be performed on individual parts at various stages of system development. It is important not to attempt to test too large a portion of the control system at one time, since trouble-shooting several untested sub-systems for one or more problems could be a very costly and time-consuming effort. As each test is completed, successfully tested sub-systems should be integrated for system tests, and after all sub-systems have been checked out, total system and acceptance tests should be conducted. Initial testing should be a stimulated system test at the systems house staging area.

These tests should be designed to detect and correct any equipment or installation problems as early as possible in the test sequence. Specific test plans should be written by the engineering staff to cover all phases of check-out. These test plans should describe the purpose of the tests, the test equipment required and the step-by-step method of implementation necessary to accomplish the specified test; define the necessary test prerequisites; and include associated test data sheets, which must be completed with test data as each progresses.

Equipment I/O Acceptance Tests

A test plan for system test at the systems house staging area should consist of complete I/O operational checks. This method of testing should result in a complete check of the system's data base and the total hardware I/O configuration. After I/O tests have been completed, the digital and analog systems should be completely operational. When this point has been reached, simulated process inputs should be connected to all the I/O terminations in a plant section and software and digital/analog hardware interface tests should be performed.

As part of the system's test plan, the systems house should generate a written software package for use during the implementation of the check-out procedures. This includes copies of the software and desired or acceptable test results. For the test, all system equipment should be interconnected. Each system module should be thoroughly checked out under simulated process conditions at the sensor. Alarm conditions should be created during check-out to verify alarm actions.

System Burn-in and Acceptance Testing

While standard module testing should include elevated temperature burn-in to eliminate excessive marginal components failures, the total system should be burned-in under elevated temperatures in excess of 100 hours. The system should be tested with all modules integrated and performing under as close to actual in-plant operation as possible. During the burn-in, the system engineers should hold training sessions for plant operational people under these simulated conditions.

The test should follow the agreed procedures with regards to both software and hardware testing. The user should understand, in detail, what the test is to accomplish and how it is to be performed. The user should examine, prior to contract award, test procedures with a previous customer contact from a test performance check. The importance of an agreed test plan has, in the past, been overlooked by the user and has created problems in system acceptance. In many cases, the acceptance of a poorly tested system in-house will cause a delay in the plant start-up in the field.

CONCLUSIONS

The successful interface to a systems house requires that the user have a basic understanding and can define the type of system required to control a specific process. The type of system must be selected with a knowledge of *resource sharing systems*, *redundancy reliability systems* and *distributed process control systems*. The type of system selected should be made after full consideration is given to *system reliability* and *system availability*. The type of system will influence the selection of the systems house. Systems houses have different experience levels relative to types of industries and types of systems.

The system specification is the basic element for a good interface with the systems house. A detailed specification will generally result in a good response, both in price and proposal.

A three party project is a good means to implement a successful project if the user does not have the necessary technical personnel to generate the system specifications and monitor the performance of the systems house. Many of the engineering contracting firms have become very successful applications organizations with expertise in specific processes and types of systems. The distributed micro based process control system, as defined in this chapter, is a good example of the type of system offered today (EMC EMCON-D) by a specific system house. Another distributed system is the TDC-2000 system offered by Honeywell.

The systems house procedures for implementation of a successful project requires *a detailed transfer report* (transfer of the system information from sales to engineering); *a user and systems house engineering kick-off meeting* (where a clear, precise understanding between the user and the systems house engineering group is established); *the establishment of monthly project meetings* (to monitor the process of the systems house and the user); and *the establishment and agreement of the system test plan for in-house acceptance of the system.*

The user must remain in control during the life of the project. This requires a good interface with the systems house project manager and a very good understanding of the project schedule. The task milestones and project progress should be reviewed in detail at least every month. The successful interface with the systems house is dependent on a very precise understanding of what was requested and what is to be delivered.

APPENDIX

Basic Specification for a Shared Display, Information and Distributed Control System

A. System overview
B. Hardware specifications
  1. Host computer system
    a. Central processing unit
    b. Bulk storage
    c. Peripherals
  2. Distributed microprocessors
    a. Communications control microprocessor
    b. Process control microprocessor
  3. Process I/O interface
C. Operating system software
  1. Host computer
  2. PCM
  3. CCM
D. On-line programming languages
  1. Data base compiler
  2. Continuous control compiler
  3. Sequential control compiler
  4. FORTRAN IV PLUS compiler
  5. Graphics compiler
E. System functions
  1. Input handling
    a. Variable identification
    b. Input data processing
    c. Performance calculations
  2. Recording
    a. Analog trending
    b. Record-oriented printing
    c. Magnetic tape mass historical data storage and retrieval
  3. CRT displays
    a. Alarm
    b. Interactive shared display system
    c. Utility
    d. Operator plant guidance
    e. Bar charts
    f. Plant pictorials

System Overview

The total plant control system will include all hardware and software for performing distributed data acquisition and process control. The design herein capitalizes on the following concepts.

Previously designed and common control software, operating systems and computing hardware shall be used when possible.

A distributed control system that is modular in design.

An interactive shared display system that will provide an efficient and informative man-machine interface. This system will provide total process visibility through color CRT's mounted in the operator's console.

A foreground/background operating system at the host level, from which the engineer can add or modify the following.

- System performance of data acquisition and control strategies.
- Optimization and performance routines.
- Logging and reporting functions.
- Color graphics displays.

Distributed DDC overcomes the effect of failures associated with centralized computer control systems. Several criteria which apply to system flexibility are listed below.

- The system must be capable of performing continuous control, sequential control and data acquisition via a distributed microprocessor network.
- The system must be upwardly compatible with new and improved types of memory, hardware and operating system software, developed by original equipment manufacturers. Specifically, alien operating system software or modification of the orginal equipment manufacturers' electronic architecture will not be acceptable.
- The system must support the following software packages which must be compiled, assembled and linked on-line in the host processor, then downloaded to the distributed microprocessors: data acquisition, continuous control, sequential control and calculation blocks. The operating system of the host must support FORTRAN IV PLUS to increase execution speed and reduce memory requirements of optimization programs and the like.
- Maintenance of the system shall include sufficient software and hardware that on-line preventative maintenance and diagnostic techniques are maximized and independent of the process control operation. Diagnostic information shall be displayed either at the operator's console, engineer's console or system typers, depending upon the type of system failure. The system architecture concept is composed of three major portions: 1) a supervisor minicomputer (HOST), with peripherals; 2) a distributed microprocessor network with process I/O; and 3) an interactive shared display system.

Host Computer

The minicomputer acts as a supervisor for process control functions that are performed by the distributed microprocessors. The host computer performs the following functions.

Download of programs to the distributed microprocessors while they are on-line.

Communication with the distributed microprocessors to gather process data.

Communication with the distributed microprocessors to establish control objectives and tuning parameters.

Data historian—storage and management of historical data.

CRT data trends for historical and real-time data.

CRT display of alarm conditions.

Print shift logs, alarm status messages, log operator actions and special logs.

Run user developed FORTRAN IV PLUS programs (i.e., optimization inventory, data analysis, etc.).

Perform system diagnostics.

Process Control Microprocessor (PCM)

The distributed microprocessors will perform continuous control, sequential control and data acquisition, concurrently, in the same microprocessor. Additionally, it will do the following.

Perform real-time data acquisition at specified scan rates.

Perform alarm limit-checking for absolute limits, deviation rates or warning limits on designated variables.

Communicate, by exception, variable

data information (i.e., current value, alarm status, set point, output control constants, etc.) to the host computer and to the shared display console.

Perform input signal smoothing, averaging or totalization, as required.

Perform all control and data acquisition functions, independent of host computer operation.

Be easily replaced by a cold spare and re-started by a technician when a malfunction occurs.

Interactive Shared Display System (ISDS)

This includes the following. Operator's console with color CRT's that interact with the PCM's through either the host computer or the redundant communications control microprocessors (CCM).

Ability by the operator/engineer to change specified operating parameters in the PCM's via the operator's console.

The normal operator/engineer interface will be through the host computer. When the host is unavailable, the interface will be through the CCM, with the sub-set of the operations available at the host level.

Ability to make a hard copy of any CRT display by using a videocopier.

Hardware Specifications

Host Computer System

Central processing unit. This unit shall include the following.

- One microsecond cycle time.
- 16 bit word length.
- Floating point processor.
- 124K words of memory (Core or MOS with battery back-up).
- ROM bootstrap loader.

The system shall be able to protect selected areas of core memory. Programs in unprotected areas of core memory shall not be able to write or branch into programs contained in protected areas. Specific usage of this capability shall be made during on-line programming, where programs being tested in background shall not be able to interfere with foreground operation.

A complete programmer's control panel, with indication of data words and active registers, shall be provided. It should allow the display of address or data words.

The architecture of a central processing unit shall contain, as a minimum, the following capabilities.

- Relative addressing (both forward and backward).
- Register to register operations.
- Interrupt structure with arm-disarm, and enable-disable.
- Power fail/auto re-start.
- Real-time clocks.
- Double precision arithmetic.
- Floating point processor.

Bulk storage. For *disk storage,* two types of storage will be required: 1) redundant disk drives with removable disk packs (the average access time must be less than 55 milliseconds and the on-line storage provided by these disks must be a minimum of 14 million words); and 2) Disk drive with a removable disk pack having a minimum of 35 million words of storage on-line (the average access time must be less than 40 milliseconds). For *magnetic tape transport,* the system shall contain the following features.

- The magnetic tape transport shall have a read/write minimum speed of 37.5 in./second, and a minimum density of 800 bits/in.
- The data transfer rate between the CPU and magnetic tape system shall be a minimum of 30,000 bytes/second.
- The tape utilized shall be 1.5 mil, /2 in. width and be IBM compatible, with nine channels of information.
- The tape system shall be of the "read-after-write" type, with parity-checking to provide high data processing integrity.

- Positioning of tape over read/write heads shall be by vacuum control, rather than by mechanical tension means, in order to provide more reliable data recording.

Peripherals. For *I/O printers,* the host contains a system terminal with a minimum speed of 120 CPS and RO devices (a character printer with minimum speed of 180 CPS and a line printer with minimum speed of 300 LPM). The CCM shall have an RO device with a minimum speed of 30 CPS. For *terminals,* the following items are required.

- CRT screens displaying a minimum of seven different colors, in order to highlight certain types of information and portray status conditions.
- A CRT screen memory separate from CPU, so that on temporary CPU outage or heavy loading of the CPU, the last displayed data remains on the screen.
- Communication between the CRT and the CPU at a minimum rate of 9600 baud.
- A screen refresh rate at least 60 times/second so that no flicker is perceivable. (The refresh rate shall not vary according to the amount of information displayed on the screen.)
- The loss of one power supply or one CRT controller shall not cause the failure of more than one CRT display.
- Ability of the CRT to display a minimum of 80 characters/line and a minimum of 48 lines/display of alphanumeric information.
- The alarm and utility CRT's screen size shall be 25 in. diagonally.

Distributed Microprocessors

The *communications control microprocessor* (CCM) shall include the following.

- Ten microsecond cycle time for CPU.
- 8 bit byte.
- MOS memory with battery back-up expandable to 64K bytes.
- Crystal controlled real-time clock.
- Power fail/auto re-start.
- Watch dog timer protection.
- Isolated power supplies.
- Parallel interface to host processor.
- Serial communications with I/O devices.
- Ability to handle priority interrupts.
- 19 in. rack (mountable).

Process control microprocessor (PCM). The PCM shall include the following.

- Ten microsecond cycle time for CPU.
- 8 bit byte.
- MOS memory with battery back-up expandable to 64K bytes.
- Crystal controlled real-time clock.
- Power fail/auto re-start.
- Watch dog timer protection.
- Isolated power supplies.
- Parallel interface to process I/O.
- A/D converter.
- Ability to handle priority interrupts.
- 19 in. rack (mountable).

Process I/O Interface

Process I/O shall be universal and consist of two basic multiplexer types: analog multiplexers and digital multiplexers.

Analog multiplexer. Each analog multiplexer must be able to handle 100 analog inputs of any type, including: thermocouples, current inputs, voltage inputs and RTD's. The analog multiplexer must allow for random assignment of all inputs, and must provide for a temperature sensor to be used for cold junction compensation for thermocouples. The printed circuit boards must be conformal coated to inhibit degradation from hostile environments, and it must be possible to replace a PC board without having to remove input sensor wiring from the terminal strip. All PC board edge connectors must be plated with gold over nickel to resist corrosion; all terminal strips must be plated with gold over nickel to resist corrosion; and it must be possible to perform open circuit detection for analog inputs.

Digital multiplexer. Each digital multiplexer must be able to handle any of the following types of signals.

- Contact input scanned or interrupt driven.
- Contact output.
- Pulse count input.
- Pulse train output.
- Momentary output.
- D/A output.

Each digital multiplexer must accept up to 16 printed circuit boards. The printed circuit boards must be conformal coated to inhibit degradation from hostile environments. It must be possible to replace a PC board without having to remove input sensor wiring from the terminal strip. All PC board edge connectors must be plated with gold over nickel to resist corrosion; all terminal strips must be plated with gold over nickel to resist corrosion.

System accuracy. The following are required.

- The A/D converter must provide a resolution of 13 bits plus sign.
- The A/D converter must perform true integration for one cycle of the power line frequency (16 2/3 milliseconds for 60 Hz) to obtain maximum rejection of normal mode noise.
- The system analog accuracy must be $\pm$ 0.1% of full range.
- Common mode rejection must be a minimum of 120 db at power line frequency.
- The input impedance of the system must be a minimum of 300 megohms.
- All inputs must be optically isolated.

Operating System Software

Host

The host computer shall have an operating system that is supported and provided by the original computer hardware manufacturer. The operating system must have the following characteristics.

Multiprogramming and multi-tasking (concurrent processing of several program images residing in memory) providing task check-pointing and priority scheduling.

An ANSI FORTRAN IV PLUS compiler with a data base access package provided by the control system vendor.

Peripheral device transparency to provide reassignment of peripheral I/O traffic.

On-line interactive text editor.

File-to-file utility programs.

On-line hardware diagnostic routines that can operate parallel to the process control programs.

Hardware interrupt priority handling.

Support for multi-systems consoles, including message communication.

PCM

The distributed PCM's executive system shall have the following characteristics similar to the host.

Multi-tasking to provide task suspension and priority scheduling of task execution.

Hardware interrupt priority handling.

Network communications to accommodate program downloading from the host and to respond with data transfers as required by the host or interactive shared display system.

Support for continuous control, sequential control and data acquisition.

CCM

The CCM executive system shall be essentially identical to the PCM units, except that extended support for display and communication functions is required.

On-line Programming Aids

Data Base Generation

An on-line data base compiler must be provided with the system to allow the engineer to build data base for the PCM's in the host. The compiler must allow the following functions.

- Define an input or output.
- Modify a point parameter.
- Report the point's parameters.
- Delete a point from the data base.
- Enable or disable scanning.
- Access real-time data values.

Continuous Control Compiler

The continuous control compiler will work in conjunction with the data base compiler to define control strategies for respective PCM's. The continuous control compiler must support the following control algorithms.

- PID (Any term may be left out): positional mode output, velocity mode output, set point output and non-linear error.
- Ratio and/or bias (may be linked to PID block).
- Lead/lag.
- Calculation blocks: add, subtract, multiply, divide, square root and weighted average.
- Selector/limiter: select maximum or minimum of up to three inputs; determine maximum difference of up to three inputs; and predetermine maximum and minimum outputs of the control signal.
- Parabolic compensator: the continuous control compiler is to be used at the host level and must be usable on-line with no interference with existing control strategies until the new strategy is ready for downloading.

Sequential Control Compiler

A sequential control compiler must be provided to allow the engineer to develop programs to perform logic sequencing and motor control in the PCM's. The programs will be developed on the host in the background mode and downloaded to the respective PCM. The compiler language will provide, as a minimum, all typical functions available on programmable controllers.

FORTRAN Compiler

An on-line compiler shall be furnished that is based on ANSI FORTRAN IV PLUS programming language. The compiler shall, as a minimum, be used for modifying and adding processing specifications for new and existing system inputs and outputs; modifying and adding performance calculations; and modifying and adding conditions from the automatic display of start-up/shut-down checklists.

The compiler shall be used to debug, assemble and compile changes for the above programs in a background mode while those programs are concurrently running in a foreground mode. The engineer shall also be able to load the debugged programs into the foreground operation and cause them to run.

Compiling, loading and executing of short changes to processing specifications of plant signals shall be done in one operation by the engineer, through the system terminal, unless errors are detected.

Condition compiling of selected statements shall be available for producing intermediate calculations and data during debugging statements. At final compilation, these statements will be deleted from the program to be run in the foreground, automatically, by the compiler.

Graphics Compiler

The graphics compiler must be a display language designed to give an instrument or process engineer the ability to create dynamic schematic CRT displays of process units. Process variables and status conditions may be displayed and updated periodically with real-time values. The compiler can also be used to generate log formats on printing devices.

The display language shall be a set of commands that allows the engineer to build virtually any display desired. The displays may be built such that process variables can be tested for specified conditions and cause the display to be modified if the proper conditions exist.

The compiler must be able to run in the background mode of the host without interfering with the real-time foreground system.

System Functions

Input Handling

System variable identification. All system variables addressable by the operator shall have a unique point identification which shall consist of a point number and English description. Point name must be nine characters in length. The English description shall be a minimum of 24 characters in length. The operator shall be able to request a printout and CRT display of the identification of all points of a particular sensor type in a plant group or system group.

Input data processing. Operator activities take place through the shared display system CRT's. Engineer activities take place through the engineer's terminal. Manual substitution of data can be made by the engineer at the system terminal. Certain selected inputs which have considerable process noise on them (such as pressure signals) shall have their signals filtered by a software filter. It shall be noted that this filtering is in addition to that required for filtering of electrical noise on signal leads. The filter shall smooth the most recently scanned value according to the values on older readings. The filter shall operate on each scanned value of selected inputs. All filtered points may be alarm-checked. Each system variable shall have a deadband that is adjustable by an engineer. The range of the deadband shall be between 0 and 15% on a per point basis. Calibration and conversion equations of sensors shall be implemented on a per point basis; the system shall have the capability of storing conversion curves which consist of up to 20 segments. The scan frequency of individual analog inputs shall be adjustable either to a faster or slower scan rate by an engineer. Scanning of analog inputs may be disabled and enabled by an engineer. Averages, integrations, running averages, load-weighted averages, maxima and minima shall be selectively initiated by an operator on field signals and other computed variables. The running averages are computed on the most recent set of scanned values, where the set may vary up to eight samples and is adjustable by the operator. Averages and integrations shall be computed on a 10 minute, hourly and daily basis. The others shall be computed at least hourly and daily. The following minimum quantities of time-type calculations are required: 240 averages, 60 load-weighted averages, 240 integrations, 20 maxima, 20 minima and 25 running averages. The engineer shall be able to modify and add extensive processing specifications for new and existing inputs through use of the on-line compiler. These processing specifications shall include all those functions available through the operator's panel, plus English description, hardware multiplexer address, measurement units, engineering unit conversion equations, special calibration equations and plant annunciator group. The engineer shall refer to the point number of the variable for making these changes; he shall not be required to refer to memory locations. Input shall be through the operator's or engineer's panels, and each and every input shall have adjustable alarm limits which allow the operator or engineer to stop or change the above on a per input basis.

Performance calculations. All calculations shall be programmed in a high-level language based on ANSI FORTRAN IV

PLUS. Calculations shall be made using floating point arithmetic. The calculations may be changed while the system is on-line. The programming language shall enable specifying of alternate inputs for selected field signals when the quality of those inputs is less than good. The calculations shall be run continuously at least every 10 minutes. The results may be calculated by the time-type calculations referred to above and output to hourly logs, trend logs and plant CRT trend groups. The results shall also be stored in a separate performance log. The log shall maintain averages of the good values only for each point on a daily and monthly basis, and the count of good values in each average shall also be stored and available for printout. The performance log will be programmed in FORTRAN IV PLUS. The engineer shall be able to compile changes to the performance calculations while the calculations are running.

Recording

Analog trend recording. To indicate that a new variable is on analog trend pen, the pen will automatically move from zero to full scale before starting a trace of the variable. Also, at midnight, all pens shall automatically swing from zero to full scale to provide a convenient time reference to the engineer reviewing records. If a trended variable reaches 5% or 95% of the chart range, the range will be automatically shifted down or up (respectively) by an amount equal to one-half of the displayed range. This will place the next current value near mid-scale and prevent the possibility of lost information. The new scale values will be logged on the designated printer, and the operator shall be able to cancel this function. The operator shall also be able to request summary of all trend pens currently in use on his utility CRT. This shall identify the pen, the variable being trended and the corresponding zero and full scale values. The CRT shall display this information within five seconds after operator request. If an automatic range change occurs, as stated above, the CRT display shall contain the current zero and full scale values.

Record-oriented printing. Logs shall not require pre-printed headings. Headings shall be generated with the appropriate log to allow the use of standard paper and to prepare logs in a more usable format. Through the pushing of a function button, the contents of a CRT display shall be printed exactly as displayed (except for color) at the time the copy displayed was requested. The operator shall be able to review or suppress a summary of all alarms, both digital and analog, that have occurred during the previous eight hour shift. In order to give the operator a presentation he can better comprehend, up to 16 successive trend group readings can be retained within computer memory for printout at one time.

Magnetic tape mass historical data storage and retrieval. The purpose of a mass storage and retrieval system is to have available detailed information at long intervals after events have occurred. To be properly utilized, it should not be necessary to detail what is to be stored; therefore, all inputs will be stored at all times, even if not in alarm, in as compact a form as possible. All alarms and returns to normal shall also be stored, as shall post trip logs. Retrieval of information stored shall be through programs which will output only information explicitly requested. This shall mean that all changes that occurred during selections sections of time can be output, or that the values of a selected input (for example, FD fan in-board bearing #1, input TT231) for a specified period of time (for example, 1:00 AM, August 4, 1973, to 2:00 AM, December 10, 1973) can be output as selected by plant engineers. Information to be stored during the changing of magnetic tapes of the retrieval of previously stored information shall not be lost. Finally, the system shall inform the operator or plant engineer when a storage tape is a stated percentage full and needs to be changed, and if an invalid storage tape has been mounted.

Glossary

The purpose of this glossary is to collect, in one reference section, definitions of some of the important terms used in this text. The terms defined are relevant to particular technologies and disciplines, including:

- Control and measurement instrumentation.
- Process control engineering.
- Digital computer hardware.
- Digital computer software.
- Communication as it applies to the linkage of groups of computers (e.g., microcomputers in a satellite system monitored by a host computer).
- Electronics.
- Microelectronics.

It is hoped that this glossary will aid the general reader who has an interest in closed-loop process computer control to achieve a more fruitful understanding of the concepts described in the text, and at the same time, be of service to the experienced practitioner in control systems or process computer control systems design.

The following legend identifies each term used in the glossary by application or discipline.

(COMP) = Computer (terminology applicable to digital and/or analog computer hardware and software).

(COMP-S) = Computer software only.

(COMP-H) = Computer hardware only.

(INSTR) = Instrumentation (definition of terminology relating to control or measurement instrumentation hardware and/or control systems engineering concepts).

(ELEC) = Electronic (generalized terminology relating to electronic hardware, wiring, grounding, filtering and systems design concepts).

(MICRO) = Microprocessor (terminology used in this handbook regarding microprocessor hardware and software concepts and microcomputer systems application).

(COMU) = Communications (terminology relevant to communications networks incorporating computer hardware and systems at the node points).

Where more than one legend code applies to a term, legends are listed in order of importance so that identification can be keyed to the category most likely to include the term in a general sense first. This was done only when it was thought that it would aid in defining the term. If no legend is shown, it is because it was deemed that none was necessary, since the definition is self-evident.

No attempt has been made for a definitive legend. Interpretation is left to the reader. The legends are used only to group some of the key terms into more than one usage where such a possibility exists and to aid further understanding of the term being defined.

Where more than one word is used in the term, the glossary is alphabetized according to a key word. For example, "absolute coding" is keyed to the word "absolute." Where the abbreviation is more common than the complete term, the entry is alphabetized according to its abbreviation.

The index cross-locates pertinent and relevant terminology for all text material, including the glossary.

References for glossary terminology are listed below. These references are the accepted and primary sources for such terminology, as they have been prepared by committees appointed by the leading technical societies covering the disciplines cited. For example, the I.S.A. (Instrument Society of America) references have been prepared by their Standards and Practices Department or by the Purdue Workshop on computer language standardization. The A.S.M.E. (American Society of Mechanical Engineers) Automatic Controls Division, also included in the references below, updated its terminology standard in 1972 from the original work of the late 1950's. Other references for terminology are S.A.M.A. (The Scientific Apparatus Makers Association), an organization of instrument manufactures, and The Foxboro Corporation and The Digital Equipment Corporation (manufacturers of instrumentation and computers).

1. Instrument Society of America, Standards and Practices Department, No. S-51.1, *Process Instrumentation Terminology,* Pittsburgh, Pennsylvania, 1976.
2. Instrument Society of America, Purdue Workshop publication, *Dictionary of Industrial Digital Computer Terminology*, prepared by Glossary Committee of Purdue Workshop on Standardization of Industrial Computer Languages, Pittsburgh, Pennsylvania, 1972.
3. American Society of Mechanical Engineers, Automatic Controls Division, *Terminology For Automatic Control,* No. MC 85.1, New York, Updated 1972.
4. Scientific Apparatus Makers Association, No. PMC 20.1, *Process Measurement and Control Terminology,* Endorsed by I.S.A., New York, 1973.
5. The Foxboro Corporation, *Glossary of Analog and Digital Process Control Terms,* Foxboro, Massachusetts, 1967.
6. Digital Equipment Corporation, *Introduction to Minicomputer Networks,* Maynard, Massachusetts, 1974.

Absolute Coding. Coding that is directly acceptable to the computer without modification in its address structure (e.g., machine language coding). (COMP-S) (MICRO)

Access Time. The time interval between that instant at which data is called from storage and the time at which delivery is completed; or the time between when data is requested to be stored and when storage is completed (e.g., read and write time). (COMP-S) (MICRO)

Accumulator. A register in which the result of an arithmetic or logic operation is accumulated; one or more registers associated with the ALU which temporarily store sums and other arithmetical and logical results from the ALU. (COMP) (MICRO)

Actuating Error Signal. A reference input signal minus a feedback signal. (INSTR)

Adaptive Control. A control action in which an automatic means is used to change the control mode so as to improve the performance of the control system (e.g., a "learning" mode). (INSTR) (COMP)

Address. An identification represented by a name, label or number for a location in storage (e.g., a portion of an instruction that specifies the location in memory for the instruction—or coded instruction—which designates the location of data program segments in storage). Address storage may exist in registers or memories. (COMP-S) (MICRO)

Algorithm. A computational procedure (e.g., a set of rules for a specific output generated from a specific input). (COMP) (MICRO)

Amplitude Modulation. A method of transmission in which the amplitude of a carrier wave is modified in accordance with the amplitude of a signal wave. (COMU)

Analog. The representation of numerical quantities by means of physical variables such as voltage, capacitance, resistance or induction (e.g., an R.L.C. electronic circuit used to represent a hydraulic, mechanical or control system). (INSTR) (COMP)

Analog Back-up. An alternate method of process control by conventional analog instrumentation in the event of a failure in the host computer. (INSTR)

Analog Computer. A computer that manipulates numerical quantities representing the physics and mathematics of a process as electrical and electronic variables for the solution of problems.

Analog/digital Converter. (A/D Converter) A device that converts analog input signals from process instrumentation into a digital code acceptable to computer (e.g., 4 to 20 milliamp DC analog converted to 0 to 10 volt pulse signals). (COMP-H)

And Gate. A logic operator having the property that if *A* is a value, *B* is a value and *C* is a value, then the AND of *A,B,C,* is true if all values are true, or false if any are false. (The "AND" gate implements "AND" logic.) (COMP) (MICRO)

Architecture. A design for the arrangement of the elements of a computer or microprocessor. (COMP) (MICRO)

Arithmetic and Logic Unit. (ALU) That portion of computer hardware in which arithmetic and logical operations are performed. (COMP-H) (MICRO)

ASCII. American Standard Code for Information Interchange. (also called USASCII) An eight-level code (seven bits plus a parity bit) to provide information code compatibility between digital devices of U.S. manufacture (established by ANSI—American National Standards Institute, formerly American Standards Association). (COMP) (MICRO) (COMU)

Assembler. A program that converts symbolic language into machine language by the substitution of operation codes for symbolic operation codes and absolute or relocatable addresses for symbolic addresses. (COMP-S)

Assembler Program. A program that translates man-readable source statements (mnemonics) into machine compatible object codes. (COMP-S)

Assembly Language. A machine-oriented language. A program written as a series of source statements using mnemonic symbols that define the instruction which is then translated into machine language. (COMP-S)

Asynchronous Computer. A computer in which an event starts as a result of a signal generated by the completion of a previous event or operation. (COMP) (COMU)

Asynchronous Transmission. Transmission in which time intervals between characters may be of unequal length and which is controlled by start and stop elements at the beginning and end of each character. (COMU)

Audio Frequencies. Frequencies which can be heard by the human ear (usually 15 to 20,000 cycles/second). (COMU)

Automatic/manual Station. (A/M Station) A device that enables the process operator to manually position one or more final control elements, and, in doing so, to bypass a control station (e.g., a switch that allows for transfer to or from automatic control to or from the manual mode). (INSTR)

Availability. The total amount of time that a computer is properly operating (e.g., the "UP" time). (COMP) (MICRO)

Background program. A program operating without regard to time and one which may be overridden by a program of greater urgency and priority. (COMP-S)

Back-up. A provision of an alternate mode of operation in case of failure in the primary code of operation (e.g., a redundancy feature, such as a manual back-up or analog back-up in a control loop). (COMP) (INSTR)

Band-width. A range of frequencies assigned to a channel or system (e.g., the difference expressed in Hz between the highest and the lowest frequencies of a band). (COMU)

Base. The number of characters employed in a numbering system: 2, binary; 8, octal; 10, decimal; 16, hexidecimal. Base is synonymous with radix. (COMP-S)

Batch Processing. A technique of data processing in which jobs are collected and grouped before processing. (COMP)

Baud. A unit of communication rate; the number of signals/second transmitted over a communications link. (One baud is equal to one bit/second in a train of binary signals.) (COMU)

Benchmark. A sample routine against which the performance of a computer is compared and evaluated. (COMP)

Binary. A condition in which there are just two possibilities (e.g., plus or minus, on or off, 0 or 1). (COMP) (MICRO)

Binary Coded Decimal (BCD). A decimal notation in which the decimal digits are represented by a group of binary digits using zeros and ones. (COMP-S) (MICRO)

Binary Digit. A character used to represent one of two digits in the binary numbering system. (COMP) (MICRO)

Bipolar. As in bipolar device (e.g., a crystal of silicon in which holes flow from base to emitter and electrons from emitter into base). (ELEC) (MICRO)

Bit. Contraction of "binary digit." (COMP)

Block. A set of words, characters or digits handled as a unit. (COMP-S)

Block Diagram. A diagram of a system, instrument, computer or program in which various functions are represented by annotated boxes and other symbols (such as triangles and circles) with interconnecting lines between the symbolism denoting information flow. (COMP) (INSTR)

Bode Diagram. A graph of log amplitude ratio and phase angle data on a log frequency base for mathematical transfer functions used for frequency-mode solutions to control problems. (INSTR)

Boolean Algebra. The processes used in the algebra formulated by George Boole for operations of formal logic. (Usually used for sequencing and simple switching systems.) (COMP)

Bootstrap. A technique designed for computer usage to bring the computer into a desired state by means of its own action (e.g., a machine routine whose first few instructions are sufficient to bring the rest of the data or instruction into the computer from an input device). (COMP-S)

Buffer. A storage device used to compensate for a difference in rate of flow of data or time of occurance of events when transmitting data from one device to another (e.g., an isolating circuit inserted between other circuits to prevent interactions, to match impedances, to supply additional drive capability or to delay the rate of information flow). (COMP) (ELEC)

Byte. A sequence of adjacent binary digits or "string" operated upon as a unit and shorter than a word. A byte is usually one-half a word (e.g., six-bit, eight-bit or nine-bit word. (COMP-S)

Call. To transfer control to a specified sub-routine. (COMP-S)

Calibrate. To verify, by comparison with a standard, the values to which scale or chart graduations should be placed in order to correspond to a series of values of the quantity which the instrument is to measure, receive or transmit. A measuring instrument is usually field calibrated at its low, mid and high points. Temperature (thermocouple) standards, for example, are maintained by the U.S. Bureau of Standards. (INSTR)

Carrier. A continuous frequency capable of being modulated by a signal. (COMU)

Cascade Control Action. Control action in which the output of one controller becomes the set point for another controller. (INSTR)

(CRT) Cathode Ray-Tube. A television-like picture tube used in visual display terminals. (COMP-H)

CCITT. Comité Consulatif Internationale de Telegraphie et Telephone. An international consultative committee that sets communications usage standards. (COMU)

Central Processor. The portion of a computer system that performs computation. It houses the arithmetic and control units and working memory. (COMP-H)

Centralized (Computer) Network. A computer network configuration in which a central node provides computing power, control or other services. (COMP) (COMU)

Channel. That part of a communications system that connects a message source to a message sink. (COMU)

Character. An elementary symbol which expresses information, usually including the decimal digits 0 to 9, the letters A through Z and as symbolism used to denote functions. (COMP)

Chip. An element, usually of silicon, housing thin layers of metal oxide semiconductors as NP or PN junctions. (ELEC) (MICRO) (COMP)

Circuit. In communications, the complete electrical path providing one or two way communication between two points comprising associated go and return channels.

Clear. To erase the information in a storage device by replacing its contents with zeros. (COMP-S) (MICRO)

Clock. A device that generates periodic signals used for synchronization (e.g., a device that measures and indicates time). (COMP)

Clock Frequency. The master periodic pulse frequency which schedules the operation of the computer. (COMP)

Closed-loop. An information or path that includes a forward path, a feedback path and a summing point and forms a closed circuit. (INSTR) (COMU)

CMOS (Complimentary MOS Device). (*See also:* PMOS and NMOS) A single chip of silicon housing both PMOS and NMOS transistor circuitry. (Note that if the substrate is N-type silicon, then the NMOS transistor portion is produced in a sub-island of P-type material. (ELEC) (MICRO)

Code. A set of rules specifying the way in which data may be represented; the set of procedures in the Standard Code for Information Interchange. In data processing, the representation of data or a computer program in a symbolic form that can be accepted by a data processor. (COMP-S) (MICRO)

Coding. A list of computer instructions used to solve a procedure, function or problem. (COMP-S)

Common Mode Interference. The form of electronic signal interference which appears between the measuring circuit terminals and electrical ground. (INSTR) (ELEC)

Common Mode Rejection. The ability of an electronic circuit to mitigate common mode voltage (usually expressed as a ratio or in decimal values). (ELEC) (INSTR)

Communications Computer. A computer that acts as the interface between another computer or terminal or a computer controlling data flow in a network. (COMU) (COMP)

Compiler. A program that translates a problem-oriented language to a machine-oriented language, such as FORTRAN or PL/M, and substitutes sub-routines and single machine instructions for symbolic inputs (e.g., a device that translates higher-level language into machine code). (COMP)

Computer Network. An interconnection of computer systems, terminals and communications facilities. (COMU)

Console. A part of a computer used for communication between the operator or engineer and the computer; a terminal providing user input and output capability. (COMP-H)

Control Algorithm. A mathematical expression for the control action to be performed. (COMP-S)

Control Block. That circuitry which performs the control functions of the CPU; it decodes microprogrammed instructions, generating the internal control signals that perform desired operations. (COMP) (MICRO)

Control Character. A character whose occurance in a particular context initiates, modifies or stops a control function. In the ASCII code, any of the 32 characters in the first two columns of the standard code table. (COMP-S) (COMU)

Control Mode. A specific type of control action such as proportional, integral or derivative. (INSTR) (COMP)

Control Program. A sequence of instructions that guide the CPU through the operations it must perform; it is usually stored permanently in ROM memory, where it can be accessed by the CPU. (COMP-S)

Control Station. The station on a network which supervises the network control procedures such as polling, selecting and recovery in computer communication. In process control, the panel-mounted receiver instrument housing the control function, setpoint and output signals. (COMU) (INSTR)

Control System. A system in which automatic guidance or manipulation are used to achieve a desired value of a variable. (INSTR) (COMP)

Control Unit. In a digital computer, the hardware that effects the gathering of instructions in proper sequence, the interpretation of these instructions and the application of this information to the arithmetic unit according to this interpretation. (COMP-H)

Controller. An instrument which reglates a controlled variable (e.g., a feedback control system that compares a controlled variable with a setpoint, and adjusts a manipulated variable as a function of the error). It may include a reference input element, a summing point, a control element, a feedback element and a sensor. (INSTR)

Conversational. A mode of processing that involves a sequential interaction between a user at a terminal, via keyboard display hardware, and a computer. (COMP)

Core Memory. A high-speed, random access storage device utilizing a matrix array of ferrite cores comprising the computer's working memory. (COMP-H)

Core Resident. A term pertaining to certain key programs permanently stored in core memory. (COMP)

Corner Frequency. The value of the abscissa of the conjunction in an asymptotic approximation of log amplitude versus log frequency graphical plot. (Used in estimating and stating frequency response characteristics for control elements.) (INSTR)

Counter. A device or memory location whose contents can be successively increased or decreased in increments. (COMP)

CPU. (Central Processing Unit) The heart of any computer system. The CPU consists of storage elements or registers, computational circuits in the ALU, the control block and I/O. (COMP) (MICRO)

CPU Time. Central processing unit time; a measure of system usage by a user, based on the total amount of computer processing time used. (COMP)

CROM (Control Read Only Memory) A ROM which has been microprogrammed to decode control logic. (MICRO) (COMP)

Cross-assembler. A program assembled by another computer. (COMP)

Cross-talk. The unwanted transfer of energy from one circuit,—the disturbing circuit— to another circuit—the disturbed circuit. (COMU)

Cybernetics. The theory of automatic control and communication in the machine and the animal (a term invented by the late Professor Norbert Weiner of M.I.T.). (INSTR)

Cycle Time. The basic unit of computer speed; the time required for a read and write operation in core memory. (COMP)

Damping. A control instrument response in which the controller output settles to its steady-state value after a change in the value of measured signal. (The time response to a step change that is as fast as possible without overshoot is called "critically damped"; "underdamped" is when overshoot occurs; and "overdamped" is when response is slower than critical.) (INSTR)

Data. A term used to denote information which can be processed or produced by a computer or control system (e.g., information in numerical code which is assigned an address in memory that the CPU uses when storing or fetching the information). (COMP)

Data Acquisition. The retrieval of data from remote sites initiated by a central computer system. (COMP)

Data Base. The sum-total of information available to a computer system (e.g., a structured collection of information as an entity or a collection of related files as an entity). In microprocessor communications, the data bus is bidirectional and can transfer data to and from CPU, memory and peripheral devices. (COMP)

DDC. (Direct Digital Control) The control of an automatic function in a process control loop by the computer. (The computer regulates the valve or final control element without any action by an intermediate analog control instrument.) (COMP)

DDCMP. (Digital Data Communications Message Protocol) A uniform method for the transmission of data between sations in a point-to-point or multi-point data communication system. The method of physical data transfer used may be parallel, serial synchronous or serial asynchronous. (COMU)

Dead Band. The value range through which an instrument input can be varied without initiating output response. (Usually expressed in percent of span of the chart or scale of that instrument.) (INSTR)

Dead Time. The interval of time between initiation of an input change and the start of an output response. (INSTR)

Debug. To detect, locate and remove errors in a program; the search for and elimination of sources of error in programming routines. (COMP)

Decentralized (Computer) Network. A computer network in which some of the network control functions are distributed over several network nodes. (COMU)

Dedicated. A device, program or function that is set aside for special use. For example, a microprocessor may be such a device that has been programmed for a single application, such as distillation column control or sequencing control of dryers. (COMP) (MICRO)

Delay Distortion. Distortion resulting from non-uniform speed of transmission of the various frequency components of a signal through a transmission medium. (COMU)

De-limiter. In computer software, a character that separates and organizes elements of data, or a device (in electronic control systems) that separates portions of a control function and thus prevents or limits interaction between process units or control functions. (COMP)

De-modulation. The process of retrieving an original signal from a modulated carrier wave. (Used in data sets to make communications signals compatible with computer signals.) (COMU)

Derivative (Rate) Control Action. Control action in which the output is proportional to the rate of change of the input. (INSTR)

Deviation. Any departure from a desired value. (INSTR)

Deviation Alarm. An alarm caused by the departure of a controlled variable from a desired value by a specified amount. (INSTR) (COMP-H)

Diagnostic Routine. A program designed to locate malfunctions in computer hardware or software. (COMP-S)

Digital. Information presented in the form of data comprising numerical digits. (COMP)

Digital Back-up. An alternate method of digital process control initiated by use of special purpose digital logic in the event of a failure in the computer system. (COMP)

Digital Computer. A computer that operates on a collection of data by the execution of arithmetic and logic processes on that data.

Direct Acting. Operation of a final control element in a manner that is directly proportional to the control input. (INSTR)

Direct Acting Controller. A controller in which the absolute value of the output signal increases in the absolute value of the input. (INSTR)

Direct Control of a System. A condition in which a process variable is manipulated or restrained by a final control element. (INSTR) (COMP)

(DMA) Direct Memory Access. A method that permits I/O transfer directly into or out of memory without passing through the CPU general registers; performed either independently of the processor or on a cycle stealing basis. (COMP)

Disc or Disk. A flat circular plate with a magnetic surface on which data can be stored by selective magnetization of portions of the surface. (COMP-H)

Distributed Network. A network configuration in which all node pairs are connected directly or through redundant paths through intermediate nodes. (COMU)

Double Precision. A programming technique in which two computer words are used to represent a number. (COMP-S)

Down time. The time interval during which a device is malfunctioning. (COMP)

Drum. An item of computer hardware used for memory storage (e.g., a right circular cylinder with a magnetic surface on which data can be stored by selectively magnetizing portions of the cylindrical surface). (COMP-H)

Dump. A method in which the contents of all or part of a storage are copied from internal storage into an external storage device. (COMP-S)

Duplex. Simultaneous, two way independent transmission in both directions. (COMU)

Dynamic Memory. A memory that is faster than a static memory and uses less power, but one which must be refreshed periodically in order to maintain a programmed state. Primarily designed into MOS technology. (COMP-H) (MICRO)

(EIA) Electronic Industries Association. A standards organization specializing in the electrical and functional characteristics of interface equipment. (ELEC)

(ESS) Electronic Switching System. A common carrier communications switching system using solid state devices and other computer related equipment and functions. (COMU)

Engineering Units. Units of measure as applied to a process variable (e.g., GPM, degrees F, BTU).

Equalization. A compensation technique used for an increase in attenuation as frequency increases; used to produce a flat frequency response. (INSTR) (ELEC)

Execution Time. A term expressed in clock cycles required to carry out computer instruction. (Given clock frequency, the actual time can be calculated.) (COMP)

Executive Program. A program that controls the execution of all other programs in the computer. (COMP-S)

Feedback Elements. Elements in a control system that change the feedback signal in response to a controlled variable signal. (INSTR) (COMP)

Feedback Signal. A return signal that results from the measurement of a controlled variable. (INSTR) (COMP)

Feedforward Control Action. A control action in which a process variable signal representing one or a number of constraints on that process variable is converted into a corrective action in order to minimize deviations of the controlled variable. (INSTR) (COMP)

Fetch. To go after and return with a desired function, instruction, program or data (etc.) within a computer software system. (COMP-S)

Fields. A source statement, usually made up of four code fields acceptable by the assembler. (The four fields usually denote Label, Operator, Operand and Comment.) (COMP-S)

Final Control Element. The last device in the control loop; it changes the value of the manipulated variable upon command from some outside element, a manual operator, a control instrument or a computer. (INSTR)

Firmware. Solid state software that is permanently fixed in PROM memory. (COMP-S)

Fixed Heads. Stationary reading and writing heads mounted on bulk memory devices. (COMP-H)

Fixed Point. A numeration method in programming a computer in which the position of a point is fixed with respect to one end of the numerals used in the program according to a pre-set rule. (COMP-S)

Flag. An indication, by the presence or absence of a signal, of hardware condition or program status. (COMP-S)

Flag Bit. An information bit which indicates a form of demarcation that has been reached, such as overflow or carry. It is also used as an indicator for interrupts. (COMP-S)

Flip-flop. A circuit or device containing active elements, capable of assuming either one of two stable states at a given time (e.g., 0 or 1, or plus or minus). (ELEC)

Flow Chart or Flow Diagram. A sequence of functions or operations drawn with the aid of symbols to indicate an executive program. Flowcharts enable the designer to visualize the procedure necessary for each item on the program. A complete flowchart leads directly to final coding for a computer. (COMP-S)

Foreground Processing. A high-priority processing method that usually results from real-time entries, which are treated preferentially, by means of interrupts, over lower priority "background" processing. (COMP-S)

Fortran (FORmula TRANslating system). A computer programming language designed for the solution of arithmetic and logical programs. (Used primarily for solving problems in science and engineering.) (COMP-S)

(FM) Frequency Modulation A method of transmission whereby the frequency of the carrier wave corresponds to changes in the signal wave. (COMU)

Front-end Processor. A communications computer associated with a host computer performing line control, message handling, code conversion, error control and applications functions, such as control and operation of special purpose terminals. (COMP)

Gate. A device with one output channel and one or more input channels, so designed that the output channel state is determined by the input channel states. (COMP-H)

Hardware. The physical equipment of a computer, as opposed to a computer program or function (i.e., mechanical, electrical, magnetic or electronic devices comprising the physical entity of a computer).

Hard-wired Logic. Logic design systems based on the interconnection of numerous integrated circuits representing logic elements. (COMP-H)

Head. A device that reads, records or erases data on a storage medium; a small electromagnet used to read, write or erase data on a magnetic drum or tape. (COMP-H)

Header. The control information prefixed in a message (e.g., the source or destination code, priority or message format). (COMU)

Heterogeneous Network. A network of dissimilar host computers, such as those of different manufacturers.

Hexadecimal. A system of whole numbers in which 16 is used as a base. Sixteen hexadecimal digits are used (0 through 15), as compared to ten numerical digits (0 through 9) in the conventional Arabic system. Therefore, an additional six digits, representing 10 through 15, are introduced. The hexadecimal code is: 0, 1, 2, 3, 4, 5, 6, 7, 8, 9, A, B, C, D, E, F. The decimal number 16 is the hexadecimal number 10. The decimal number 26 is the hexadecimal number 1A. (COMP-S)

Hierarchical Network. A computer network in which processing and control functions are performed at different levels by several computers especially suited to the functions performed (e.g., a plant host computer operating in the background on complex optimization routines while also monitoring, in the foreground, various distributed micros).

High-level Language. A problem-oriented programming language (as opposed to a machine-oriented programming language) in which the instruction format is more suitable to the problems to be handled than the language of the machine on which they are to be implemented. (COMP-S)

Hollerith. A system of encoding alphanumeric information onto cards. These cards were first used in 1890 for the U.S. Census and were named after Herman Hollerith, the inventor of the system. (COMP)

Homogeneous Network. A network of similar host computers which may be comprised of one model from one manufacturer.

Host Computer. (*See also:* Hierarchical Network) A computer attached to a network providing functions, such as computation, data base access or special programs or programming languages.

Host Interface. The interface between a communications processor and a host computer. (COMU) (COMP)

HZ (Hertz) A unit of frequency equal to 1 cycle/second. Cycles are referred to as Hertz in honor of Heinrich Hertz. (ELEC)

Identification. A method providing individual equipment or organizational characteristics or codes to gain access to computer programs, processes, files or data. (COMU) (COMP)

Information Bit. A bit generated by a data source and which is not used for error by the data transmission system. (COMP-S)

Information Path. The functional route by which information is transferred in a one way direction from a single data source to a single data sink. (COMP)

Initialize. A procedure in which various programming devices in a computer, such as counters, switches and addresses, are set to zero or other starting values at the beginning of, or at prescribed points in, a computer program (e.g., the re-setting of all hardware controls to starting values at the beginning of each new program). (COMP)

Input. The data to be processed (or a device used for conveying data) into another device. (COMP-H)

Input Signal. A signal applied to a device, element, system or instrument.

Instruction. A statement that specifies an operation and values or locations of its operands (e.g., a computer operation, stored in memory, that commands the CPU to perform arithmetic and logic functions, control peripheral devices or indicate subsequent instructions). (COMP-S)

Instruction Set. The list of instructions executed by a computer to provide the basic information necessary to assemble a program. (COMP-S)

Interactive. The exchange of information and control between a user and a computer process, between analog controllers or between process units.

Integrated Circuit. A circuit element incorporating transistor, diode and resistor elements in a semiconductor chip. (MICRO)

Integrated Injection Logic (I^2L) A logic circuit incorporating two transistor types in a single unit, where the substrate serves as an emitter of an npn transistor with an upward current flow. This chip substrate is also the base of another transistor, a pnp, in which current flow is lateral; a technique for increased packing density of logic elements, more efficient than older bipolar techniques. (ELEC) (MICRO)

Interface. A shared boundary defined by common physical interconnection characteristics, signal characteristics and/or interchanged signals. A device or equipment enabling operation between two systems or a shared logical boundary between two software blocks (e.g., a common boundary between adjacent components, circuits or systems, enabling the devices to acquire information from one another). Buffer, "handshake" and adapter are terms synonymous with interface.

Interference (Electrical) An intermittent or undesirable voltage or current from an external source or sources affecting the circuitry of an instrument or instrument system.

Interpreter. A method or program that translates a high-level language into machine language. (COMP-S)

Interrupt. The suspension of normal programming routine in order to handle a sudden request for service. The importance of the interrupt capability of a computer depends on the applications to which it will be exposed. A control system for a continuous chemical process may more easily accept interrupt functions than one designed for a sequential or batch process.

I/O (Input/Output) In large computers, the front-end hardware that enables interface to the outside world. In microprocessors, the package pins that are tied directly to the internal bus network to enable an interface between the microprocessor and sensing instrumentation. (COMP-H) (MICRO)

Jump. The jump operation, like the branch operation, is used to control the transfer of operations from one point to a more distant point in the control program. Jumps differ from "branching" in that the relative addressing mode is not used. (COMP)

Label. A numerical value or a memory location in the programmable system. The absolute address is not necessary, since the intent of the label is a general destination. Labels are a requisite for jump and branch instructions. (COMP-S)

Large Scale Integrated Circuit (LSI) A device on one chip of silicon substrate, housing a series of electronic circuits or gates exceeding 100 such gates per chip. (Currently, the state-of-the-art technology exceeds 3000 gates per chip.) (ELEC) (MICRO)

Library. A collection of programs written for a particular computer, minicomputer or microprocessor; a collection of standard programs with which problems or parts of problems may be solved. (Runge-Kutte numerical methods solution to higher ordinary differential equations in mathematics is an example.) (COMP-S)

Line. A portion of a circuit external to the computer, consisting of the conductors connecting a communications hardware or computer to the host computer (e.g., a group of conductors on the same route in the same cable). (COMU)

Linear Programming. An analysis or solution of problems in which the linear function of a number of variables is to be maximized or minimized. The solution requires that the variables be subject to a number of constraints in the form of linear inequalities. (COMP-S)

Link. Any specified relationship between two nodes in a network. (COMU)

Load Sharing. The distribution of a given load among several computers on a network. (COMU)

Logic. In computer or control systems terminology, the hardware and software implementation of Boolean algebra principles; the mathematical treatment of formal logic in which a system of symbols is used to represent quantities and relationships. Typical symbols or logical functions are AND, OR, and NOT. When translated into a switching circuit, this is referred to as a "gate." A switch has only two states—open or closed—thus making possible the application of binary numbers to the solution of mathematical problems.

Machine Language. (*See also:* Binary) A binary language; coding written in a language acceptable to the computer without further modification.

Manual Back-up. An alternative method of process control by manual adjustment of the final control elements in the control loop, in case of failure of the computer. (INSTR)

Mark. The presence of a signal. In communications, a mark represents the closed condition or current flowing and is equal to a binary one condition. (COMP) (COMU)

Mask. A pattern of bits used to extract selected bits from a string. Also a method of manufacture of a microprocessor chip to perform a particular process control, or some other, function. (ELEC) (MICRO)

Mathematical Model. A mathematical representation of the physics of a process, device or system.

Matrix. A two-dimensional rectangular array of quantities. Matrices are manipulated in accordance with the rules of matrix algebra in mathematics. In computer hardware, a logic network is designed in the form of an array of input leads and output leads.

Medium Scale Integrated Circuit. (MSI) A solid-state, metal oxide semiconductor device housing the equivalent of less than 100 transistor gates. (ELEC) (MICRO)

Memory. That portion of computer system hardware into which information can be inserted and held for future use. (Storage and memory are interchangeable terms.) Memories accept and hold information only in the form of binary numbers. Storage can be core, disk, drum or semiconductor based hardware.

Memory Protect. A means of protecting the contents of memory from alteration by preventing the execution of a memory modification via detection of a guard bit associated with the accessed location in memory. (COMP)

Microcomputer. (*See:* Microprocessor)

Microprocessor. The microprocessor is a central processing unit fabricated on one or two chips with arithmetic and logic unit, control block and register array. If joined to memory storage and peripheral I/O's, the combination is called a microcomputer.

Microinstruction. An instruction formed by combining two or more other instructions. (MICRO)

Microprogram. Computer instructions which do not refer to main memory storage; a computer technique which performs sub-routines by manipulating basic computer hardware—sometimes referred to as a sub-computer within the computer (e.g., instructions stored in ROM, a portion of which can implement a higher-level language program). (COMP) (MICRO)

Microprogramming. A programming procedure in which several instruction operations are combined in one instruction for greater speed and efficient use of memory. (MICRO)

Mnemonic. An alphanumeric code designed to be easy to remember and used to designate a memory location or computer operation, or a code to assist the human memory. (The binary numbered codes assign groups of letters, or mnemonic symbols, suggesting the definition of the instruction. For example, LDA for load accumulator.) Source statements may be written in mnemonic symbolism and translated into machine language. (COMP-S)

Modem. A modulator/de-modulator. A device that modulates signals transmitted over communications circuits. (COMU)

Modulation. A method in which a characteristic of one wave is varied in accordance with a characteristic of another wave. (COMU)

MOS. (Metal Oxide Semiconductor) The physical structure of an MOS field effect transistor (FET) is metal over silicon oxide over silicon. The metal electrode is a gate; the silicon oxide is an insulator; and carrier-doped regions in the silicon substrate are the drain and source. A device thus produced is similar to a capacitor. (MOS is slower than bipolar since the "capacitor" must charge up before current can flow.) Three advantages of MOS are process simplicity because of reduced fabrication; savings in space, enabling functional density; and ease of interconnection on chip. (ELEC) (MICRO)

Movable Heads. Hardware designed to enable the use of movable reading and writing heads on bulk memory devices. (COMP-H)

Multi-element Control System. A control system using input signals from two or more process variables for the purpose of changing the action of the control system based on the given multi-process variables. (INSTR) (COMP)

Multiplex. To mix, or simultaneously transmit, two or more messages or signals on a single channel. In communications, a process of transmitting more than one signal at a time over a single link (e.g., frequency sharing or time sharing signals). (COMP) (COMU)

Multiplexer. A device used to enable multiplexing. It may or may not be a stored program computer, as in an application connecting field instrument signals to a communications link. (COMP-H) (INSTR)

Multiprogramming. A method for the time shared control of two or more programs by a computer.

Nesting. A method to enclose a sub-routine in a larger routine, but not necessarily as part of the larger routine. (COMP)

Network. An interconnected or interrelated group of nodes in communication between computers. (COMU)

Network Control Program. That portion of an operating system in a host computer that establishes and breaks logical connections and communicates with the network and with user processes within the host computer. (COMP) (COMU)

NMOS. *(See also:* PMOS) A field effect metal oxide semiconductor based technology in which only electron charges are active. An NMOS device consists of two islands of N-type silicon, formed in a P-type silicon substrate. (ELEC) (MICRO)

Node. An end-point of any branch of a network, or a junction common to two or more branches of a network. (COMU)

Noise. An unwanted component of a signal or variable which masks its information content. (COMU) (INSTR) (ELEC)

Non-linear system. A system whose function cannot be represented by a first-order mathematical differential equation.

Object Program. A code which is the output of an automatic code translation program such as an assembler or compiler; the end result of the source language program after it has been translated into machine language. (COMP-S)

Octal. The selection or condition in which there are eight possibilities; the numeration system with a radix of eight. (COMP-S)

Off-line. Equipment or software not under control of the central processing unit. (COMP)

Offset. A steady-state deviation of the controlled variable when the set point is fixed. (INSTR)

On-line. Devices or programs under direct control of the central processor unit in a computer (e.g., a computer that is monitoring or controlling a process or operation, or a user's ability to interact with a computer). (COMU)

Operand. A quantity on which a mathematical operation is performed. One of the instruction fields in an addressing statement. (COMP-S)

Operating Code. A source statement which generates machine codes after assembly (e.g., software that controls the execution of computer programs and controls scheduling, debugging, I/O control, accounting, storage and data manipulation). The part of a computer instruction which specifies the operation to be performed. (COMP-S)

Optimize. A system or method for the establishment of control in order to make control effective within a system of constraints.

Optimizing Control Action. Control action that automatically adopts and maintains the most advantageous value of a process variable, rather than holding it at one fixed set point value.

OR. A logic operator having the property that if A is a value, B is a value and C is a value, then the OR of $A, B, C,$ is true if at least one value is true, and false if all statements are false.

Overflow. Overflow results when an arithmetic operation generates a quantity beyond the capacity of the register. (COMP-S)

Pack. To include several discrete items of information in one unit of information, or the relocation of data to make efficient use of available storage capacity. (COMP-S)

Packet. A group of bits, including data and control elements, which is switched and transmitted as a composite whole. The data and control elements and, possibly, error control information are arranged in a specified format. (COMP-S)

Packing Density. The number of useful storage cells per unit of dimension; the number of bits per inch stored on a magnetic tape or drum; or, in a microprocessor chip, the number of transistor-equivalents housed on the one chip. (COMP-S) (MICRO)

Parallel. Pertaining to the simultaneous transfer and processing of all bits in a word or other unit of information. (COMP-S)

Parallel Operation. Processing all the digits of a word or byte simultaneously by transmitting each digit on a separate channel or bus. (COMU)

Parity Check. A method of checking the validity of information, word by word. The total number of binary ones and zeros in a word is always even or odd, and each word contains a parity bit whose value is a function of the number of digits. (The parity bit checks whether the word is true or false.) (COMP-S)

Password. A word or string of characters that is recognizable by automatic means and that permits a user access to protected storage, files or input or output devies. (COMP-S)

Patch. A temporary section of coding into a routine to correct or alter the routine. (COMP-S)

Peripheral. Devices for entering data into—or receiving data from—the computer. (COMP-H)

Phase Modulation. A method of transmission whereby the phase angle of the carrier wave is varied in accordance with the signal. (COMU)

PMOS. (*See also:* NMOS) A field effect metal oxide semiconductor technology in which islands of P-type material in an N-type substrate of silicon produce only holes as charge carriers. (ELEC) (MICRO)

Pointer. A register which serves as a reference point to a memory location.

Point to Point Connection. A network configuration in which a connection is established between two terminal installations (e.g., a circuit connecting two points without the use of any intermediate terminal or computer). (COMU)

Polling. The process of inviting another station to transmit data (in communications between computers); a method used to identify the source of interrupt requests. (When several interrupts occur at one time, the control program decides which one to service first.)

Port. A device which provides electrical access to a system or circuit. The point at which the I/O is in contact with peripheral devices. (COMP)

Priority. Level of importance of a program or device. (COMP-S)

Priority Interrupt. The temporary suspension of a program currently being executed in order to execute a program of higher priority. (COMP-S)

Problem-oriented Language. A programming language designed for an easily acceptable or understandable expression of a given class of problems (e.g., FORTRAN, for scientific and engineering problems). (COMP-S)

Procedure-oriented language. A programming language designed for an easily understood expression of procedures used in the solution of a wide class of problems (e.g., ALGOL). (COMP-S)

Process Control Loop. A system of control devices connected together via signal linkages to control one sub-unit operation of a process. (INSTR)

Process. A systematic sequence of production operations to create a specified result or product.

Program. A computer operation procedure (e.g., a set of instructions to allow computer operation in a pre-designed pattern). Software.

Programmer. A person who prepares computer operation procedures by means of flowcharts and coding.

PROM. (Programmable Read-only Memory) A block of information within a memory bank that can be altered by reprogramming in the field with the aid of programming equipment. Program data stored in ROM's are called firmware because they cannot be altered in "EPROM" eraseable microprocessors; reprogramming is accomplished via portable ultraviolet light devices for erasure of existing information. (COMP-S)

Proportional Band. The change in input required to produce a full range in output, due to proportional control action; the percent of instrument chart or scale over which the final control element is moved from a minimum position to a maximum position (or *vice versa*); usually stated in input units or percent of the input span. PB = 1/gain. (a 20% PB denotes a 20% chart range for full valve travel). (see Sensitivity) (INSTR)

Proportional with Integral Control. (Reset) A control result in which the output is proportional to the input plus the time integral of the input. (INSTR)

Proportional Plus Derivative Control. (Rate Action) A control result in which the output is proportional to the input plus the time of rate of change of the input. (INSTR)

Proportional with Integral and Derivative Control (P.I.D.) A control result in which the output is proportional to the input plus the time integral of the input and the time rate of change of input. (INSTR)

Protocol. A formal set of rules controlling the format and timing of data storage and processing. (COMP-S)

Pulse. A sudden and recognizable change of short duration in the level of an electrical signal.

Pulse Code Modulation. Modulation of a pulse train in accordance with a given code. (COMU)

RADIX. Synonymous with the BASE of a number. (COMP-S)

RAM. A term denoting random access memory in a computer; rapid access to any storage location point in the memory by means of geometric coordinates to allow information to be "written" into or "read" out of memory. (COMP)

Ramp Response. The total (transient plus steady-state) time response resulting from a sudden increase in the rate of change in input from zero to a finite value. (INSTR)

Random Access. The process of obtaining data from (or placing data into) storage, where the time required for such access is independent of the location of the data most recently obtained or placed into that storage. (COMP-S)

Ratio Controller. A controller that maintains a predetermined ratio between two or more process control variables. (INSTR)

Read. To acquire data from a source. (COMP-S)

Real-time Clock. A register and/or circuitry which automatically maintains time in conventional units for use in program execution. (COMP)

Real-time Operation. A technique that allows the computer to utilize information at the time that it becomes available, as opposed to batch processing, which occurs at the time the information was generated. (COMP)

Real-time Program. A program that operates concurrently with an external process which it is monitoring or controlling in respect to time.

Real-time System. A system performing computation during the time the related physical process runs so that the results of the computation can be used in controlling the process. (COMP)

Recursive. The iterative use of a sub-routine in completing the solution of a problem. (COMP-S)

Redundancy. A computer protocol in which a portion of the total characters or bits are eliminated without loss of information. (COMP-S)

Refresh. The packaging of memory circuits to keep voltage intact when electric charges (i.e., data) leak from these dynamic memory elements. (COMP)

Register. A memory in small scale; words stored are arithmetical, logical or transferral operations; storage in registers can be temporary. (COMP-S)

Relocate. To move a routine from one portion of storage to another and to adjust the necessary address references so that the routine, in its new location, can be executed. (COMP-S)

Remote Job Entry. (R.J.E.) The submission of jobs through an input device that is connected to a computer through a communications link (e.g., a mode of operation allowing input of a batch job by a card reader at a remote site, and receipt of the result via a line printer or card punch).

Repeatability. The precision among repeated measurements of output for given values of input under given operating conditions over a period of time from either direction of the instrument scale. (INSTR)

Resolution. The smallest interval between two

sequentially displaced points or events which can be distinguished, one from the other. (INSTR)

Resource. Computational power, programs, data files, storage capacity or a combination of these in a computer system.

Response Time. The elapsed time between the generation of the last character of data and receipt of the first character of the reply. In control or measurement systems, the elapsed time between the appearance of an output signal for a given input stimulus.

ROM. A term denoting read-only memory in a computer; a memory so constructed that data within a block of information may only be written or added to; a fixed block of information or a matrix of undifferentiated cells or bits in memory.

Routine. The instructions that perform a specific task in a computer. (COMP-S)

Scale. A mechanical arrangement on an instrument used to indicate values of set point, process variable or manual settings. To change a variable by a factor in order to bring its range within desired limits. (INSTR)

Scale Factor. A number used as a multiplier, chosen so that it will cause a set of quantities to fall within a given range of values. (INSTR)

Scan. A gathering of data from sensor instrumentation by a computer, for use in calculations through a multiplexer (e.g., a sequential interrogation of devices under programmed control).

Sense. To detect the presence of a process condition. (INSTR)

Sensing Element. A device responsive to a value of a measured physical quantity. (INSTR)

Sensitivity. The ratio of a change in output to the change in input upon upset and return to the steady-state; usually expressed as a numerical ratio. (synonymous with GAIN.) (INSTR)

Serial. (*See also:* Parallel) The sequential bit transfer and processing of each bit in a word or unit of information. (COMP-S)

Serial Transmission. A method of transmission in which each bit of information is sent sequentially on a single channel. (COMU)

Service Routine. A routine in general support of the operation of a computer (e.g., an input/output, diagnostic, tracing or monitoring function).

Set Point Control (SPC) A control mode in which the computer supplies a calculated set point to a conventional analog instrument control loop. (INSTR)

Shared Time Control. One control action divided between several loops.

Shift. The movement of information to the right or left in an arithmetic register of a computer. Information leaving a register at one end does not re-enter at the other end. (COMP-S)

Signal to Noise Ratio. A ratio of signal amplitude to noise amplitude. (INSTR)

Significant Bits. The most significant bit (MSB) in a byte of eight numbers is on the extreme left; the least significant bit (LSB) is on the extreme right; and remaining bits are weighted according to their position in the byte between the MSB and the LSB. (COMP-S)

Simplex Mode. Operation of a channel in computer communication in one direction only, with no reversal. (COMU)

Simulation. A method utilizing mathematics and physics in which the representation of physical systems within the computer signify process variables; the operations then accomplished by the computer represent the process itself, and information thus produced by the computer represents the results of the process. (COMP-S)

Simulator. A device or computer program that operates on a model of a process; a program that emulates logical operation and is designed to execute object programs generated by a cross-assembler on a machine other than the one being worked on. (Used to check and debug programs prior to commitment to memory.) (COMP-H, S)

Slave. A remote system or instrument whose functions are controlled by a central "master" system or control instrument. (INSTR)

Slice. A type of chip architecture permitting stacking of devices to increase word bit size. (MICRO)

Smooth. An operation that decreases or eliminates rapid fluctuations in data. (INSTR)

Software. The totality of programs and routines

associated with a computer; compilers; the documentation associated with a computer, such as manuals and circuit diagrams. Software is the computer's instruction manual. The term was probably chosen to contrast with hardware and it is the language used by a programmer to communicate with the computer. This language in the memory of the computer is mathematical; therefore, the programmer must convert instructions into digits or bits of information in order to process data.

Solid State Software. In microprocessors, this software is permanently fixed in PROM memory. (COMP-S)

Source Language. A program code used as input to a translation program, such as an assembler or a compiler. (COMP-S)

Source Statement. A program written in other than machine language, usually in three-letter mnemonic symbols giving the definition of the instruction (e.g., the "executive," which translates into operating machine code, and the "assembly," which documents the source program but generates no code). (COMP-S)

Span. The algebraic difference between the upper and lower values of an instrument measuring range. (INSTR)

Stack. A block of successive memory locations accessible from one end on a last-in-first-out basis (LIFO) and coordinated with the stack pointer, which keeps track of storage and retrieval of each byte of information. (COMP-S)

Star Network. A computer network with peripheral nodes all connected to one or more centrally located computers. (COMU)

Static Memory. Static memories retain their programmed state as long as power is applied, as in MOS and bipolar devices. (COMP-H)

Steady-state. The characteristic of a process condition in which values such as rate or amplitude exhibit very small changes over long periods of time. (INSTR)

Step Response. The time response of an instrument upon being subjected to an instantaneous change in its input from one steady-state level to another. (INSTR)

Storage. (Memory) A device in which information can be entered or in which it can be held, and from which it can be called at a later time.

Sub-routine. A series of computer instructions that performs a specific task or function with a location specifying where to return in the main program after the task has been completed; a part of a master routine that may be called at will to mesh with a variety of master routines. (COMP-S)

Supervisory. A process computer application in which the computer performs higher-level process calculations but does not actuate the final element. The computer may reorganize the mathematical calculations for the process model in the background mode while it also resets control instrument set points in the foreground mode.

Supervisory Control. (*See also:* Cascade Control) A control action in which a control loop operates independently subject to corrective action from an external source. A set point change can be made manually or by another analog controller or by computer.

Supervisory Programs. Computer programs that have the primary functions of scheduling, allocating and controlling system resources rather than processing data.

Symbolic Coding. Symbolism in which other than binary machine language is used.

Synchronous Computer. A computer in which each event starts as a result of a signal generated by a clock.

Synchronous Transmission. The data and/or bits are transmitted at a fixed rate with the transmitter and receiver synchronized, eliminating the need for start-stop elements. (COMU)

Syntax. The structure of expressions in a computer language; the rules governing the structure of a language.

System. A collection of hardware organized in such a way as to achieve a process objective.

Systems Analysis. The study and definition of a control problem and the steps leading to a solution of the problem (i.e., "scientific method"—cause and effect data analysis; problem is defined; a solution is proposed and documented).

Systems Engineering. The specification, design and implementation of a control system resulting from the analysis of a control problem.

Table. A block of information in memory, used as data by a program. (COMP-S)

Tag. Information used as an identifier or label for other information purposes. (COMP-S)

Teletype. A trademark of the Teletype Corporation, referring to one of their series of teleprinters.

Terminal. A device or computer which may be connected to a local or remote host system, and for which the host system provides computational and data access services. (COMP-H)

Text. A sequence of characters forming part of a transmission that is sent from the data source to the data sink and contains the information to be conveyed. (COMU)

Throughput. The speed with which problems are performed. (COMP-S)

Time Constant. For a first-order system, the time required for the output to complete 63.2% of the total rise or decay as a result of a step change of the input. (INSTR)

Time Sharing. A method of operation in which a computer facility is shared by several users for different purposes at the same time. The computer services each user in sequence, but at so high a speed it appears that all users are handled simultaneously. (COMU)

Transducer. (*See also:* Transmitter) A device that receives a signal in one form and converts it to another form of the same or other physical quantity. (INSTR)

Transfer Function. The operational equivalent of a complex mathematical function that permits solution by simple arithmetic.

Transmitter. A device that converts a measured variable, by means of a sensing element, to a standardized transmission signal. Electronic signals are usually 4 to 20 milliamps; pneumatic signals are usually 3 to 15 psig. (INSTR)

Truncate. To delete the least significant digits of a measurement or a number, thereby decreasing precision. (COMP-S)

Truth Table. A table based on a logic function (listing all possible combinations of input data) and indicating, for each combination thereof, true output values to that logic. (COMP-S)

TTL. (Transistor-transistor logic) A form of bipolar microelectronic digital logic circuit technology that couples transistor elements to achieve devices not possible with discrete components (e.g., a multiple-emitter transmitter). (ELEC) (MICRO)

Tuning. The adjustment of control constants in algorithms or analog controllers to produce a desired control effect (e.g., the tuning of a proportional band value to achieve a desired control effect based on proportional-only control). (INSTR)

Turnaround Time. The elapsed time between submission of a job to a computing center and the return of results, or the time required to reverse the direction of transmission from sender to receiver (or *vice versa*) when using a two way alternate circuit in communication between computers.

Two's Complement Numbers. The ALU in a computer performs standard binary addition using the two's complement numbering system to represent both positive and negative numbers. The positive numbers in two's complement representation are identical to the positive numbers in standard binary logic. (COMP-S)

Update. To modify a program according to current information. (COMP-S)

Virtual Circuit. A connection between a source and a sink in a network. This may be achieved by different circuit configurations during transmission of a message. (COMU)

Volatile. Semiconductor read/write memories are volatile and cannot retain a program unless power is continually applied. Power failure will wipe out the contents of read/write memories. (This does not apply to PROMs.) (COMP)

Watch Dog Timer. An electronic interval timer that generates a priority interrupt unless periodically recycled by a computer; used to detect hardware or software failure conditions. (COMP)

Word. A sequence of bits or characters treated as a unit and capable of being stored in one computer location; a group of "characters" treated as a unit and given a single location in computer memory. While a byte is a group of bits, a word is a group of numeric and/or alphabetic characters and symbols (the two terms are sometimes used interchangeably). (COMP-S)

Word-time. The data transfer rate (words per second) between a peripheral device and the computer. (COMP-S)

Write. To deliver data to storage. (COMP-S)

Bibliography

Books

1. Soucek, Branko, *Microprocessors and Microcomputers*. New York: John Wiley & Sons, 1976.
2. Rao, Guthikonda V., *Microprocessors and Microcomputer Systems*. New York: Van Nostrand Reinhold Co., 1978.
3. Franks, G. E., *Modeling and Simulation in Chemical Engineering*. New York: Wiley-Interscience, 1972.
4. McCabe, Warren L. and Smith, Julian C., *Unit Operations of Chemical Engineering,* 3rd Edition, 1976.
5. Shinskey, F. G., *Process Control Systems*. New York: McGraw-Hill Book Co., 1967.
6. Bibbero, Robert J. (Process Control Div. Honeywell, Inc.), *Microprocessors in Instrumentation and Control*. New York: John Wiley & Sons, 1977.
7. Torrero, Edward A., *Microprocessors: New Directions for Designers*. Hayden Book Co., Inc., 1977.
8. Elphick, Michael S. *Microprocessor Basics*. Hayden Book Co., Inc., 1977.
9. Altman, Lawrence (Editor), *Microprocessors 1975*. McGraw-Hill Electronic Book Services.

Articles

A. Noyce, Robert N., Microelectronics, *Scientific American*, p. 62 (Sept. 1977).

B. Meindl, James B., Microelectronic Circuit Elements, *Scientific American* (Sept. 1977).

C. Holton, William C., The Large Scale Integration of Microelectronics, *Scientific American* (Sept. 1977).

D. Hodges, David A., Microelectronic Memories, *Scientific American* (Sept. 1977).

E. Toon, Hoo-Min D., Microprocessors, *Scientific American* (Sept. 1977).

F. Oliver, Bernard M., The Role of Microelectronics in Instrumentation and Control, *Scientific American* (Sept. 1977).

G. Mead, Carver A., Microelectronics and Computer Science, *Scientific American* (Sept. 1977).

H. Skrokov, M. Robert, Microprocessor Control Benefits on Olefins Plant Design, *A.S.M.E. Talk and Paper*, Mexico City (Sept. 19–24, 1976)

I. Skrokov, M. Robert, The Benefits of Microprocessor Control, *Chemical Engineering Magazine* (Oct. 11, 1976).

J. Harrison, T. J., I.B.M. *Micros, Minis and Multiprocessing, Instrumentation Technology* (Feb. 1978).

Index